NETWORK
FLOW
PROGRAMMING

BOARD OF ADVISORS, ENGINEERING

NETWORK
FLOW
PROGRAMMING

Paul A. Jensen
J. Wesley Barnes

Operations Research Group
Department of Mechanical Engineering
University of Texas at Austin

John Wiley & Sons
New York Chichester Brisbane Toronto

Library of Congress Cataloging in Publication Data:

Jensen, Paul A   1936–
   Network flow programming.

   Includes index.
   1. Network analysis (Planning) 2. Programming
(Mathematics) I. Barnes, J. Wesley, joint author.
II. Title.

T57.85.J39   658.4′032   79-26939
ISBN 0-471-04471-7

Printed in the United States of America

10 9 8 7 6 5 4 3 2 1

To our wives,
Margaret Eastlack Jensen
and Susan Geron Barnes

PREFACE

Today, both theoretical and applied interests in the use of network flow programming are experiencing an expansion unrivaled by that of any other optimization technique. Perhaps the most important reasons for this expansion are the fresh advances in network computational methods that allow analysts to solve problems of enormous size formerly unassailable by any other approach. Networks have been used in innumerable applications to represent such things as inventory systems, river systems, distribution systems, precedence ordering of events, flowcharts, and organization charts. In fact, the network representation is such a valuable visual and conceptual aid to the understanding of the relationships between events and objects that it is used in virtually every scientific, social, and economic field.

In this book, we present a synthesis of the more important techniques, both recent and traditional, that are related to network flow programming. The level of discussion is easily within the reach of first-year graduate or advanced undergraduate students. The two background requirements are a reasonable facility with a general-purpose programming language, such as Fortran IV, and an understanding of Linear Programming that is consistent with a good undergraduate text in Operations Research. The population possessing this background includes most persons holding undergraduate degrees in Computer Science, Engineering, Operations Research, Applied Mathematics, and Business Administration. However, the most important and largest segment of our intended audience is the practitioner.

With this audience in mind, we do three things. First, we emphasize concepts rather than theoretical proofs. Such proofs abound in the literature, and at the end of the book an extensive bibliography is provided which covers the theoretical underpinnings of the subject. Second, we include in the text a coherent computational package of efficient algorithms that will solve any of the various network programming problems addressed in the book. Third, we take great care to present both the relationships between Linear Programming and Network Flow Programming and the interrelationships that join the various network flow programming problems and their suggested techniques of solution.

In the process of achieving these three objectives, we have formulated a reasonably complete exposition of the total spectrum of single commodity network flow programming problems. This presentation ranges from basic topics such as network storage techniques and handling representations to such advanced topics such as stochastic and generalized networks.

The principal strengths of this book can be summarized as follows:

The description of the various network flow programming problems in a consistent and unified notation. This makes evident the strong relationships between the problems and the relationships to the underpinning linear programming theory.

The modular nature of the manipulation and optimization algorithms and programs. Algorithms are presented in small packages that are easily understood by the student. Complex operations are performed by collections of modules. Modularization also provides for easy substitution of algorithms for experiments that test alternative computational approaches.

Emphasis on the computational aspects of algorithms. The recent emergence of network flow programming is primarily due to the computational procedures used to store and manipulate networks and network components in the computer. The serious student or user of network programming cannot neglect this aspect.

The method of presentation of the algorithms. We present algorithms at three levels: a rigorous flowchart using notation related to the computer programs (the form of the flowcharts is particularly compact and easily understood), a parallel English language description, and an example.

The scope of the material covered. We cover all of the important single commodity network flow programming problems. This includes extensive discussion of generalized networks and networks with separable nonlinear cost functions. In particular, this gathering of subjects is an especially useful one not currently available in existing texts.

Although most of the material in this book derives from previously published work, some appears here for the first time. The results and algorithms given here are immediately applicable to a wide variety of "real-world" problems. The computer codes, which we designed to be especially useful for teaching, can be obtained from us for a nominal handling charge. In addition, a solutions manual containing detailed answers to all exercises in the book is available to instructors and practitioners from John Wiley on written request.

There is ample material for a two-semester course, with some augmentation—perhaps with case studies. The first semester would comprise the material of Chapters 1 through 7. The second semester would deal with the more advanced material in the remaining chapters and the augmentations selected by the instructor. A particularly able class should be able to cover Chapters 1 through 10 in a single concentrated semester. These comments are based on our more than five years of experience teaching this material.

We express our thanks to the great number of graduate students in the Mechanical Engineering Department's Operations Research graduate program at the University of Texas at Austin who gave us assistance in the development of this book. Finally, special thanks are due to Margaret Jensen, who typed the numerous revisions of the manuscript.

Paul A. Jensen
J. Wesley Barnes

CONTENTS

ALGORITHMS

NETWORK
FLOW
PROGRAMMING

CHAPTER 1

NETWORK FLOW MODELS

1.1 INTRODUCTION

As illustrated in Figure 1.1, a *network* is a collection of *nodes* and *arcs*. This representation is useful for modeling a wide range of physical and conceptual situations. Networks have been used in innumerable applications to represent such things as inventory systems, river systems, distribution systems, precedence ordering of events, flowcharts, and organization charts. The network representation is such a valuable visual and conceptual aid to the understanding of the relationships between events and objects that it is used in virtually every field of scientific, social, and economic endeavor.

Some practical situations that can be represented by a network also have the characteristic of flow; that is, water may flow in a pipe network, traffic may flow in a street network and products may flow in a distribution network. Models of such situations are called *network flow models*. In this book, we restrict our attention to models of this type and, as we will see, many problems not obviously in this class can be represented by network flow models.

Further, we consider network flow models in which the amount of flow in each arc is controllable and the objective is to choose values for the arc flows that optimize some measure of effectiveness. To illustrate, suppose Figure 1.2 defines a network flow model. Each arc in the network has flow directed as specified by the arrow head of the arc. In this model, each arc has been assigned three parameters: a *lower bound*, which is the minimal amount that can flow over the arc; a *capacity*, which is the maximum amount of flow that the arc can carry; and a *cost* for each unit of flow that passes through the arc.

Since a time period is implied in most network formulations, flows and capacities are usually stated in terms of "flow-per-unit-time." If no lower bound parameter is present on an arc, it is assumed that the lower bound is zero. In

1

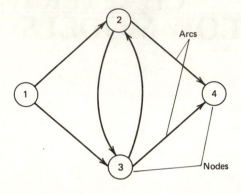

Figure 1.1
Network

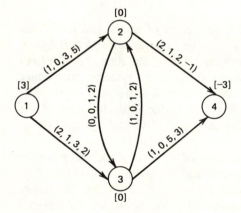

Figure 1.2
Network Flow Problem with Solution

addition lower bounds present no practical problems since a very simple transformation, discussed in Chapter 3, may be used to remove nonzero lower bounds from any network. The required quantities of flow entering or leaving the network at each node are also specified. These parameters of the nodes are called *external flows*. A positive external flow enters the network at a node and a negative external flow leaves the network. Flow is conserved at each node. Thus, the flow entering a node from the arcs of the network plus the external flow at the node must equal the flow leaving the node on the arcs of the network.

The flows on the arcs are controllable within the limits, or constraints, set by arc capacities, conservation of flow, and node external flows. Clearly, these arc flows are the decision variables of an optimization problem. The optimization problem is to choose the arc flows, within the above restrictions, to minimize the total cost of the flow. As the reader may easily verify, the flows shown in Figure 1.2 are optimal for the given parameter values.

1.2   RELATIONSHIPS BETWEEN NETWORK FLOW PROGRAMMING PROBLEMS

The problem of optimizing some objective subject to constraints is called a *mathematical programming problem*. Because all the problems considered in this text are defined by a network that carries flow, we use the term *network flow programming problem*. The problem of Figure 1.2 is a specific example of the pure, linear, minimum cost flow problem. The schematic relationships joining this basic problem to other network flow programming problems that are considered in this text are shown in Figure 1.3. The central point in this figure is the pure, linear, minimum cost flow problem. The problems listed to the left are less general in the sense that they are specializations of this basic problem. Problems listed to the right are more general in that this basic problem is in some way a specialization of each of these problems. The general linear programming problem is also shown in this figure to indicate its relationship to the network programming problems. Algorithms have been identified to solve each of the problem classes in Figure 1.3. Algorithms for the less general problems are more efficient, in computational time and memory requirements, than those for the more general problems. Algorithms defined for a more general problem can solve a less general form of that problem, while the converse is rarely true.

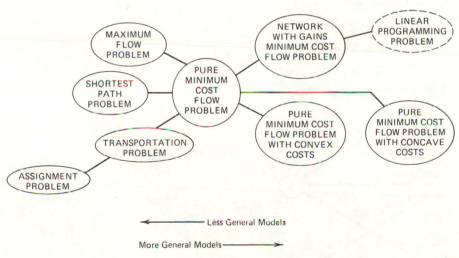

Figure 1.3
Network Flow Programming Problems Relationships

1.3 SPECIALIZATIONS OF THE PURE, LINEAR, MINIMUM COST FLOW PROBLEM

Transportation Problem

Of frequent use in practice, the *transportation problem* is a special case of the minimum cost flow problem where the network representation has a distinct form: the nodes can be partitioned into two sets N_1 and N_2 such that all arcs originate in N_1 and terminate in N_2. Three other special properties are:

1. All arcs have infinite capacities, which allows omission of the capacity parameter from the arc parameter list.
2. All nodes have nonzero fixed external flows.
3. The sum of the external flows over all nodes is zero.

An example transportation problem network model, with its associated optimal flows, is presented in Figure 1.4.

Assignment Problem

An important specialization of the transportation problem is the *assignment problem*, in which both $|N_1| = |N_2|$ and all demands and supplies are unity. Given the cost associated with pairing any object $i \epsilon N_1$ with any object $j \epsilon N_2$, the problem is to find an exhaustive one-to-one pairing of the two sets' elements that minimizes the sum of the pairing costs. This problem is illustrated by the network of Figure 1.5. Again the flows present in Figure 1.5 are optimal and

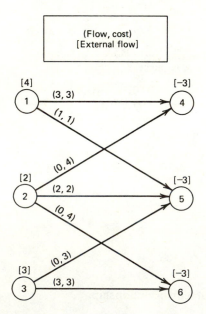

Figure 1.4
Example Transportation Problem

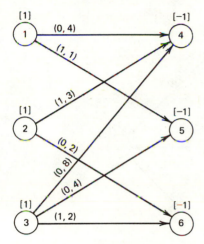

Figure 1.5
Example Assignment Problem

imply the following optimal pairings: node 1 and node 5, node 2 and node 4, and node 3 and node 6.

Shortest Path Problem

In the *shortest path problem*, two nodes are designated as the *source* and the *sink*. The arc cost is commonly given the physical interpretation of *arc length*. The *optimal path* is that sequence of arcs connecting the source to the sink such that the sum of the arc costs on the path is minimized. An example shortest path problem appears in Figure 1.6. The optimal flow pattern in Figure 1.6 implies the shortest path consists of arcs (1, 3), (3, 4), and (4, 5).

Since nodes that are neither source nor sink will always have a fixed external flow of zero, no external flow designation is given for nodes 2, 3, and 4. This convention will be followed consistently throughout the book; that is, the absence of fixed external flow parameters will be interpreted as zero external flow.

Maximum Flow Problem

For the *maximum flow problem*, arc capacity is the only relevant parameter. Once a source node and a sink node are identified, the problem is to maximize the flow passing from source to sink. An example maximum flow problem is shown in Figure 1.7, where node 1 is the source and node 6 is the sink. One of the several alternate optimal flow patterns is also given in Figure 1.7.

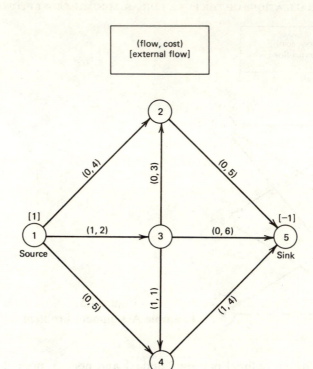

Figure 1.6
Example Shortest Path Problem

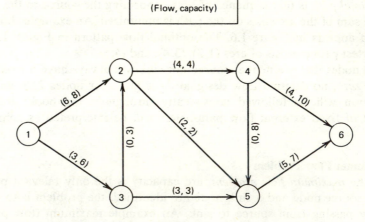

Figure 1.7
Example Maximum Flow Problem

6

As we will see in Chapter 7, the *minimal cut problem* is equivalent to the maximum flow problem and is solved concurrently. The coobjective is to find the set of arcs of minimal total capacity that when removed from the network will prohibit all flow from the source to the sink. The minimal cut for the example of Figure 1.7 consists of arcs (2,4), (2,5), and (3,5). Note that the total capacity of these arcs is 9, which is equal to the value of the maximum flow.

1.4   THE NETWORK-WITH-GAINS MODEL

One of the generalizations of the pure minimum cost flow problem is the "network-with-gains problem." We will restrict our attention to this generalization for the present and defer discussion of networks with convex and concave arc costs until Chapter 3.

The *network-with-gains model* or *generalized network model* is an important class of network programming models for which efficient solution algorithms exist. For these models, flow is not necessarily conserved as it passes through an arc but rather it is multiplied by the arc's *gain parameter*. For example, consider arc (1,3) in Figure 1.8, which receives two of the three units of fixed external flow available at node 1. Because arc (1,3) has a gain parameter of 0.5 only 0.5*2 or 1 unit of flow arrives at node 3. Similarly, 1.5 units of flow leave node 2 by means of arc (2,4) but 3 units of flow arrive at node 4 because arc (2,4) has a

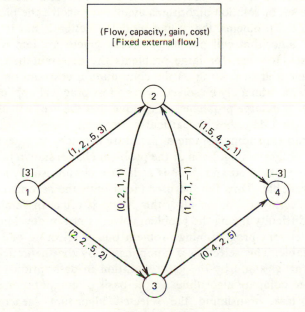

Figure 1.8
Example Network-with-Gains Problem with Optimal Solution

gain parameter equal to 2. Clearly any pure network model is a special case of the network-with-gains model where all gain parameters equal 1.

As we see in later chapters, the inclusion of the gain parameter adds greatly to the modeling power of network approaches. As might be expected, however, this additional capability is accompanied by increased complexity and some loss of solution efficiency.

1.5   A PREVIEW OF THINGS TO COME

The prospective user of network flow programming faces three tasks.

1. Specifying his problem as a network flow problem.
2. Choosing an algorithm to solve the problem.
3. Obtaining a reliable computer program of the chosen algorithm that can solve his problem efficiently.

The first of the above jobs is the modeling task. Although we address methods of appropriately modeling physical systems as network flow programming problems throughout the book, we give that subject a particularly concentrated discussion in the following chapter. In general, a particular problem may fall in several classes of the network flow programming models. For instance, a shortest path problem may be stated as a linear program, as a pure minimum cost flow problem, or as a shortest path problem. The model selected will determine the set of solution algorithms available to solve the problem.

For small-sized problems, the proper model selection is not too important as any available algorithm will probably solve a given problem with acceptable computer cost. However, for large problems the determination of the proper model will have a great bearing on the computation cost and on the size of the problem that can ultimately be solved. Since most practical applications of these techniques result in large problems, we assume that the user will choose the least general model that describes his particular problem.

The large and expanding volume of literature provides an extensive theory for network flow programming. Each of the problem classes shown in Figure 1.3 has been extensively explored and there is at least one and usually several solution algorithms available. Thus for the user faced with the second task above, the problem of selecting a solution algorithm, there is often an "embarrassment of riches." The difficulty facing the problem solver is not in developing algorithms to solve his network programming problem but in selecting one of the several that are available. This selection is complicated by the theoretical emphasis of most journal articles and by the great variation in descriptional notation. It is also difficult to compare algorithms on the basis of computational efficiency.

The third task, translating the selected algorithm—generally stated in prose—to a computer program, is a very necessary step. Problems of sufficient size to warrant solution by network programming methods are too large for hand calculations. The translation may take many forms and involve some important programming decisions. One important decision regards the representation of the network and its parameters in computer memory. This decision, made early in the programming process, has a major effect on the computer memory space and computation time requirements for the solution of the network flow programming problem. Even when the problem solver chooses to use an independently written computer coding of his selected algorithm, an intimate knowledge of the implications of the programming approaches used is very important for the appropriate selection of computing codes.

In the chapters to follow, we attempt to aid in the three tasks by giving clear and detailed descriptions of the various problem classes and by presenting their solution algorithms in a common notation. While describing a number of solution algorithms that have appeared in the literature, we note strong similarities by showing that the algorithms are all based on the same underlying theory. We describe the algorithms in both prose and flowchart format. The former aids in understanding the process; the latter aids in the translation to a computer program. The computer storage representations of the networks and the algorithms described are, in our opinion, an excellent compromise between memory requirements, efficiency, and ease of conceptual understanding. The algorithms included in this book allow the solution of any network flow programming problem shown in Figure 1.3.

1.6 HISTORICAL PERSPECTIVE

The pioneering works describing an optimization problem involving flows in networks are generally attributed to Hitchcock (1941), Kantorovich (1942), and Koopmans (1947), who described what is now referred to as the transportation problem. In the years since that time, hundreds of articles and several books have been written, some introducing new problem classes structured on networks, many suggesting new algorithms and modifications of algorithms to solve these classes, and some surveying the work that went before.

A great deal of the algorithmic development for the assignment, transportation, maximum flow, and the pure minimum cost flow problem occurred during the 1950s. Dantzig (1951), Flood (1953), Charnes and Cooper (1954), Gleyzal (1955), Ford and Fulkerson (1956), Munkres (1957), Eisemann (1957), and Dennis (1958) provided solutions to the transportation problem during this period. The assignment problem was specifically considered by Kuhn (1955, 1956), Motzkin (1956), and Munkres (1957). The maximum flow problem was considered in papers by Dantzig (1956), Ford and Fulkerson (1957a), and Fulkerson and Dantzig (1955). Algorithms for the pure minimum cost flow problem, or as it is sometimes called *the capacitated transshipment problem*, appear in papers by Fulkerson (1961b), and Orden (1956).

Early consideration of capacitated problems were by Ford and Fulkerson (1957b, 1962). Beale (1959) studied the convex cost transportation problem.

Books by Dantzig (1963), Charnes and Cooper (1961), and Ford and Fulkerson (1962) provide descriptions of much of the work that occurred in these exciting years. The Charnes and Cooper and Dantzig books are general linear programming texts and provide much of the ground work for the algorithms discussed in this text. The book by Ford and Fulkerson is a classic of network flow programming and for many years was the only book in the field. It looks at network flow problems from a graphical, rather than algebraic, point of view. Many of the Ford and Fulkerson algorithms, particularly the out-of-kilter algorithm, are in use to this day.

Many variations and improvements in algorithms and several books have appeared during the 1960s and 1970s. We reserve consideration of the literature of this period until specific classes of problems are discussed in later chapters.

The continuing evolution of computers during the years since the early developments has been a major impetus to the application of network flow programming and the continued algorithmic development. Most of the algorithms to be described in this book are too complex for hand calculations for any but the smallest problems. Thus, the computer is, in a practical sense, a necessary condition to the usefulness of network flow programming.

EXERCISES

1. Suppose a problem has been modeled as a transportation problem except that all arcs have flow capacities associated with them. How does this model differ from the pure, linear, minimum cost flow problem?
2. Suppose an arc from node 3 to node 4 is added in Figure 1.4 and the cost associated with the arc is 1 unit of cost per unit of flow. How does this change the optimal flow pattern in the network?
3. The absence of an arc connecting nodes 2 and 5 in Figure 1.5 implies that pairing nodes 2 and 5 is either impossible or prohibitively expensive. How would the optimal flow pattern change if an arc (2,5) with zero cost were added to Figure 1.5?
4. Suppose arc (1,3) in Figure 1.6 is, for some reason, no longer available. What is the new optimal flow pattern? Is there more than one alternate optimal flow pattern?
5. Find two other alternate optimal flow patterns for the network of Figure 1.7.
6. Suppose the capacity of arc (2,4) in Figure 1.7 is increased to 9 units of flow. What is the new optimal pattern of flow?
7. Why are there no stipulated fixed external flows in Figure 1.7?
8. What would be the effect of changing the gain parameter of arc (2,4) in Figure 1.8 to a new value of 3?

CHAPTER 2
MODELING APPLICATIONS OF
NETWORK PROGRAMMING

2.1 INTRODUCTION

In this chapter, we use the graphical network model and specific examples of physical applications to illustrate how the various classes of network flow programming may be applied to the "real world." After discussing the concept of slack node parameters, we discuss applications of the pure minimum cost flow problem. We then present example applications of the less general models discussed in Chapter 1. Finally, we consider applications of networks with arc gain parameters. Consideration of networks with convex and concave costs is deferred until Chapter 3.

For the reader's convenience, let us recall some major points of the previous chapter.

All the network models are described by arc and node parameters. For the pure minimum cost flow problem, arcs may have three parameters: a lower bound on flow, an upper bound on flow, and a cost-per-unit flow. A network-with-gains problem will have an additional arc parameter, the arc gain. Associated with each node is a fixed external flow parameter that is positive if flow is to enter the network at the node, negative if flow is to leave the network at the node, and equal to zero if flow neither enters nor leaves the network at the node.

2.2 SLACK NODE PARAMETERS

It is natural at this point to add two additional node parameters, the slack external flow and the cost-per-unit slack external flow. Slack external flow is *not* required by the network. However, it may be achieved up to a specified maximum amount independent of a node's required fixed external flow. One

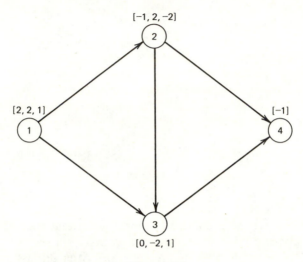

[Fixed external flow, maximum slack external flow, slack cost]

Figure 2.1
Network with Slack Node Parameters

need only pay the cost-per-unit flow that is associated with that "slack flow." Just as with fixed external flows, the maximal slack external flow is positive (negative) if the slack external flow is to enter (leave) the node. For example, in Figure 2.1, at least two units of flow enter the network at node 1. If costs are in dollars per unit flow, up to as many as two more units of flow may enter at node 1 at a cost of $1 per unit of flow. A required one unit of flow must leave at node 2 and as many as two units of flow at a price of *minus* $2 per unit flow (actually a profit of $2!) may enter at node 2. Interpretation of nodes 3 and 4 follows in a similar fashion.

2.3   THE PURE, LINEAR, MINIMUM COST FLOW PROBLEM— EXAMPLE APPLICATIONS

The Tanglewood Chair Manufacturing Co.

The Tanglewood Chair Manufacturing Co. has four plants located around the country. The fabrication and assembly cost per chair and the minimum and maximum monthly production for each plant are shown below.

Plant	Cost per Chair	Maximum Production	Minimum Production
1	$5	500	0
2	7	750	400
3	3	1000	500
4	4	250	250

The company obtains the twenty pounds of wood required to make each chair from two suppliers who have agreed to supply any amount ordered. In return, the company guarantees purchase of at least eight tons of wood per month from each supplier. The cost of wood is $0.10/lb from Supplier 1 and $0.075/lb from Supplier 2. The shipping cost in dollars per pound of wood from each supplier to each plant is shown below.

	Plant 1	Plant 2	Plant 3	Plant 4
Wood 1	0.01	0.02	0.04	0.04
Source 2	0.04	0.03	0.02	0.02

The chairs are sold in New York, Houston, San Francisco, and Chicago. Transportation costs in dollars per chair between the plants and the cities are:

		City NY	City H	City SF	City C
	1	1	1	2	0
	2	3	6	7	3
Plant	3	3	1	5	3
	4	8	2	1	4

The following table shows the minimum demand that must be satisfied, the maximum demand, and the selling price for chairs in each city.

City	Selling Price	Maximum Demand	Minimum Demand
NY	$20	2000	500
H	15	400	100
SF	20	1500	500
C	18	1500	500

We desire to answer the following four questions consistent with minimizing the total cost while satisfying all of the restrictions cited above.

1. Where should each plant buy its raw materials?
2. How many chairs should be made at each plant?
3. How many chairs should be sold at each city?
4. Where should each plant ship its product?

The network model for this problem is shown in Figure 2.2. Since only one kind of flow is allowed, we must express all flows either in pounds of wood or in

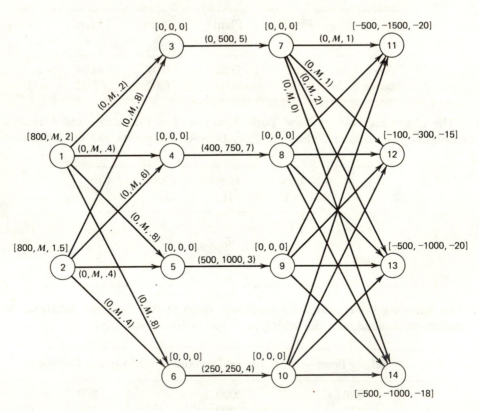

Figure 2.2
Chair Manufacture Model

chairs. For this problem, we have arbitrarily decided to express the flows in terms of chairs. Necessarily, wood-related costs and bounds in Figure 2.2 are expressed in terms of equivalent chairs. Note that M, denoting a very large number, has been used as the upper bound on arcs that represent unbounded paths for flow.

When the minimal cost flows are obtained, the arc flows will correspond to specific shipments of chairs through the system. Clearly, the flow on arc $(5, 9)$ will represent the production in plant 3, and the quantity of chairs shipped from plant 1 to Houston will be represented by the flow on arc $(7, 12)$.

As we will see in Chapter 3, a characteristic of the solution of this type of problem is that, when all arc capacities and all fixed external flows at the nodes are integer, the optimum solution will be integer. Thus, for this example problem we know that any solution will only prescribe movements of completely assembled chairs.

Each physical element of the problem is represented by a node or arc: because of modeling considerations, the converse is not necessarily true. Thus nodes 1 and 2 represent wood sources 1 and 2, respectively. Nodes 3 through 10 are present only for modeling convenience. Thus, the arcs joining nodes 1 or 2 to nodes 3, 4, 5, and 6 depict transport from the wood sources to the plants. Similarly, arcs $(3, 7)$, $(4, 8)$, $(5, 9)$, and $(6, 10)$ represent the four plants and the arcs connecting nodes 7 through 10 to nodes 11 through 14 depict transport from the plants to the cities. Nodes 11, 12, 13, and 14 represent the four cities.

One recommended modeling rule is that parameters of each arc or node should relate to only the physical subsystem represented by that arc or node. Thus arc $(7, 11)$ represents the transportation link from plant 1 to New York. The optimal solution would not change if we violated this "rule" by setting the capacity for this arc to plant 1's production capacity of 500 (rather than M). However, the conceptual clarity and dynamic flexibility of the network model would be unnecessarily obscured and constrained. For instance, if the production capacity of plant 1 were increased above 500, we would have to remember to increase the capacity of arc $(7, 11)$ by the same amount.

Notice how the node parameters have been used in the model to easily incorporate the minimal wood purchases at nodes 1 and 2 and the minimal demands at nodes 11, 12, 13, and 14 into the model. At nodes 1 and 2, the company guarantee of eight tons of wood or, equivalently, 800 chairs is required by the nodes' fixed external flows. Since no upper limit on wood supply is present, $+M$ is assigned as slack external flow for nodes 1 and 2. In a like manner, at nodes 11 through 14, the minimal demand for chairs at each city is required by the fixed external flows. The maximum slack external flows are assigned the magnitude of the *difference* between the maximum and minimum demand and a negative sign is present to stipulate that the slack flow leaves the network.

Hickory–Dickory Loading, Inc.

The subject of conveyor loading gives rise to our next example. Suppose the units produced by a plant's three manufacturing machines are moved by conveyor to shipping docks. There are three separate conveyor belts: belt A, which can carry output from machines 1 and 3 to docks 4 and 5; belt B, which can carry the output of machine 2 to docks 4 and 5, and belt C, which can carry the output from machines 2 and 3 to dock 6. The belts' directional movements are shown in Figure 2.3. Past experience has shown that a belt can move no more than 50 units per hour and a dock can ship no more than 60 units per hour.

If we desire to minimize the total distance traveled by the produced units, a minimum cost network model for this problem is shown in Figure 2.4, where P_1, P_2, and P_3 are the given production rates in units per hour, respectively. In this model, the manufacturing machines are represented by nodes 1, 2, and 3. Arcs $(1,7)$ and $(3,8)$ are dummy arcs representing the act of placing each machine's production on the associated conveyor belt wherein zero distance of travel is

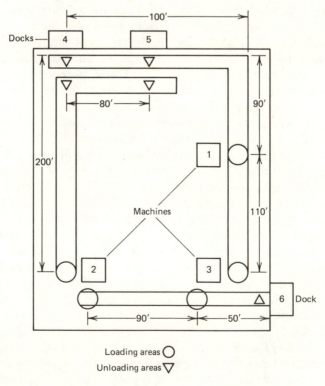

Figure 2.3
Schematic Diagram of the Conveyor System

assumed. Nodes 7 and 8 are the merge points for production from machines 1 and 3 and machines 2 and 3, respectively. Nodes 4, 5, and 6 are the shipping docks. All arcs except (1,7) and (3,8) are segments of conveyor belt with the stipulated capacities and unit costs. The solution for this model will indicate the optimal flows on the various conveyor belts represented by the arcs.

Hirem-Firem Temporaries, Inc.

The Hirem-Firem company supplies clerical employees upon demand to the various business concerns in Houston, Texas. The best available estimates of demand for the next four months are:

Month	1	2	3	4
Demand	22	15	32	8

The costs to hire an employee at the start of each month and the costs to lay off an employee at the end of each month are:

Month	1	2	3	4
Dollar/hire	170	290	150	80
Dollar/fire	220	230	110	120

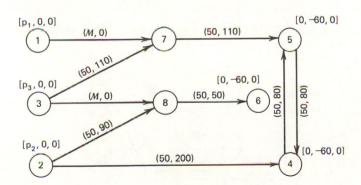

[Fixed flow, maximum slack flow, slack cost]
(Capacity, cost)

Figure 2.4
Conveyor Loading Network Model

All employees are paid at the same rate of $250 per month. A history of bankruptcy in similar companies that have not satisfied customer demands makes it mandatory that the estimated customer demands be satisfied. Temporary clerical help is at times hard to find. Once employees are hired, they will work until laid off and currently employed individuals have no discernible effect on the number of "new hires" available. The best estimates of possible new hires in the four months are 25, 10, 24, and 10 for months 1 through 4, respectively. It is assumed that all employees hired during this period will be laid off at the end of the fourth month.

The company naturally desires to obtain the optimal policy of hirings and layoffs at the beginning and end of each month such that their total cost of providing employees to their various customers is minimized. The network formulation of this problem appears in Figure 2.5.

In Figure 2.5, nodes 1 through 4 represent the beginning of each month. Positive slack external flows provide the possibility of hiring at these points in

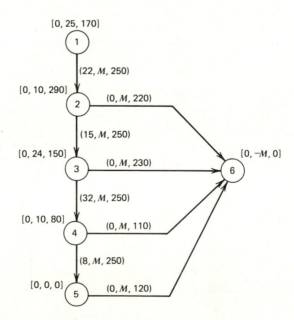

Figure 2.5
Network for Hirem-Firem Temporaries, Inc.

time. The flow on arcs $(1,2)$, $(2,3)$, $(3,4)$, and $(4,5)$ indicates the number employed during the four months. The lower bounds enforce the requirement that demand be met. The flows on arcs entering node 6 represent the employees laid off at the end of each month.

Bromwich Restaurant Services

The Bromwich company, Helmut Bromwich—President, is in the business of catering to restaurants all over the city. The business consists mostly of providing customers with a daily supply of tableclothes, uniforms, and napkins. However, Bromwich also provides a silverware–china service and has a service and repair unit for maintenance of restaurant machinery, (i.e., stoves, dishwashers, etc.).

Since napkins are one of the "big runners" in terms of numbers and expense, the optimal napkin handling policy is of great interest. Since Sunday, Monday, and Tuesday are slow days, we can restrict our attention to the remaining days of the week. Past history has shown that typical daily demands for napkins are:

Day	W	T	F	S
Demand	450	650	975	850

Bromwich can provide napkins through new purchases or by sending them to the Pathway Industrial Laundry. Pathway provides quick overnight service or slow one-day service. For various reasons such as competitive demand for services and napkins, the purchase price and laundry costs vary according to the day of the week. The costs are given below in terms of dollars per napkin.

Day	Purchase	Fast Laundry	Slow Laundry
W	1.00	0.75	0.60
T	1.25	1.35	0.95
F	3.75	1.75	1.75
S	2.25	1.25	1.10

The problem is further complicated, from Bromwich's viewpoint, by the OSHA ruling that all napkins used in any one week must be destroyed at the end of that week so as to minimize the spread of communicable diseases.

The network for the solution of this problem is given in Figure 2.6. In Figure 2.6, node 1 represents the store that supplies new napkins, nodes 2 through 4 represent the laundry, nodes 6 through 9 represent the customers, and node 5 represents the dumping point for surplus napkins not used on Saturday. All arcs originating at node 1 represent the flow of new napkins to customers on each of the four days. Arcs (2,7), (3,8), and (4,9) represent the flow of quick laundered napkins and arcs (2,8) and (3,9) represent the flow of slow laundered napkins. Arcs (2,3), (3,4), and (4,5) represent dirty napkins that are held out of the cleaning process until the next day. Note that no provision is made for stocking new or laundered napkins for use on later days. Bromwich does not have storage space for this purpose.

Amalgamated Grain Export, Limited

AGE is an established agricultural export firm centered in London, England, with branch offices in every major grain-producing country in the world. Recent

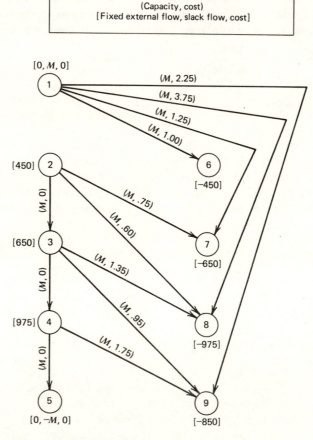

Figure 2.6
Bromwich–Pathway Napkin-Handling Network

crop failures in the USSR have given rise to an unusual opportunity for humanitarian service on the part of AGE (and also, incidentally, for large profits!).

Since it is probable that the next wheat harvest will be more successful, this opportunity will exist only for the next five months. The world grain commodity market is highly variable and competitive. This causes month-to-month price changes and also limits the amount of warehouse space available for storage of grain.

The company's senior bookkeeper has compiled the following table of information.

Month	Purchase Cost ($/Ton)	Selling Price ($/Ton)	Holding Cost ($/Ton)	Purchase Availability (Tons)	Warehouse Availability (Tons)	Demand (Tons)
1	85	95	2	6000	5000	3500
2	70	90	4	7000	4000	2500
3	105	100	6	5000	4000	4000
4	95	100	4	3000	6000	6000
5	103	110	4	3000	7000	4500

Figure 2.7 shows the network that AGE used to minimize the cost of providing the wheat to the USSR. In Figure 2.7, nodes 1 through 5 represent the supply points for wheat (warehouses and wheat suppliers) in each month. Nodes 6 through 10 represent the demand points in the USSR for each month. Arcs (1,2), (2,3), (3,4), and (4,5) reflect the cost of storing surplus grain in each month and, finally, arcs (1,6), (2,7), (3,8), (4,9), and (5,10) reflect the flow of grain from the supply points to the demand points.

Joe's Tavern

Joseph R. Smithey owns and runs a tavern on the "main drag" adjacent to the Big State University campus. For this reason, the majority of his clientele are university students. Joe has an understanding with the University Police Department. Since the officers have been around a little while, they know that college students have a tendency to be somewhat more rowdy than ordinary bar customers. However, the police require that Joe not let his customers get so drunk that they fall off their bar stools.

Joe has somewhat of a dilemma. His continued economic livelihood depends upon both happy customers and happy policemen.

Here is a situation typical of most any night at the tavern.

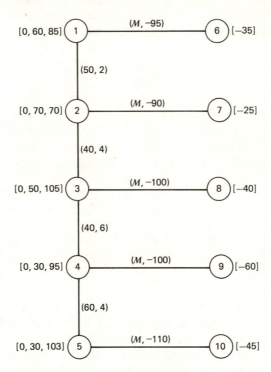

Figure 2.7
AGE Wheat Network

Four thirsty customers from the Brazos Coop are sitting at the bar with 40 oz of Jack
   Beem to divide between them. (Needless to say, none will be left when the evening is
   over.) Since Joe will be pouring, he largely controls the amount of Jack Beem that each
   customer will receive.

Being a good host, Joe knows each customer should receive at least 3 oz of the liquor.
   However, since customer 1 bought the bottle it is certain he will complain unless he
   receives at least 15 oz. Further, customer 3 has liver problems and should receive no
   more than 5 oz.

The probability that customer i will *not* fall off his stool after drinking x_i
ounces of Beem is

$$P_i = e^{-\lambda_i x_i}$$

where

$$\lambda_1 = .01, \qquad \lambda_2 = .05, \qquad \lambda_3 = .10, \quad \text{and} \quad \lambda_4 = .01$$

How many ounces of Beem should Joe give each customer in order to maximize the probability that all four customers stay on their stools?

Clearly the probability that all four customers stay on their stools is $P = P_1 * P_2 * P_3 * P_4$ and we want to maximize the value of P. This *nonlinear* "objective function" is not amenable to network flow programming techniques. However, recalling that a product of variables may be expressed as the sum of the logarithms of the variables suggests a transformation that will make our objective function suitable for solution by network techniques.

$$\ln P = \ln P_1 + \ln P_2 + \ln P_3 + \ln P_4$$

Therefore,

$$\ln P = -0.01 x_1 - 0.05 x_2 - 0.10 x_3 - 0.01 x_4$$

Maximizing P is equivalent to maximizing $\ln P$, which is equivalent to *minimizing* $-\ln P$. Therefore if Joe allots his four customers amounts of liquor such that the weighted sum

$$0.01 x_1 + 0.05 x_2 + 0.10 x_3 + 0.01 x_4$$

is minimized, he will maximize the probability that no one falls off a stool.

Figure 2.8 illustrates the network that Joe used to arrive at the optimal distribution of Jack Beem. Nodes 1 through 4 correspond to the customers and node 5 corresponds to the bottle of Beem. The four arcs represent the weighted "pouring operations" by Joe to each of the four customers.

2.4 THE TRANSPORTATION PROBLEM—EXAMPLE APPLICATIONS

Generous Electric—Southwest Refrigerator Division

Generous Electric owns primary warehouses located near its production plants and secondary warehouses in most major cities in the United States. As stocks of refrigerators are depleted in the secondary warehouses, a demand is created for replenishment from the primary warehouses. Company economists have arrived at the unit costs associated with transporting refrigerators from any primary warehouse to any secondary warehouse. Using the information below, the company wants to derive a replenishment policy that will minimize the cost of updating the secondary warehouse inventories.

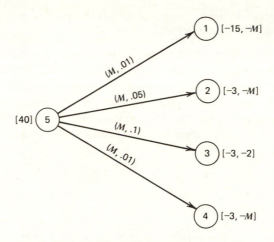

[Fixed external flow, slack external flow]
(Capacity, cost)

1 [−15, −M]

(M, .01)

(M, .05)

2 [−3, −M]

[40] 5

(M, .1)

3 [−3, −2]

(M, .01)

4 [−3, −M]

Figure 2.8
Joe's Apportionment Network

TRANSPORT COST, DOLLARS PER UNIT

Secondary Warehouses	Primary Warehouses		
	Houston	Dallas	Lufkin
Austin	4	7	6
Corpus Christi	6	9	7
Oklahoma City	6	4	5
Houston	2	5	5
San Antonio	4	9	7
Nacogdoches	3	6	7

City	Demand	City	Availability
Austin	49	Houston	173
Corpus Christi	38	Dallas	68
Oklahoma City	17	Lufkin	44
Houston	64		
San Antonio	57		
Nacogdoches	9		
Total	234	Total	285

The network used for obtaining the optimal replenishment policy is given in Figure 2.9. In Figure 2.9, nodes 1 through 3 are the primary warehouses and nodes 4 through 9 are the secondary warehouses. All arcs simply represent the transportation links between the primary and secondary warehouses that they connect.

Node 10 is a "dummy" secondary warehouse. Recall that we stated in Chapter 1 that "the sum of the external flows over all nodes is zero." Since the number of refrigerators available at the primary warehouses exceeds those needed by the secondary warehouses, we must provide a dummy secondary warehouse with a fixed external flow of -51 refrigerators; that is, total demand equals total supply.

The Battle of Clesius

General Ozymandias has just received word that the Visigoths are on a forced march and will attack the northern provinces in three days unless a sufficient force can be gathered in time to repel the invaders.

The general has three message runners in his camp and can communicate by means of signal flags to three other nearby encampments that have two, one, and two runners. No other encampments can be contacted except by message runners. If the other camps receive the message, they will send troops as quickly as possible. Of course, the sooner the message is received the more troops any camp can get to the battlefield on time. The following table gives the expected number of troops that each camp can provide dependent upon the message source.

EXPECTED NUMBER OF TROOPS

Message Source	Frontier Camps											
	1	2	3	4	5	6	7	8	9	10	11	12
1	225	200	210	150	60	175	0	10	25	25	0	190
2	200	210	210	25	90	90	22	0	100	100	0	120
3	210	220	225	75	50	160	75	65	50	160	35	65
4	75	90	95	100	75	135	90	50	75	200	25	175

Figure 2.10 gives the transportation network that General Ozymandias used to maximize the expected number of troops present at the battlefield in time to

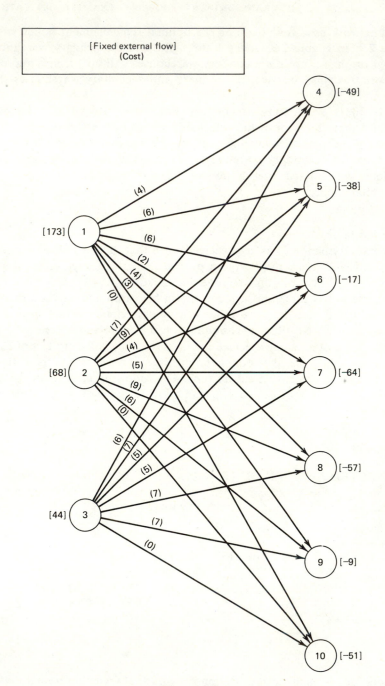

Figure 2.9
Generous Electric Distribution Network

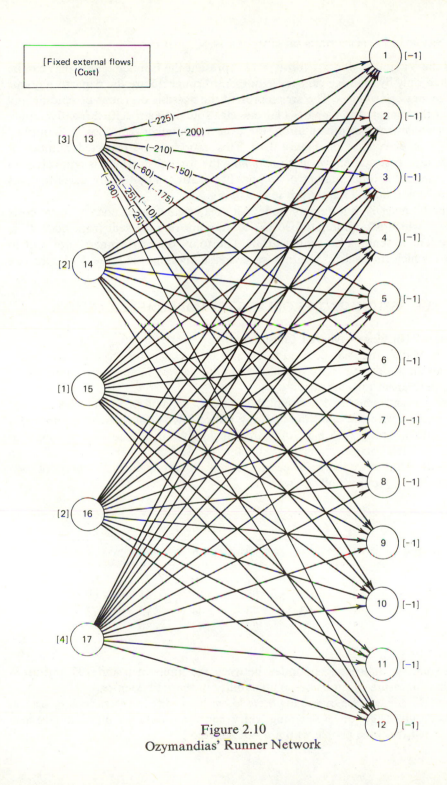

Figure 2.10
Ozymandias' Runner Network

27

fight the Visigoths. Nodes 1 through 12 represent the frontier camps that can be reached only by message runners. Nodes 13 through 16 are the departure points for the message runners. The arcs represent the possible decisions of sending any one of the runners from a particular message source to a specified frontier camp.

For consistency of presentation, the expected number of troops is multiplied by -1 in every case on Figure 2.10. This expresses the expected number of troops as a cost and allows us to *minimize* the total cost, which is equivalent to maximizing the total number of troops. For clarity, only the arc costs for arcs originating at node 13 are given in the figure.

Finally, node 17 is a dummy node. All arcs originating at node 17 have costs of zero, which is appropriate because a "dummy runner" sent from node 17 is equivalent to sending no runner at all. The four "dummy runners" are sent to camps which, in reality, will not receive the message that the Visigoths are attacking.

2.5 THE ASSIGNMENT PROBLEM—EXAMPLE APPLICATIONS

Samuel Hitchings—Marital Consultant

Date-Match, Inc., has decided to expand their capabilities to offer not only a computerized dating service but also a similar service for persons interested in more permanent attachments.

Using slightly modified personal data forms, the data processing department has come up with the following "pairing indices" for a four-person group and a seven-person group who have indicated an interest in this service. In order to avoid possible biases, knowledge of the gender of either group has been withheld. All that we know for certain is that one group consists wholly of men and the other group wholly of women.

Group 2	\multicolumn{7}{c}{Group 1}
	1	2	3	4	5	6	7
1	35	95	87	0	72	63	10
2	53	90	85	15	65	69	75
3	91	93	73	44	75	73	68
4	15	20	22	45	30	42	12

The pairing indices are bounded between the limits of 0 and 100 and are a weighted measure of probable compatability in married couples.

Sam Hitchings, fresh out of Big State U. with an M.S. in Operations Research, has been given the job of deciding which couples should get married. (He also gets to tell the three people who are not selected.)

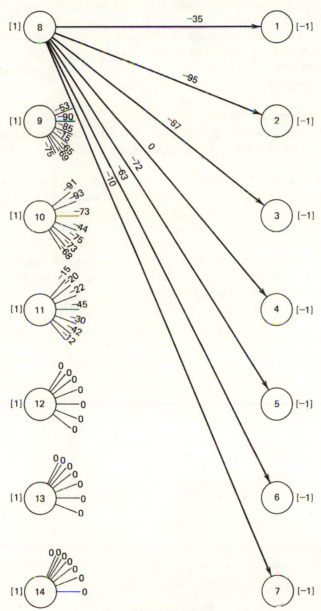

Figure 2.11
Marryin' Sam's Pairing Network

29

Among other approaches, one method of selecting the pairings would be to maximize the sum of the pairing indices of the four selected pairs. Figure 2.11 pictures an abbreviated version of the network that Sam used for that purpose. Again, the negatives of the indices are used as costs whose total is to be minimized.

In Figure 2.11, nodes 1 through 7 represent the clients in Group 1, nodes 8 through 11 represent the clients in Group 2, and nodes 12 through 14 represent "dummy" members of Group 2 who will be paired with the members of Group 1 who are not selected.

Richardson's Machine Shop

Richardson's Machine Shop makes, by special order, tools, dies, and prototypes from design blueprints. Two machines are available to perform the shop's work.

Today, six jobs are to be performed. Since the jobs are performed by contract for a set amount of payment, the shop supervisor wants to minimize the average completion time for all the jobs. (This is equivalent to minimizing the sum of all completion times.)

The machines have different capabilities. This results in some jobs' requiring more time on one machine than on the other. In fact, job number 4 cannot be performed on machine number 2. The times required by each machine to perform any of the jobs are given below.

Machine	Job					
	1	2	3	4	5	6
1	3	7	9	2	9	4
2	7	12	10	M	6	6

Since there are more jobs than machines and all jobs must be performed, some of the jobs will have to wait on other jobs.

One possible but nonoptimal assignment of the jobs to the machines is

The sum of the completion times for this assignment is:

$$3*2+2*9+1*4$$
$$+3*7+2*12+1*10=83$$

Observe that if p_{ij} is the time required by job i on machine j and job i is the lth job among the K_j jobs assigned to machine j, the contribution of job i to the sum of the completion times is

$$(K_j - l + 1)p_{ij}$$

The maximum value of K_j for any machine j is the number of jobs, n.

These observations lead to the network formulation presented in Figure 2.12. Nodes $11, 21, \ldots, 42$, and 52, which appear on the right side of Figure 2.12, represent the various possible scheduled positions on machines 1 and 2. For example, node 41 represents the fourth position on machine 1 or, in general, node lj represents the lth position on machine j. Nodes 1 through 6 represent the six jobs and nodes 7 through 10 and 13 represent dummy jobs that take no time and fill machine positions not filled by real jobs.

The arcs of Figure 2.12 represent the possible assignments of each job to each position on the machines. Most of the arcs are omitted from Figure 2.12. However, with one exception, the network model it describes has arcs from each node on the left to every node on the right. The exception is that *no* arcs extend from node 4 to nodes 12, 22, 32, 42, or 52.

A complete table of the arc costs associated with Figure 2.12 is given below.

						Machine Position Nodes					
		11	21	31	41	51	61	12	22	32	42	52
	1	18	15	12	9	6	3	35	28	21	14	7
	2	42	35	28	21	14	7	60	48	36	24	12
	3	54	45	36	27	18	9	50	40	30	20	10
	4	12	10	8	6	4	2	M	M	M	M	M
Job	5	54	45	36	27	18	9	30	24	18	12	6
nodes	6	24	20	16	12	8	4	30	24	18	12	6
	7	0	0	0	0	0	0	0	0	0	0	0
	8	0	0	0	0	0	0	0	0	0	0	0
	9	0	0	0	0	0	0	0	0	0	0	0
	10	0	0	0	0	0	0	0	0	0	0	0
	13	0	0	0	0	0	0	0	0	0	0	0

In general, the arc cost for arc (i, lj) may be computed as $(n - l + 1)p_{ij}$.

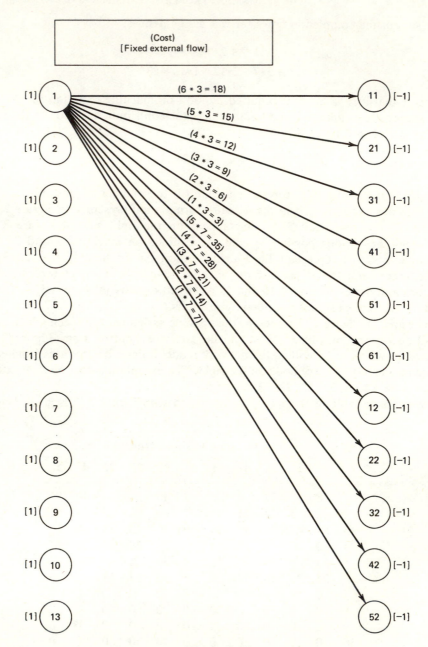

Figure 2.12
Machine Shop Scheduling Network

32

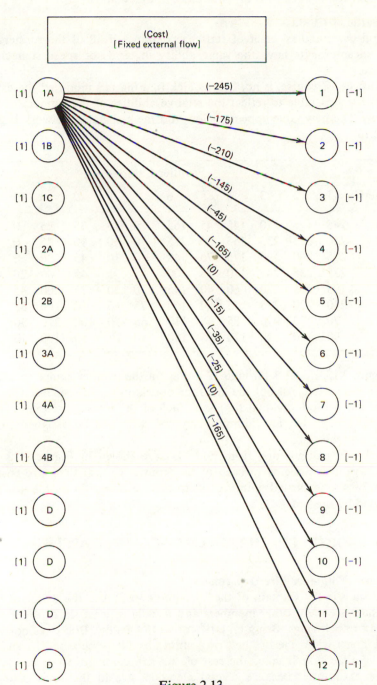

(Cost)
[Fixed external flow]

[1] 1A   (−245)   1 [−1]

[1] 1B   (−175)   2 [−1]

[1] 1C   (−210)   3 [−1]

[1] 2A   (−145)   4 [−1]

[1] 2B   (−45)   5 [−1]

[1] 3A   (−165)   6 [−1]

[1] 4A   (0)   7 [−1]

[1] 4B   (−15)   8 [−1]

[1] D   (−35)   9 [−1]

[1] D   (−25)   10 [−1]

[1] D   (0)   11 [−1]

[1] D   (−165)   12 [−1]

Figure 2.13
Ozymandias' Improved Runner Network

33

The Battle of Clesius—Revisited

General Ozymandias' chief of staff has inquired if all of the runners at the various encampments have the same ability in terms of speed, tenacity, and endurance.

The answer, of course, is no. Some quick figuring has modified the expected-number-of-troops table to reflect the relative abilities of the runners at each of the camps. The new table appears below and the associated network is given in Figure 2.13.

Runners	Frontier Camp											
	1	2	3	4	5	6	7	8	9	10	11	12
1A	245	175	210	145	45	165	0	15	35	25	0	165
1B	220	225	235	155	30	225	0	10	10	25	0	190
1C	205	200	195	175	75	140	0	10	45	20	0	235
2A	235	235	245	110	75	200	85	95	45	60	20	35
2B	195	205	220	60	40	145	70	50	75	200	45	90
3A	210	220	225	75	50	160	75	65	50	160	35	65
4A	70	70	70	85	65	110	65	90	100	210	30	190
4B	90	110	120	120	95	150	110	65	50	150	10	140

In Figure 2.13, nodes 1 through 12 are again the frontier camps to which the runners are directed. All but the last four nodes on the left side of Figure 2.13 represent a specific individual runner at each of the message sources. The final four nodes on the left are "dummy runners" who will be assigned to camps receiving no message.

Only the arcs originating from node 1A are shown in Figure 2.13 but all runners have arcs directed to each of the frontier camps. Each arc cost is the negative of the expected number of troops resulting from the assignment represented by the arc.

2.6 THE SHORTEST PATH PROBLEM—EXAMPLE APPLICATIONS

Wemperly Volunteer Fire Department

Bob Bishop is the Captain of the Wemperly VFD and the driver of the fire truck. There are eight other members and usually at least six of them respond when the fire bell rings. Being an efficient sort of fellow, Bob does not want to leave his home when the fire bell rings until absolutely necessary. Usually, the amount of time that it takes the rest of the fire company to assemble at the station is 15 minutes because several members live in the surrounding rural country about Wemperly.

The question Bob must answer is, "How many minutes after hearing the fire bell can he wait before leaving his house if he is to arrive no later than the last member of the company?"

Using the map that the local county agricultural agent gave him, Bob constructed a network representation of the roads that he could use to drive from his house to the fire station. Since Bob was very familiar with the roads, he could estimate quite closely the amount of time it would take to pass over any segment of road in the area. With this knowledge, he could use the network to find the path that would take him the least total time and thus answer his question.

The network he used appears in Figure 2.14, where node 1 is Bob's house, node 14 is the fire station, nodes 2 through 13 are road intersections, and the arcs are road segments between the intersections. Each arc cost is Bob's estimate of the time it would take to traverse that segment of road.

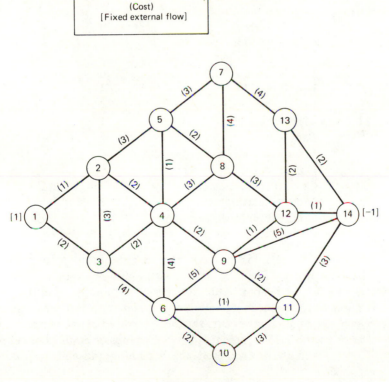

Figure 2.14
Wemperly VFD Network

Cooperative Construction, Inc.

Coop Construction, one of the newer firms in town, has just been awarded a contract to build a new field house at the Haayes Consolidated High School. Because of rather severe penalties for failure to meet the contractual completion date, the company president has instructed the Planning Department to determine which parts of the construction project might be particularly important to completing the project on time.

As a preliminary step, the Planning Department has broken down the project into mutually exclusive subprojects and has recorded not only the time required for each subproject but also the precedence relationships joining the subprojects. This information is given below. Figure 2.15 presents a network representation of that information.

Subproject	Description	Days Required	Predecessor Subprojects
A	Land survey	1	—
B	Laying foundation	2	A
C	Erect walls	4	B
D	Construct roof	7	C
E	Lay wood floor	5	B
F	Finish interior walls	6	C
G	Install plumbing	4	C
H	Install electricity	4	D
I	Mount basketball standards	1	D
J	Paint court boundaries	2	E
K	Install heating/cooling system	6	G, H
L	Construct indoor track	6	E
M	Construct stadium seats	4	E, F
N	Mount electronic scoreboard	2	F
O	Construct refreshment bar	2	E, F, G

As indicated in Figure 2.15, the arcs represent the various subprojects that make up the total project. The arcs labeled with d and having a cost of zero are dummy arcs which have been inserted to remove any ambiguity in the graphical characterization of the precedence relationships. The arc costs are expressed as the negative of the time requirements for the associated subprojects.

The nodes of Figure 2.15 represent either beginnings or endings of subprojects or both. Nodes 1 and 14 also represent the beginning and end, respectively, of the entire project.

To assist in answering the president's questions, the Planning Department found the shortest path in the network of Figure 2.15. This corresponds to the

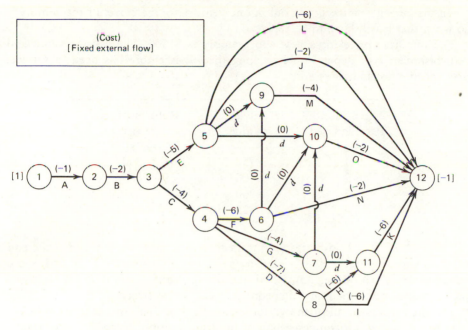

Figure 2.15
Haayes Field House Construction Network

longest path in the construction project. Any subproject on this path can tolerate *no delay* without delaying the project as a whole. The Planning Department also conducted a complete sensitivity analysis of the network to obtain the amount of delay that each subproject not on the longest path could tolerate without delaying the total project completion.

Mid-States Engineering Company

Mid-States is a general civil engineering and construction company. One of its speciality operations is to cut "rain grooves" in heavily traveled streets in two large cities. The rain grooves are cut parallel to the major axis of the roadway and assist traction in wet weather by providing conduits to carry away the rain water while allowing most of the road surface to remain relatively dry.

Naturally, the large subassembly used for this cutting operation dulls and wears down on a regular basis, about every three months. When this happens, a decision must be made. The company can either purchase a new subassembly or refurbish the old one. If a new subassembly is purchased, the old one is sold to a junk metal company for its salvage value.

Because of new competitors, this specialty service will no longer return a sufficiently high profit under a new contract. Therefore, it will be terminated when the current contract runs out in one year. Also, the current subassembly is so worn that it will have to be replaced.

The purchase price for a new subassembly is $1500. The salvage value and refurbishment cost depend on how long the subassembly has been in service. The specific dollar figures are:

Three-Month Periods Used	Salvage Value	Refurbishment Cost
1	750	300
2	500	600
3	200	875
4	25	1000

The network that Mid-States used to decide what decision to make at the end of each three-month period is shown in Figure 2.16. Nodes 1, 2, and 3 are simultaneously period beginning and end points while nodes 0 and 4 represent the start and end of the remaining contract year, respectively.

The arcs represent all of the various sequences of decisions that are possible over the planning horizon consisting of the four 3-month periods. For example, arc (0,2) represents buying a new subassembly at the start of the first period, refurbishing it at the start of the second period, and finally selling it for its salvage value at the end of the second period.

The cost associated with arc (0,2) is the sum of the purchase cost of $1500, the one period refurbishment cost of $300, and the salvage "cost" for a two-period usage of −$500. Thus $1500+$300−$500=$1300. The optimal sequence of decisions is that sequence associated with the shortest path through the network of Figure 2.16.

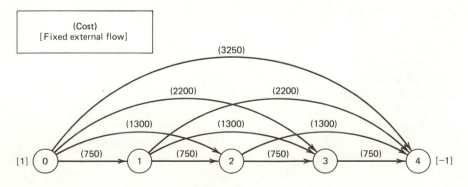

Figure 2.16
Mid-States Rain Groover Network

2.7   THE MAXIMUM FLOW PROBLEM—EXAMPLE APPLICATIONS

Krumley Painting and Wallpapering

The Krumley family has painted almost every house in Bula, Texas (Pop. 498). The family is large enough that there are three crews of workers available throughout the spring and summer months. Edward Krumley, the family patriarch and head of the company, has contracted for five jobs during this season. Each job may be started on or after various dates and must be completed on or before agreed-upon dates in the future. Mr. Krumley has been in the business long enough to estimate, to the nearest week, the required time to perform any of the jobs.

The data for each of the jobs is given below.

Job	Start Date	Completion Date	Required Time (Weeks)
A	April 1	April 28	2
B	April 14	May 12	4
C	April 1	May 26	7
D	April 21	May 26	3
E	April 1	May 26	6

Jobs need not be worked on continuously from start to finish. However, Mr. Krumley makes it a practice to assign his crews in such a way that in any given week they will be working on only one job. Further, no two crews ever work on the same job at the same time.

Using this data, Mr. Krumley wants to know if it is possible to finish all the jobs by their completion dates. Figure 2.17 gives one kind of network that could be used to answer Mr. Krumley's question. Nodes 1 through 8 represent each of the eight weeks occurring between April 1 and May 26. Nodes 9 through 13 represent jobs A through E, respectively. Nodes 0 and 14 are the source and sink for the maximum flow problem model.

The arcs originating at node zero all have a capacity of 3, reflecting the fact that three crew-weeks of work are available in each week. The arcs terminating at node 14 all have capacities equal to the number of crew-weeks required to complete the job associated with the origin node of each arc.

All other arcs have a capacity of one unit and represent the possible allocation of one crew-week from a particular week-long period to a specific job. Note that weeks 1 and 2 are linked only to jobs A, C, and E, week 3 is linked to all but job D, and weeks 4 through 8 are linked to all of the jobs.

If it is possible to finish all the jobs on time, the maximal flow in the network will equal 22; that is, all the arcs terminating at node 14 will have flows equal to their capacities.

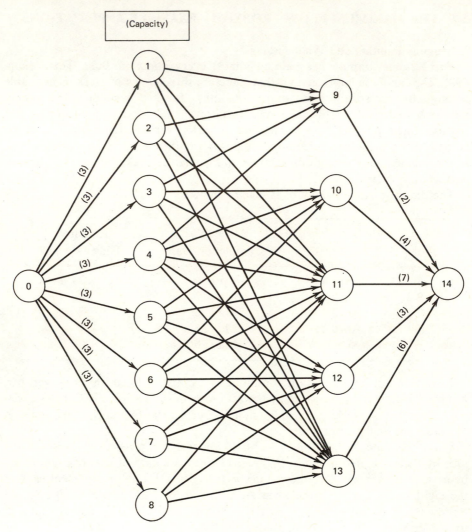

Figure 2.17
Krumley Scheduling Problem

The ACME Bookbindery

The president of the ACME Bookbindery is concerned with productivity in his Manor Road plant. There are nine steps in binding a book: (1) open boxes, (2) collate pages, (3) trim pages, (4) glue backs, (5) attach hardback binding, (6) emboss binding with title, (7) box books, (8) attach mailing labels, and (9) load on trucks.

Step	Number of Stations	Capacity of Station (Books/Day)
1	1	76
2	2	10, 12
3	2	10, 13
4	4	7, 9, 3, 2
5	2	22, 23
6	1	42
7	3	15, 6, 4
8	2	10, 15
9	1	105

Using the information above, the supervisor of the Manor Road plant has been charged with the responsibility of both determining the maximum number of books that can be bound in a day and making recommendations that will increase this output at as small a cost as possible.

Figure 2.18 gives the network the supervisor used to arrive at the suggestions he made to his boss. The nodes represent the nine steps in binding a book and the arcs are work stations associated with their origin nodes. The capacities of the arcs are the capacities of the work stations in books per day.

Operation Quarantine

The U.S. Pest Service has determined that the annual migration of the flying corn borer is about to begin again this year. Unless exhaustive measures are taken, farmers are expected to suffer severe economic loss. Over the past few years, the paths of migration of the corn borer have been extensively mapped. Using this information, the service would like to determine the least costly way to assure that every corn borer will undergo an application of insecticide.

Since the cost of "dusting" any migration path is directly proportional to the maximum number of borers that might use the path, the schematic representation of the migrating paths given in Figure 2.19 was used to answer the question

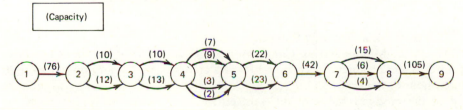

Figure 2.18
ACME Bookbindery Network

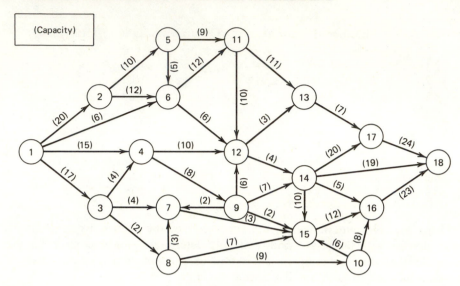

Figure 2.19
Corn Borer Migration Paths

asked above. In Figure 2.19, the nodes are junction points for the arcs, which represent the migration paths. Nodes 1 and 18 are the starting and ending points of the migration, respectively. The arc capacities are the maximum number of borers (in thousands) expected to use a particular path. These capacities are based largely on historical data.

The Pest Service formulated this problem as a maximum flow problem that upon solution yielded a minimal cut, that is, a set of migration paths of minimal total capacity that if removed from the network of Figure 2.19 would prohibit any corn borer from reaching node 18.

All migration paths in this minimal cut received extensive insecticide dusting by the service's crop duster aircraft.

2.8 NETWORKS WITH GAINS—EXAMPLE APPLICATIONS

Muskogee Manufacturing Company

Muskogee Manufacturing Company has five machine centers and produces four different products. According to union contract, the factory operates for 20 eight-hour days each month. Each machine center has the capability to produce all of the products but with different efficiencies. The table below shows the number of hours required to produce one unit of product on each machine.

Products	Machines				
	1	2	3	4	5
1	2	1.3	4	5.2	3
2	1	0.7	2.3	3	2.1
3	7	4.2	5	4	3
4	0.7	1.2	5	3	2

Product and machine information is shown below.

Product	Maximum Sales in Month	Revenue per Unit
1	100	$30
2	150	15
3	125	35
4	20	20

Machine	Production (Cost/Hour)
1	$10
2	8
3	7
4	9
5	11

Figure 2.20 presents the network Muskogee Manufacturing Company used to arrive at the monthly optimal assignment of products to machines. Nodes 1 through 4 represent factory outlet points for products 1 through 4, respectively. Notice that revenue is expressed as a negative cost and the maximum number of sales is expressed as a slack external flow.

Nodes 5 through 9 represent machines 1 through 5, respectively. It is not required that all 160 hours per month be used at any machine center. Thus, the available work time is expressed as a slack external flow with an associated cost equal to the production cost per hour.

Finally, the arcs are present not only to show the products that a machine center can produce but also to convert the flow of hours of production into units of product. The latter operation is implied by the gain parameter for each arc (i,j), which is equal in each case to the reciprocal of the number of hours required for machine i to produce one unit of product j.

The Tanglewood Chair Manufacturing Company—Revisited

It has always been assumed that each plant takes 20 lb of wood to make chairs. However, recent efficiency studies have shown that the four production plants take different amounts of wood to produce a chair: respectively, plants 1, 2, 3, and 4 really take 25, 20, 18, and 23 lb per chair.

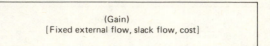

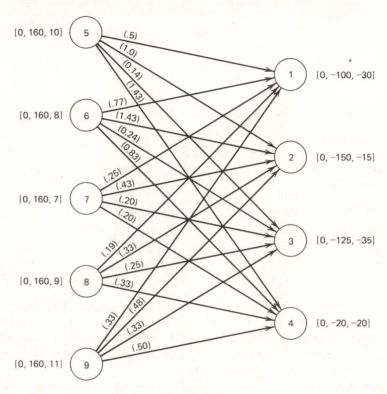

Figure 2.20
M. M. Co. Machine Assignment Network

These relative efficiencies must be taken into account to assure a truly optimal operation policy.

A modified network, using arc gain parameters to include the differing efficiencies, is shown in Figure 2.21. Figure 2.21 differs from Figure 2.2 only at nodes 1 and 2 and at the arcs originating at nodes 1 and 2. These nodes and arcs now have all of their parameters expressed in terms of tons of wood. The flows, in tons of wood, are converted to chairs by the gain parameters of the arcs originating at nodes 1 and 2.

Northwest Power and Light

Northwest Power and Light has two fields from which they obtain the fuel that powers their two electricity generating plants. Because the ore, as it comes

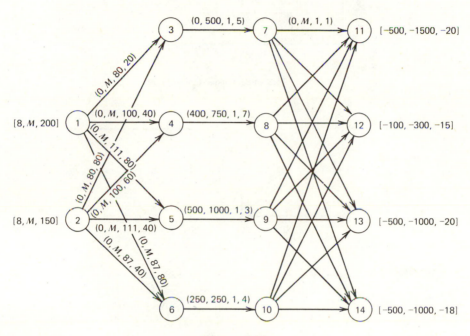

[Fixed external flow, slack flow, cost]
(Lower bound, upper bound, gain, cost)

Figure 2.21
Tanglewood Chair Network—Revisited

from the mine, is mixed with a high percentage of other impurities, it is sent to one of two locations for separation and refining before being sent to the generating plants. The costs and amounts of ore available at the two fields are shown below.

The two refineries process the mineral ore. Refinery A requires 2 units of mineral ore for each unit of refined fuel and has a capacity of 50 units of ore per year. Refinery B requires 3 units of mineral ore for each unit of refined fuel and has a capacity of 75 units of ore per year. The output of refinery B is of somewhat better quality than that of refinery A as described below. Refinery cost is one dollar per unit of ore for each refinery.

The power generating stations burn the refined fuel to produce electric power. Both stations are identical except in capacity. Each unit of the refined fuel produced by refinery B produces one unit of electric power. Each unit of refined fuel produced at refinery A produces 0.8 units of electric power. Station capacities and generating cost per unit of output power are as follows.

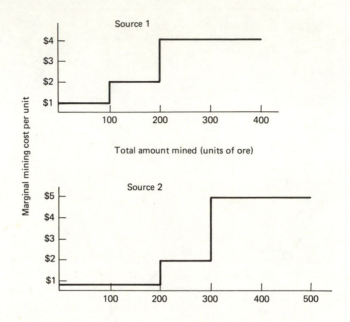

Station	Cost ($/power unit)	Capacity (power units/year)
A	1.50	10
B	1.75	30

Power demand for each of the next 5 years has been estimated at 22 and 17 units for stations 5 and 6, respectively. Distances between ore fields and refineries in miles are:

Field	Refinery A	Refinery B
1	100	200
2	80	70

Shipping cost for mineral ore is $0.01 per unit per mile. Distances in miles between refineries and power plants are as follows.

	Station	
Refinery	A	B
A	100	120
B	80	50

Shipping cost for refined fuel is $0.01 per unit per mile.

There is a transmission line that can be used to move power between the two generating stations. Power may be transmitted in either direction on this line. The length of the line is 100 miles. Transmission costs for power on this line is $0.005 per unit per mile. Each mile of transmission results in a 0.1% loss in power transmitted.

Northwest Power and Light used the network in Figure 2.22 to determine the optimal policy of mining, ore usage, and power transmission for the next five-year period. The ore fields, refineries, and power stations are partially identified on Figure 2.22. The fact that ore field 1 is represented by nodes 11 and 12, refinery A is represented by nodes 31 and 32, and station A is represented by nodes 51 and 52 completes that identification.

For each ore field, the limitation on the amount of ore extraction is given as a maximum slack external flow. The different marginal mining costs are repre-

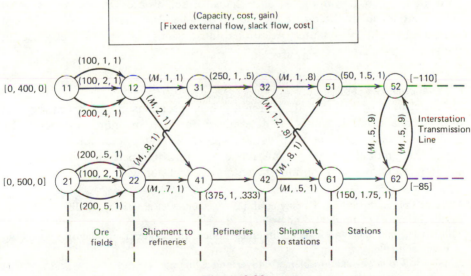

Figure 2.22
NWP & L Energy Problem

sented by multiple arcs where each different value of marginal costs has its own associated arc. The increasing nature of the marginal (arc) costs will assure that the arcs are used in the proper temporal order. The arcs representing shipment of raw ore to the refinery simply charge for that shipment by setting each arc cost equal to the product of the shipping cost per unit per mile times the number of miles covered.

The arcs joining the refinery nodes, (31,32) and (41,42), do two things.

1. Change the flow from units of raw ore to units of refined fuel by multiplying the amount of flow by the reciprocal of the amount of raw ore required to produce one unit of refined fuel.
2. Limit the amount of raw ore that is refined. The arcs associated with shipment of the refined fuel to the stations exact the appropriate cost per unit shipped and model fuel efficiency.

The station arcs, (51,52) and (61,62), limit the amount of power produced and charge for each unit of power that is produced. Finally, the interstation transmission line not only exacts its direct monetary charge but also causes approximately a 10% power degradation over its 100-mile length (because $0.9 \simeq (0.999)^{100}$).

2.9   HISTORICAL PERSPECTIVE

The literature contains a number of descriptions of applied problems for network flow programming. A nonexhaustive set of problem classes is listed next; citations to the literature are included. Several of these classes have been illustrated within this chapter. Additional illustrations appear in the exercises at the end of the chapter.

Production Scheduling

This class of problems deals with multiperiod problems in which a production or purchase schedule is to be determined for one or more products. The problems usually involve time variable costs, selling prices, and production capacities. Product inventories interrelate the individual period models, and the solution is to provide an optimal inventory control policy.

Multifacility, multiproduct: Dorsey et al. (1974), Dorsey et al. (1975), Ratliff (1976).
Multiechelon, single product: Zangwill (1969).
Multiperiod purchase and reconditioning policy (the caterer problem): Ford and Fulkerson (1962).
Application to large-scale problems: Glover and Klingman (1975).
Scheduling one product on several facilities: Bowman (1956), Prager (1957).
Multiperiod single product: Charnes and Cooper (1961).

Sequencing

Certain sequencing problems or problems of scheduling events in time are amenable to network flow programming solutions. Usually these applications are very dependent on specific characteristics of the problem under consideration. Slight variations of these characteristics may make the efficient network flow techniques not applicable and indeed most sequencing problems are computationally very difficult. Even for the difficult problems, however, network flow algorithms are often used as a subroutine in a branch and bound or dynamic programming approach.

Finding the minimum number of docking facilities to meet a given schedule of arrivals and departures: Dantzig and Fulkerson (1954).
Multivehicle tanker scheduling: Bellmore et al. (1971).
Parallel processor scheduling problems: Bartholdi et al. (1976).
Sequencing jobs on machines: Veinott and Wagner (1962), Lawler (1964).
Critical path scheduling: Kelley (1961).
Crash scheduling on project networks: Fulkerson (1961a), Falk and Horowitz (1972).

Assignment Problems

If a one-to-one assignment from one group of items to another group of items is required, an assignment problem may be applicable.

Assignment of tasks to machines to minimize total setup time: Dantzig (1963).
Optimal marriage assignment to maximize happiness: Halmos and Vaughan (1950).

Transportation

Survey of transportation applications: Potts and Oliver (1972).
Traffic assignment: Hershdorfer (1966), Jewell (1967).
Location-allocation: Cooper (1972).

Miscellaneous Applications

Bid evaluation and selection: Stanley et al. (1954), Waggener and Suzuki (1967).
Equipment replacement: Dreyfus (1960), Bennington (1974).
Application to combinatorial problems: Fulkerson (1966).
Racially balancing schools: Belford and Ratliff (1972).
Rectilinear distance facility location: Francis and White (1974).
Assignment of production of multiple products to machines of differing efficiencies (machine loading problem): Charnes and Cooper (1961).
Water resources planning: Texas Water Development Board (1972), Texas Water Development Board (1975), Bhaumik and Jensen (1974).
Physical network natural flow problems (i.e., electricity or water): Birkhoff and Diaz (1956), Minty (1960), Hu (1966), Hu (1967).

Electric power planning: Malek-Zavarei and Aggarwal (1972).
Cash flows: Charnes and Cooper (1961).

EXERCISES

1. For disposal of 210 gallons of a highly corrosive liquid, the Cochard Chemical Company has available only seven 50-gallon containers, each with a different corrosion resistance. The probability of container i leaking its contents is $P_i = 1 - e^{-r_i v_i}$, where r_i is the resistance coefficient and v_i is the volume of the corrosive liquid in gallons. Formulate a linear network flow model that can be used to determine the amount of liquid to put into each container such that the probability that none of the containers fail is maximized.   .

2. Eight tankers are scheduled to arrive and depart from a port at the times given in the table below. Set up a minimum cost flow model that will find the smallest number of docks required to service the tankers within the arrival and departure times and the schedule of the tankers at the docks.

Tanker	Arrival Date	Departure Date
1	0	3
2	3	5
3	4	5
4	7	9
5	2	5
6	7	10
7	4	6
8	6	9

3. The Smallville School District has been ordered to desegregate its elementary schools effective beginning the fall semester. In addition, it must present an implementation plan to the federal district court in six weeks. Smallville has five subdistricts and three elementary schools. The school board has compiled the following information.

Subdistrict	Number of Whites	Number of Blacks	Total
1	35	15	50
2	25	10	35
3	90	5	95
4	10	115	125
5	15	30	45
Total	175	175	350

School	Capacity
1	170
2	80
3	100

Distance (Miles)

	School		
Subdistrict	1	2	3
1	3	7	9
2	4	4	12
3	6	4	10
4	7	5	4
5	4	7	3

While the federal court would like a perfect 50%-50% racial split at each of the three schools, any split between 40%-60% to 60%-40% will be acceptable. Formulate a network flow model that may be used to minimize the total student-miles traveled daily while assuring that:

(a) All students go to school.
(b) No school is overloaded.
(c) Every school has between 40% and 60% whites in its student population.

4. Over the next three months, 20,000 boxcars of Wyoming coal must be transported from Butte to distribution points at St. Louis, Houston, and New Orleans. The three cities have contracted for 4000, 10,000, and 6000 boxcars, respectively. Because of different rail carriers' practices, three different routes are possible.

(a) Direct transport to St. Louis, Houston, and New Orleans at a cost of $1200, $1400, and $1500 per boxcar, respectively.
(b) Transport first to Wichita at a cost of $600 per boxcar; then, a decoupling–coupling charge of $30 per car; finally, a charge of $600, $400, and $400 per car for final transport to St. Louis, Houston, and New Orleans.
(c) Transport first to St. Louis at a cost of $1200 per car; then by river barge to New Orleans at $250 per car (equivalent); finally, by boat across the Gulf of Mexico to the Port of Houston at a cost of $150 per car (equivalent). Various government regulations stipulate that no more than 20% of the 20,000 cars may be sent to Wichita. Because of other committments, the direct rail lines from Butte to Houston and New Orleans are limited to 1000 and 3000 cars per month, respectively. Formulate a network flow model to minimize the total costs of transporting the 20,000 cars of coal.

5. Huey P. Short, the senior senator from Louisiana, is sponsoring a bill whose sole
    purpose is to classify the Bayou Stickleback as an endangered species. There is one
    major Eastern senator who could provide a significant stumbling block to the
    passage of the Short bill, Hiram P. Crane. Senators Short and Crane have long been
    in separate camps. Therefore, *very little* direct influence is possible. However, by
    exercising "senatorial courtesy" and "calling in" some "favors," Senator Short can
    bring indirect pressure to bear upon Senator Crane. Short's senatorial staff has
    quantified the amount of relevant influence that Short can place on key senators
    who either have Senator Crane's "ear" or know someone who does.

Senator	Maximum Influence from Short
Buttle	20
Bellemay	15
Martinez	9
Carter	4

Obviously, the four direct contacts listed above will only pass on pressure that is
applied to them and they can only do so through their personal contacts.

Senator	Maximum Influence from Buttle
Bellemay	5
Jones	6
Crane	4

Senator	Maximum Influence from Bellemay
Martinez	3
Jones	4
Crane	5

Senator	Maximum Influence from Martinez
Jones	4
Carter	5
Crane	7

Senator	Maximum Influence from Carter
Jones	1
Crane	3

Senator	Maximum Influence from Jones
Crane	8

The staff feels that 20 units of pressure on Crane should assure his support for the Stickleback. Formulate a network model that can be used to determine if Short can count on Crane's support.

6. The diagram below pictures a production process. The product is produced by machines M1, M2, and M3. These machines produce at a rate p_1, p_2, and p_3 units per hour, respectively. The product is loaded on forklift trucks at the machines and moved to the paint booths P1 and P2. Paint booth P1 can process 100 units per hour and P2 can process 50 units per hour. Each forklift truck can carry 2 units. While the products are being painted, the trucks drive around the paint booths on the aisles provided. The painted products are reloaded on the trucks and moved to the packaging machines M4, M5, and M6. These can handle 30, 60, and 80 units per hour, respectively. After being reloaded on the trucks, the product is moved to the docks D1 and D2 and shipped. The docks can each ship 100 units per hour. The trucks are then driven around the outside of the building back to the machines to be reloaded. Each of the aisles inside the plant can pass at most 40 trucks per hour in either direction. The road outside the plant can handle all traffic. It costs 1 cent to move a truck one foot. Set up the network whose minimal cost circulation gives the optimum routing of trucks to move all of the production.
7. A high-priority job is given to a machine shop. The manager wants to get started on the job right away by assigning the five parts in the job to five different machines of the seven in the shop. The table below shows the time (in minutes) required to change over the several machines to start on the new job. Find an assignment of parts to machines to minimize total machine changeover time.

	Machines						
Parts	1	2	3	4	5	6	7
1	22	28	34	41	45	22	52
2	19	27	45	52	44	28	21
3	52	45	37	28	21	17	22
4	37	55	40	45	55	39	31
5	56	25	33	54	25	31	56

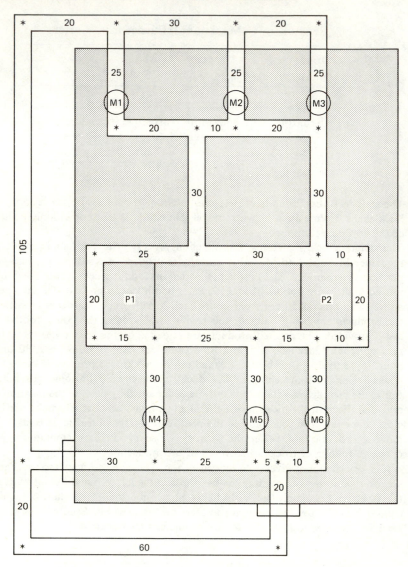

8. Mary Peters is being promoted in three months and would like to finish her current job in the style that got her promoted. She is currently responsible for sales and inventory of a particular kind of air bearing. The highly competitive market causes marked fluctuations in purchase cost, selling price, and inventory costs. She has a fixed amount of storage available to her that will house C bearings. The following dollar figures are the best estimates available for the next three months.

Month	Purchase Cost (per Bearing)	Selling Price (per Bearing)	Storage Cost (per Bearing per Month)
1	P_1	S_1	h_1
2	P_2	S_2	h_2
3	P_3	S_3	h_3

The current market is such that Mary can obtain as many bearings and sell as many bearings as she desires. There is only one limitation. She must store the bearings after purchase, to protect them from the elements, even if she is selling part or all of her inventory during this month. Formulate a network model that will enable Mary to maximize her profit over the next three months for any values of the P_i, S_i, and h_i. (Assume the current inventory is empty.)

9. Four power plants are located at points A, B, C, and D along a cool Colorado stream. Each plant alone generates enough waste heat to increase the temperature of the river water by 10°. The government wants to control the amount of heat dumped in the river by the four plants so that at no point in the river should the temperature exceed 60°. At its source, the river temperature is 50°. The river itself provides cooling with heated water cooling by 3° from A to B, 5° from B to C, 4° from C to D, and 2° from D to the mouth of the river. These reductions hold as long as the water is above 50°. The water never cools to below 50°. As an alternative to using the river, the plants can cool by other means. The cost per degree of cooling outside the plant is shown below with x_i the amount (in degrees) of nonriver cooling provided for plant i.

Plant	Cost of Cooling
A	$10x_A$
B	$15x_B$
C	$20x_C$
D	$15x_D$

Formulate a network model that the company can use to minimize the cost of meeting the government restrictions.

10. The Helvetian Navy has the task of supplying the infantry in the current conflict. In order to do this, four supply ports on the coast of Helvetia may be used to load the ships and four ports on the coast of the enemy territory may be used to unload the ships that survive the crossing of the Dante Straits. Because the enemy fleet of submarines is operating in the straits, some losses will be incurred. The probabilities that a ship traveling from a particular supply port to a specific destination port will make the crossing successfully are given below.

		Supply Port			
		1	2	3	4
	1	0.9	0.7	0.5	0.5
Destination	2	0.8	0.5	0.6	0.8
port	3	0.4	0.6	0.8	0.9
	4	0.6	0.9	0.8	0.6

There are two additional problems: The enemy has placed floating proximity mines on the boundary of each loading port and destination port. Therefore, some losses will also be felt that are independent of the risks involved in crossing the straits. The probabilities of successfully exiting (or entering) a port are given below.

	Ports			
	1	2	3	4
Supply	0.7	0.9	0.8	0.9
Destination	0.6	0.7	0.5	0.7

Times required to transport supplies (including loading time) between the various supply and destination ports are (in days):

		Supply Port			
		1	2	3	4
	1	2	5	8	10
Destination	2	4	3	6	7
port	3	9	5	3	6
	4	6	4	3	1

The Logistics Staff has reduced the requirements of the fighting forces supplied by each of the destination ports to an equivalent number of shiploads. In addition, they have conferred with each of the infantry commanders and have determined when the supplies should arrive at each destination port. These requirements and elapsed days before they are needed are given below.

	Destination Port			
	1	2	3	4
Requirement	35	50	20	15
Elapsed days	6	5	6	4

Because of the difficulty of protecting stockpiles of unloaded supplies for more than short periods of time, the ships should not arrive any more than three days before the supplies are needed. Of course, supplies that arrive more than one day after the time stipulated by the military commander are of no value. The number of ships available at each port and the costs per loaded ship (10^3) are given below.

	Supply Port			
	1	2	3	4
Ships	160	75	75	65
Cost	75	65	90	85

Develop a network flow programming model that will assist the Admiral of the Helvetian Navy in:

(a) Determining whether it is possible to supply the infantry's needs.
(b) If so, determining the shipping policy that will minimize the cost of supplying those needs.

11. The Generous Electric economists (Section 2.4) have discovered that it is possible to use intermediate shipping points to achieve replenishment of the secondary warehouse inventories. For example, it may save money to ship, say, from Dallas to Austin, unload Austin's requirement, and then proceed to Corpus Christi. This indirect shipping, because of economies of scale, could cost less than *direct* shipments to the two cities. The economists have identified the Austin secondary warehouse, and the Houston and Dallas primary warehouses as possible intermediate shipping points to the other cities. By the end of the week, they will also have the additional costs for shipments between points that were not considered possible in the past. Your job is to reformulate the network model so that the new optimal shipping pattern can be found once those costs are known.

12. Northwest Power and Light has just received estimates of the power demand for the period beginning 6 years from now and ending 20 years from now. Because of uncertainties in predicting future demands, the estimates are given in terms of total demands for 5-year increments of time.

	Station	
Time	5	6
6–10	100	95
11–15	90	110
16–20	85	120

Modify the network model given in Figure 2.22 to reflect this new information so that the optimal power distribution policy for the next 20 years may be obtained.

CHAPTER 3
FORMALIZATION OF
NETWORK MODELS

3.1 NETWORK NOTATION

Structure

The structure of a network model is defined by nodes and arcs. A node i is an element of the list of nodes, $N = [1, 2, \ldots, i, \ldots, n]$. An arc may be defined by an ordered pair of nodes (i,j) or as an element, say element k, of the list of arcs, $M = [1, 2, \ldots, k, \ldots, m]$. Thus, an arc may be identified as arc k, arc $k(i,j)$, or (i,j). An arc $k(i,j)$ is said to originate at node i and terminate at node j. Alternatively i is called the *origin node* and j is called the *terminal node* of arc k. We identify the *origin* and *terminal lists*

$$\mathbf{O} = \left[o_1, o_2, \ldots, o_m \right]$$
$$\mathbf{T} = \left[t_1, t_2, \ldots, t_m \right]$$

where o_k and t_k are the origin and terminal nodes, respectively, of arc k. The collection of nodes and arcs is a directed network (or simply network) written $D = [N, M]$, where the values n and m and the sets $\mathbf{O}$ and $\mathbf{T}$ are sufficient to define the connections present in the network.

Additional structural notation that may be useful defines the list of arcs that originate at node i, $M_{Oi} = [k | o_k = i]$, and terminate at node i, $M_{Ti} = [k | t_k = i]$. The number of arcs in the list M_{Oi} is the quantity m_{Oi} and the number of arcs in the list M_{Ti} is the quantity m_{Ti}.

The Graphical Network Model

This representation provides a visual display of the network flow model. Nodes are symbolized by circles containing the node index; node variables and parameters are shown is *square brackets* adjacent to the node. An arc is shown by a directed line segment connecting its origin and terminal nodes; the arc index appears adjacent to the arc. Arc variables and parameters are shown with *parentheses* adjacent to the arc. Since the variable and parameter lists will differ between the several problem classes to be considered, variable and parameter list definitions will appear with each graphical model. Figure 3.1 is a graphical model showing the structural characteristics of an example network, where the variable and parameter lists are left empty.

Flow

A variable quantity that characterizes most of the problems to be considered is arc flow. Written as either f_k or $f(i,j)$, arc flow usually models some physical flow quantity such as flow of fluid, flow of persons, or flow of money. A principal characteristic of arc flow is that it is conserved at nodes; that is, the flow into a node equals the flow out of a node.

Although there are many network flow applications that involve multicommodity flows, we will limit our attention to single commodity flows where flows on two or more arcs leaving a node are indistinguishable as to type.

We can partition network problems according to whether flow is conserved in the arcs. For the pure network problem, flow is conserved in an arc. Thus, if f_k is the flow in arc k at its origin and f'_k is the flow in the arc at its terminal, a pure network problem has

$$f'_k = f_k$$

For a network with gains, flow is not conserved in the arc. Here a *positive gain parameter* a_k is defined such that

$$f'_k = a_k f_k$$

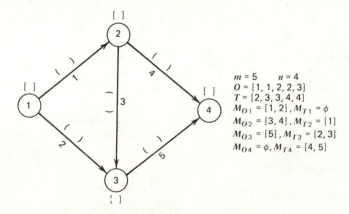

$m = 5 \qquad n = 4$
$O = [1, 1, 2, 2, 3]$
$T = [2, 3, 3, 4, 4]$
$M_{O1} = [1, 2], M_{T1} = \phi$
$M_{O2} = [3, 4], M_{T2} = [1]$
$M_{O3} = [5], M_{T3} = [2, 3]$
$M_{O4} = \phi, M_{T4} = [4, 5]$

Figure 3.1
Graphical Representation of the Network

If $a_k = 1$ flow is conserved; if $a_k < 1$ flow is lost in the arc; if $a_k > 1$ flow is gained in the arc.

The *flow vector* is defined

$$\mathbf{f} = [\, f_1, f_2, \ldots, f_m \,]'$$

where the prime symbol indicates the defining row vector is transposed; that is **f** is a column vector.

Cost

A cost may be associated with the flow in an arc. The *arc cost* $h_k(f_k)$ is a function only of the flow in arc k and is independent of flow in other arcs. Our objective will be to minimize the *network cost*, which is the sum of the arc costs.

$$H(\mathbf{f}) = \sum_{k=1}^{m} h_k(f_k)$$

We identify three classes of cost functions: linear, convex, and concave. The linear cost function is

$$h_k(f_k) = h_k f_k$$

where h_k is an arc parameter giving the arc cost per unit flow, and the vector of arc costs is

$$\mathbf{h} = [\, h_1, h_2, \ldots, h_m \,]$$

It follows that the linear network cost is

$$H(\mathbf{f}) = \sum_{k=1}^{m} h_k f_k = \mathbf{hf}$$

Figures 3.2*a* and 3.2*b* show several representative convex and concave cost functions, respectively.

Capacity

The parameter defining an upper bound for flow on each arc is the arc capacity c_k.

$$f_k \leqslant c_k$$

The arc capacity vector

$$\mathbf{c} = [\, c_1, c_2, \ldots, c_m \,]'$$

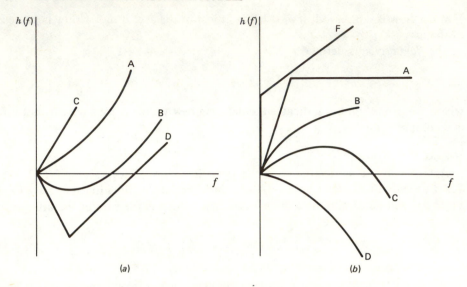

Figure 3.2
Representative Nonlinear Arc Costs (*a*) Convex Costs
(*b*) Concave Costs

may be used to write the set of arc capacity constraints

$$\mathbf{f} \leqslant \mathbf{c}$$

Lower Bounds

A common requirement in many modeling situations is that arc flow be at least as great as some lower bound. Therefore, let us define the lower bound parameter for each arc $\underline{c}_k$ to constrain the flow f_k such that

$$f_k \geqslant \underline{c}_k$$

The arc lower bound vector

$$\underline{c} = \left[\underline{c}_1, \underline{c}_2, \dots, \underline{c}_m\right]'$$

may be used to write the set of arc lower bound constraints

$$\mathbf{f} \geqslant \underline{c}.$$

External Flows

External flows enter or leave the network at the nodes. For most network models, the external flows represent connections to the world outside the system being modeled and hence are very important to an accurate representation.

There are two kinds of external flows for a node, the *fixed flow* and the *slack flow*. The fixed flow at node i, b_i, enters the network if $b_i > 0$ and leaves the network if $b_i < 0$. Any feasible solution stipulates network flows such that all fixed external flows are accommodated.

The value of the slack external flow at node i, f_{si}, is determined as part of the optimization procedure. The direction of this flow and the bound on its value is specified by the node parameter b_{si}, the *slack flow capacity*. If b_{si} is positive (negative) f_{si} enters (leaves) the network and is bounded by $0 \leqslant f_{si} \leqslant |b_{si}|$.

The cost per unit of external slack flow, h_{si}, is a node parameter used to add the cost of external flow for all nodes to the arc costs yielding the network objective function.

$$\mathbf{H(f)} = \sum_{k=1}^{m} h_k(f_k) + \sum_{i=1}^{n} h_{si} f_{si}$$

Figure 3.3*a* illustrates a network with slack external flow parameters.

An alternative representation of slack flows that uses arcs rather than the node parameters will be used for later algorithms. For this representation, a slack node that does not require conservation of flow is designated as node n. For a positive (negative) slack flow f_{si}, an arc is constructed from (to) the slack node to (from) node i. The capacity of the arc is $|b_{si}|$ and the cost on the arc is h_{si}.

Conservation of Flow

A feasible **f** conserves flow at all nodes of the network, excepting the slack node. With slack arcs and slack node defined as above, all flows except fixed external flows are arc flows. Thus, conservation of flow implies for each node:

Total arc flow leaving the node-Total arc flow entering the node

= Fixed external flow at the node

Algebraically, the conservation-of-flow constraint for a general node i is

$$\sum_{k \in M_{Oi}} f_k - \sum_{k \in M_{Ti}} f_k = b_i$$

for the pure network problem and

$$\sum_{k \in M_{Oi}} f_k - \sum_{k \in M_{Ti}} a_k f_k = b_i$$

for the network-with-gains problem.

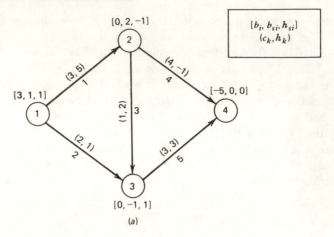

(a)

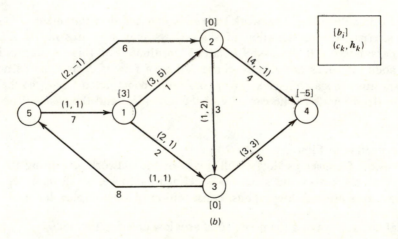

(b)

Figure 3.3
Representation of Slack External Flows (a) With Node Parameters
(b) With a Slack Node

For the example problem of Figure 3.3b, the complete set of conservation-of-flow constraints is

$$
\begin{aligned}
\text{node 1:} \quad & f_1 + f_2 & & & - f_7 & & = 3 \\
\text{node 2:} \quad & -f_1 & + f_3 + f_4 & - f_6 & & & = 0 \\
\text{node 3:} \quad & -f_2 - f_3 & + f_5 & & + f_8 & = 0 \\
\text{node 4:} \quad & & -f_4 - f_5 & & & = -5
\end{aligned}
$$

Note that each column representing a nonslack arc has exactly two nonzero entries, one with coefficient $+1$, and one with coefficient -1. Slack arcs have only one nonzero entry because no conservation-of-flow constraint is written for the slack node.

This characteristic is true for any pure network problem since each nonslack arc flow is represented in exactly two conservation-of-flow equations. The flow variable $f_k(i,j)$ appears with a $+1$ coefficient in row i and a -1 coefficient in row j. Later in this chapter, we discuss this property's implications on the character of the optimal solution to the network flow problem. If we denote the matrix of coefficients associated with the conservation-of-flow equations as $\mathbf{A}$ and the column vector of fixed external flows as $\mathbf{b}$, the conservation of flow equations may be written as $\mathbf{Af} = \mathbf{b}$.

For the network-with-gains problem the gain factor a_k multiplies the flow variable $f_k(i,j)$ in row j of the constraint set. Thus, if the network structure of Figure 3.3b also included arc gains, the $\mathbf{A}$ matrix would appear:

$$\mathbf{A} = \begin{bmatrix} 1 & 1 & 0 & 0 & 0 & 0 & -1 & 0 \\ -a_1 & 0 & 1 & 1 & 0 & -1 & 0 & 0 \\ 0 & -a_2 & -a_3 & 0 & 1 & 0 & 0 & 1 \\ 0 & 0 & 0 & -a_4 & -a_5 & 0 & 0 & 0 \end{bmatrix}$$

For the network arcs each column of $\mathbf{A}$ still contains two nonzero entries, but the negative entry is $-a_k$ rather than -1.

3.2   TWO USEFUL TRANSFORMATIONS

Transformation for Zero Lower Bounds

Available core storage will limit the size of problem that can be solved. Since each parameter requires a fixed amount of computer storage, it is important to minimize the number of parameters used to define a network.

It is possible to express any network flow problem with nonzero arc lower bounds as an equivalent problem with all zero lower bounds. This allows one to delete the lower bound from the parameter list by making the following transformation for each arc $k(i,j)$ with a nonzero lower bound $\underline{c}_k$:

(a) Replace $\underline{c}_k$ by zero.
(b) replace c_k by $c'_k = c_k - \underline{c}_k$.
(c) replace f_k by $f'_k = f_k - \underline{c}_k$.
(d) replace b_i by $b'_i = b_i - \underline{c}_k$.
(e) replace b_j by $b'_j = b_j + a_k \underline{c}_k$.

This transformation is performed one arc at a time for all arcs with nonzero lower bounds. Each new transformation uses the updated parameter values that resulted from the previous step.

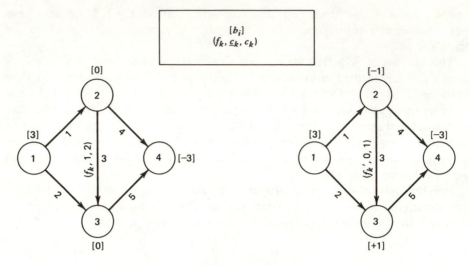

Figure 3.4
Transformation to Delete Nonzero Lower Bound

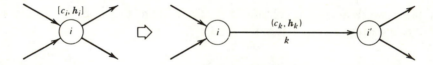

Figure 3.5
Translation of Node to Arc Parameters

This procedure is illustrated in Figure 3.4 where arc 3 with a lower bound of 1 is transformed to an arc with a lower bound of zero.

In subsequent discussions, we assume that all arc lower bounds are zero. The parameter $\underline{c}_k$ will therefore not appear explicitly in the algorithms or arc parameter lists.

Node Parameters and Variables

It may be convenient in the initial modeling process to associate such parameters as capacity and cost with one or more nodes. These node parameters can easily be translated to arc parameters by replacing such a node with two nodes and a connecting arc as in Figure 3.5. The arc then appropriates the node parameters.

3.3 THE ALGEBRAIC NETWORK MODEL

The graphical representation plus the node and arc parameters are sufficient to define a network flow model. The characteristics of the slack node, the objective function, and the familiar concept of conservation of flow are implicit in the definition. Indeed, the simple interpretation of the network flow model is a primary strength of this representation.

The algebraic representation of the model is more rigorous because it explicitly specifies all the model assumptions. Although we prefer the graphical representation for examples, the algebraic representation is important for theoretical developments. In terms of the parameters and variables already defined, the general single commodity linear minimum cost flow problem may be expressed as the linear programming problem (LP) given below.

$$\text{Min.} \sum_{k=1}^{m} h_k f_k$$

$$\text{s.t.} \quad \sum_{k \in M_{Oi}} f_k - \sum_{k \in M_{Ti}} a_k f_k = b_i \qquad i = 1, \ldots, n-1$$

$$f_k \leqslant c_k \qquad k = 1, \ldots, m$$
$$f_k \geqslant 0 \qquad k = 1, \ldots, m$$

where node n is the slack node.

In matrix notation, the linear programming model is

$$\text{Min.} \quad \mathbf{hf} \qquad (1)$$

$$\text{s.t.} \quad \mathbf{Af} = \mathbf{b} \qquad (2)$$

$$\mathbf{f} \leqslant \mathbf{c} \qquad (3)$$

$$\mathbf{f} \geqslant \mathbf{0} \qquad (4)$$

We now summarize some important results from linear programming theory that will be important to later developments. These results are well known and will be stated here without proof.

Let $\mathbf{A}$ be partitioned so that $\mathbf{A} = [\mathbf{B}, \mathbf{R}]$. Any matrix $\mathbf{B}$ formed by a set of $n-1$ linearly independent columns of $\mathbf{A}$ is called a *basis* and the variables corresponding to these columns are basic *variables*. The vector of basic variables, implicitly written in the same order as the columns in $\mathbf{B}$, is $\mathbf{f_B}$.

The collection of nonbasic variables associated with $\mathbf{R}$ is $\mathbf{f_R}$. The LP constraints can now be written

$$[\mathbf{B}, \mathbf{R}] \begin{bmatrix} \mathbf{f_B} \\ \mathbf{f_R} \end{bmatrix} = \mathbf{b} \qquad (5)$$

Solving for $\mathbf{f_B}$

$$\mathbf{f_B} = \mathbf{B}^{-1} [\mathbf{b} - \mathbf{R} \mathbf{f_R}] \qquad (6)$$

To define a basic solution, we must first assign each nonbasic variable the value zero or c_k. Now equation (6) above yields the *basic solution* $\mathbf{f_B}$. Note that since rank $(\mathbf{A}) = n - 1, \mathbf{f_B}$ is unique. At least one optimal solution to the LP is a basic solution.

The dual to the LP may be written, after some manipulation, as

$$\text{Min.} \quad \sum_{i=1}^{n-1} \pi_i b_i + \sum_{k=1}^{m} \delta_k c_k$$

$$\text{s.t.} \quad \pi_i - a_k \pi_j + \delta_k \geqslant -h_k \quad \begin{array}{l} k = 1, \dots, m \\ i = o(k), j = t(k) \end{array}$$

$$\pi_i \text{ unrestricted} \quad i = 1, \dots, n-1$$

$$\delta_k \geqslant 0$$

In matrix notation

$$\begin{array}{lll} \text{Min.} & \pi\mathbf{b} + \delta\mathbf{c} & \qquad (7) \\ \text{s.t.} & \pi\mathbf{A} + \delta\mathbf{I} \geqslant -\mathbf{h} & \qquad (8) \\ & \pi \text{ unrestricted} & \qquad (9) \\ & \delta \geqslant 0 & \qquad (10) \end{array}$$

The primal–dual conditions for an optimal solution of the linear program have been specialized to the minimal cost network flow problem in the following three theorems.

THEOREM 1

Given a solution $\mathbf{f} = [\mathbf{f_B}, \mathbf{f_R}]^\mathrm{T}$ to the *primal* problem and a solution $[\pi, \delta]$ to the dual problem, the solutions are optimal if and only if:

1. $\mathbf{f}$ is feasible for the primal problem; that is, (2), (3), and (4) are satisfied.
2. $[\pi, \delta]$ is feasible for the dual problem; that is, (8) and (10) are satisfied.
3. Complementary slackness is satisfied; that is,

    (a) If $f_k(i,j) > 0, \pi_i - a_k \pi_j + \delta_k = -h_k$.
    (b) If $f_k < c_k, \delta_k = 0$.
    (c) If $\pi_i - a_k \pi_j + \delta_k > -h_k, f_k(i,j) = 0$.
    (d) If $\delta_k > 0, f_k = c_k$.

Because $\mathbf{c} \geqslant 0$, a restricted condition of dual feasibility may be succinctly stated in mathematical terms.

THEOREM 2

If $[\pi, \delta]$ is an optimal solution to the dual problem,
then $\delta_k = \max[0, -h_k - \pi_i + a_k \pi_j]$
Theorem 2 allows the optimality for the problem to be written as follows.

THEOREM 3

Given a solution $\mathbf{f}$ to the primal problem and the partial solution π to the dual problem, the solutions are optimal to their respective problems if and only if the following considerations are satisfied:

1. Primal feasibility.
2. $\delta_k = \text{Max}[0, -h_k - \pi_i + a_k \pi_j]$ (restricted dual feasibility).
3. Complementary slackness

    (a) $\pi_i - a_k \pi_j = -h_k$ for $0 < f_k < c_k$

    (b) $f_k = 0$ for $\pi_i - a_k \pi_j > -h_k$

    (c) $f_k = c_k$ for $\pi_i - a_k \pi_j < -h_k$ where $i = o(k)$ and $j = t(k)$.

3.4 THE LINEAR PROGRAMMING MODEL FOR THE PURE MINIMUM COST PROBLEM

We consider in this section the pure minimum cost problem and leave the more complex case of the network with gains to Chapter 10. Since the basic conceptual material concerning this problem type has already been presented, perhaps the simplest way to show the direct relationship between equations (1) to (4) of this chapter and the pure network flow problem is by means of a specific example.

The linear programming formulation, consistent with equations (1) to (4), of the network presented in Figure 3.6 is given below:

$$
\begin{aligned}
\text{Min. } & 5f_1 + 1f_2 + 2f_3 - 1f_4 + 3f_5 - f_6 + f_7 + 0f_8 \\
\text{s.t. } & f_1 + f_2   -f_7 = 3 \\
& -f_1 +f_3 +f_4 -f_6 = 0 \\
& -f_2 -f_3 +f_5 = 0 \\
& -f_4 -f_5 = -5 \\
& f_6 + f_7 -f_8 = 0
\end{aligned}
$$

$$0 \leqslant f_1 \leqslant 3, 0 \leqslant f_2 \leqslant 3, 0 \leqslant f_3 \leqslant 1, 0 \leqslant f_4 \leqslant 2$$
$$0 \leqslant f_5 \leqslant 5, 0 \leqslant f_6 \leqslant 1, 0 \leqslant f_7 \leqslant 2$$
$$f_8 \text{ unrestricted}$$

In this example, the first four constraints reflect the conservation-of-flow requirement at nodes 1 through 4. The fifth constraint is included for mathematical and notational convenience to define the external flow into the slack node. The remaining $2m - 14$ constraints reflect both the flow capacity restrictions and the nonnegativity restrictions for the amount of flow on each arc. There are $m + 1 = 8$ columns with the first 7 columns representing the 7 arcs and the last column representing the external flow into the slack node n(node 5). The slack node has the characteristic of a variable external flow and we symbolize this flow by $f_{m+1}(f_8)$.

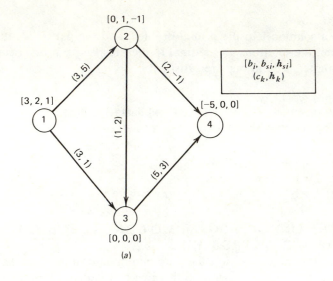

$$[b_i, b_{si}, h_{si}]$$
$$(c_k, h_k)$$

(a)

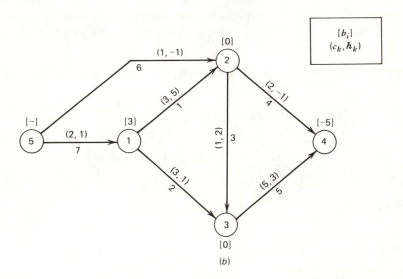

$$[b_i]$$
$$(c_k, h_k)$$

(b)

Figure 3.6
Example Pure Network Flow Problem (a) With Node Parameters
(b) With a Slack Node

Note that the **A** matrix for this problem has two entries in each column, excepting the last, which are $+1$ and -1. The last column has a single -1 in the last row.

A number of useful results are available concerning this LP structure.

It is informative to note that **A** is totally unimodular. Total unimodularity means that every subdeterminant of **A** has a value of 1, -1, or 0. Thus every basis, **B**, formed from **A** has its determinant, $|\mathbf{B}|$, equal to $+1$ or -1. If **b** is integer, the relationship $\mathbf{Bf_B} = \mathbf{b} - \mathbf{Rf_R}$ and the fact that all nonbasic flows are set at a value of 0 or c_k are sufficient to show, using Cramer's rule, that $\mathbf{f_B}$ is integer.

A basis for this problem includes column $m+1$ and $n-1$ columns from **A**, chosen so that all n columns are independent. A basic solution to the primal problem is determined by setting each nonbasic flow to either c_k or zero and assigning flows to the basic arcs in order to satisfy the required node external flows (b_i). This flow may be infeasible with some basic arc flow negative or greater than c_k. Figure 3.7 illustrates several basic solutions for the network of Figure 3.6. Figures 3.7a and 3.7b have the same basis but different flows because nonbasic arcs have been assigned different flows. The two assignments of flows to the arcs are infeasible. Figure 3.7c shows a basic feasible solution.

The dual to this pure network problem is

$$\text{Min.} \sum_{i=1}^{n} \pi_i b_i + \sum_{k=1}^{m} \delta_k c_k$$

$$\text{s.t.} \quad \pi_i - \pi_j + \delta_k \geq -h_k \qquad \text{for arc } k(i,j)\varepsilon \mathbf{M}$$

$$\pi_n = 0$$

$$\pi_i \text{ unrestricted} \qquad i\varepsilon \mathbf{N}$$

$$\delta_k \geq 0 \qquad k\varepsilon \mathbf{M}$$

The dual of the example problem is

$$\text{Min.} \quad 3\pi_1 + 0\pi_2 + 0\pi_3 - 5\pi_4 + 0\pi_5 + 3\delta_1 + 3\delta_2 + \delta_3 + 2\delta_4 + 5\delta_5 + \delta_6 + 2\delta_7$$

$$
\begin{array}{llllllll}
\text{s.t.} & \pi_1 - \pi_2 & & & +\delta_1 & & & \geq -5 \\
& \pi_1 & -\pi_3 & & +\delta_2 & & & \geq -1 \\
& \pi_2 - \pi_3 & & & +\delta_3 & & & \geq -2 \\
& \pi_2 & -\pi_4 & & & +\delta_4 & & \geq +1 \\
& \pi_3 -\pi_4 & & & & +\delta_5 & & \geq -3 \\
& -\pi_2 & +\pi_5 & & & & +\delta_6 & \geq +1 \\
& -\pi_1 & +\pi_5 & & & & +\delta_7 & \geq -1 \\
& & -\pi_5 & & & & & = 0 \\
\end{array}
$$

$$\pi_i \text{ unrestricted} \qquad i = 1,\dots,5$$
$$\delta_k > 0 \qquad k = 1,\dots,7$$

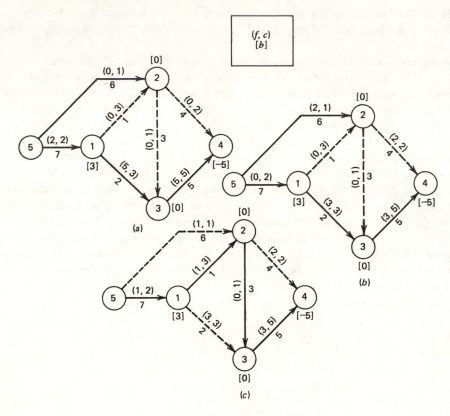

Figure 3.7
Basic Solutions for the Example Problem

The optimality conditions can be written directly from Theorem 3. For simplicity define

$$d_k = \pi_i - \pi_j + h_k \qquad \text{for arc } k(i,j)$$

1. Primal feasibility.

   (a) Conservation of flow satisfied at each node.
   (b) $0 \leqslant f_k \leqslant c_k$ for all arcs k.

2. Restricted dual feasibility.

$$\delta_k = \text{Max}[0, -d_k]$$

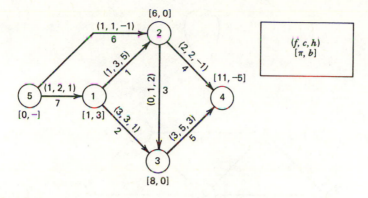

Figure 3.8

Optimal Solution for the Example Problem

3.   Complementary slackness

   (a) $d_k = 0$ for $0 < f_k < c_k$
   (b) $f_k = 0$ for $d_k > 0$
   (c) $f_k = c_k$ for $d_k < 0$

The dual feasibility condition is trivial since, given π_k and h_k, δ_k can always be chosen to satisfy that condition. Thus the primal problem and dual problems are solved if one can find flows **f** and node potentials π that satisfy primal feasibility and complementary slackness. Figure 3.8 shows an assignment of flows and potentials for the example problem. Since this assignment satisfies conditions 1 and 3 above, the solution must be optimal.

3.5   TERMINOLOGY FROM GRAPH THEORY

The reader may have noticed the rather special structure that each basic solution exhibited in Figure 3.7. This kind of special structure, to be fully described below, is one of the dominant reasons that network flow programming problems may be solved with very great efficiency. Therefore, it is appropriate that a section of this chapter be devoted to defining several graph theoretical concepts and terms that will be used throughout the remainder of this book.

A *graph* G is a finite nonempty set of nodes N with a prescribed set X of unordered pairs of distinct nodes from N. Each element of X is a *line*. A *loop* is formed by a line joining a node to itself. A *graph* neither allows loops nor multiple lines, that is, two or more lines that connect the same pair of nodes. A *multigraph* does not allow loops but does allow multiple lines. Figure 3.9 presents examples of a loop, graph, and multigraph.

Although we will refer explicitly only to graphs, the definitions given below

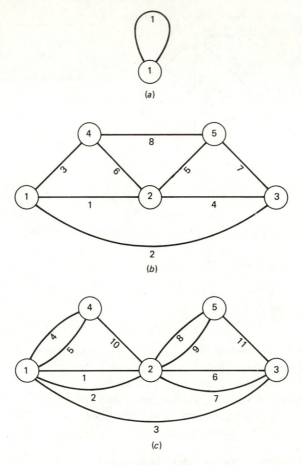

Figure 3.9
Examples of Loop, Graph, and Multigraph (*a*) Loop (*b*) Graph
(*c*) Multigraph

apply to multigraphs as well. A *walk* of a graph *G* is an alternating sequence of nodes and lines, beginning and ending with nodes, in which each line is incident with the two nodes immediately preceding and following it. If the same node begins and ends the walk, it is a *closed walk*; otherwise it is an *open walk*. A *trail* is a walk with no line repeated. A *path* is a walk with no node (and therefore no line) repeated. A *cycle* is a closed walk with no node but the beginning node repeated.

Examples of each of the terms given in the paragraph above may be easily extracted from Figure 3.9*b*. For example, one walk in Figure 3.9*b* is node 1, line 1, node 2, line 5, node 5, line 5, node 2, line 4, node 3. Equivalently, a walk may be defined by listing, in order, only the nodes or only the lines. Whichever is not listed is implicit by context. Therefore, the walk given above could also be described as nodes $1,2,5,2,3$ or lines $1,5,5,4$.

A closed walk would be formed by nodes $1,2,4,5,2,1$; a trail would be formed by nodes $1,2,5,4,2,3$; an example path is nodes $1,2,4,5,3$; and an example cycle is nodes $1,2,4,1$.

A *connected graph* has every pair of nodes joined by a path. A graph that is not a connected graph is a *disconnected graph*. The order of a walk is the number of lines in it. A *tree* is an acyclic connected graph. A *forest* is an acyclic graph. Thus all components of a forest are trees.

A *directed graph* or *digraph* is a finite nonempty set of nodes N with a prescribed set M of *ordered* pairs of distinct nodes. The elements of M are *directed lines* or *arcs*. The definitions of the terms *directed walk, directed trail, directed path, directed cycle, directed tree*, and *directed forest* follow directly from the above discussion.

A *directed network* $D = [N, M]$ is a directed graph together with one or more functions, each of which assigns a real number to each arc or to each node of the directed graph. A *subnetwork* of $D, D_s = [N_s, M_s]$, is a directed network having $N_s \subset N$, and $M_s \subset M$. A directed tree, $D_T = [N_T, M_T]$, is a subnetwork that defines a unique directed path from some specified node, called the *root node*, to each of the other nodes in N_T. A *spanning subnetwork* of D is a subnetwork of D having $N_s = N$. Therefore, a *spanning tree* of D has $N_T = N$. A *spanning forest* of D is a directed disconnected acyclic spanning subnetwork of D.

Trees have a number of characteristics.

- The number of nodes in a tree is one greater than the number of arcs. Thus a spanning tree has $n - 1$ arcs.
- For a directed spanning tree one arc terminates at each node except the root node.
- Any number of arcs may originate at a node of a directed tree.
- For any pair of nodes in a tree there is a unique path that originates at one of the nodes and terminates at the other.
- The $n - 1$ columns of the matrix A that help form the basis of the pure minimum cost problem correspond to the arcs in the set M_T of a spanning tree.
- The arcs not included in M_T are called *chords* of the tree. If a chord is added to a tree, a unique cycle is formed and the resulting network is no longer a tree.
- If there exists a directed path in a tree from node i to node j then node i precedes node j and node j succeeds node i. Similarly arc k precedes arc k' if there is a directed path that contains both arc k and k' and arc k appears first. Arc k' succeeds arc k.

Figure 3.10 illustrates four trees obtained from Figure 3.6*b*. Note that Figure 3.10*b* is a spanning tree of Figure 3.6*b* and Figures 3.10*c* and 3.10*d* taken together are a spanning forest of Figure 3.6*b*.

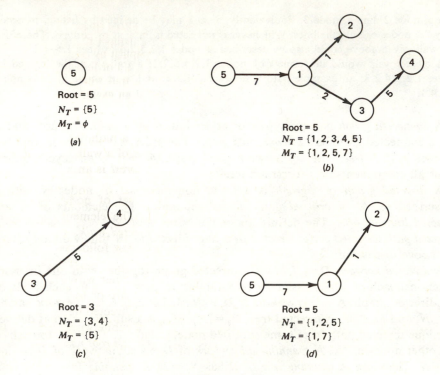

Figure 3.10
Example Network Trees

3.6   THE EXPANDED AND MARGINAL NETWORKS

Expanded Networks

As has been seen above, network models can be naturally based on directed graphs. Figure 3.11a illustrates such a network model. It is convenient from the viewpoint of obtaining optimal solutions to network flow programming problems to define an *expanded network*, which may be obtained directly from the original network.

Let us define the existence of a *mirror arc*, $-k$, for each arc $k \in M$ such that arc $-k$ connects the same nodes as arc k but has opposite direction.

$$o(-k) = t(k)$$
$$t(-k) = o(k)$$

The expanded network, $D_E = [N, M_E]$, has the same node set as D and its arc set contains not only the arcs of D but also all of the mirror arcs as well; that is, if

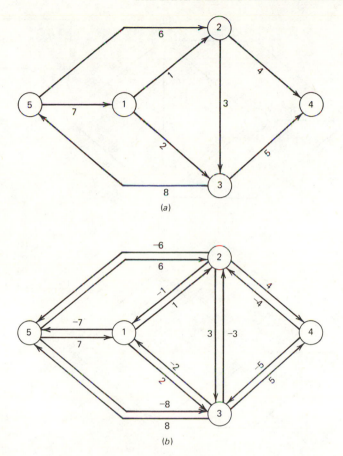

Figure 3.11
Expanded Network Example (*a*) Directed Network (*b*) Associated
Expanded Network

$M = [1, 2, \ldots, m]$ then $M_E = [1, 2, \ldots, m, -1, -2, \ldots, -m]$. Figure 3.11 shows a directed network in part (*a*) and its associated expanded network in part (*b*).

For the pure network, a cost parameter is easily provided for each mirror arc in D_E. If h_k is the cost for forward arc k, then $-h_k$ is the cost for mirror arc $-k$.

Whenever it is not obvious from context, the arcs of the original network will be called *forward arcs* to clearly differentiate them from their counterparts in the expanded network. Notice that if the original network is connected, then a *directed* path exists between every pair of nodes in the expanded network. It follows directly that the mirror arc of a mirror arc is the associated forward arc.

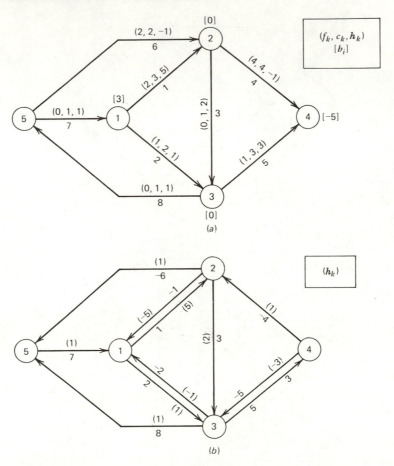

Figure 3.12
Marginal Network Example (*a*) Network with Flows Assigned
(*b*) Corresponding Marginal Network

Marginal Network

The *marginal network*, $D^* = [N, M^*]$ has the same node set as D and D_E. The arc set of D^* consists of that subset of M_E that are *admissible arcs*. A forward arc is admissible if the flow is not already equal to the arc capacity; that is, if flow can still be increased on that arc in the original network. A mirror arc is admissible if flow on the corresponding forward arc is not equal to zero; that is, if flow can be decreased on the associated arc in the original network.

The criteria for determining whether flows may be increased or decreased may differ between applications. We therefore define the function $A_d(k)$, which indicates the admissibility of the arcs of the expanded network. Thus

$$A_d(k) = 1 \qquad \text{if arc } k \text{ is admissible}$$
$$A_d(k) = 0 \qquad \text{if arc } k \text{ is inadmissible}$$

The usual definition of the admissibility function will be

$$A_d(k) = 1 \begin{cases} \text{if } k > 0 & \text{and} & f_k < c_k \\ \text{if } k < 0 & \text{and} & f_{-k} > 0 \end{cases}$$

$$A_d(k) = 0 \begin{cases} \text{if } k > 0 & \text{and} & f_k \geqslant c_k \\ \text{if } k < 0 & \text{and} & f_{-k} \leqslant 0 \end{cases}$$

We will define the function anew whenever it is used. Figure 3.12*b* shows the marginal network associated with Figure 3.12*a* when the above admissibility function is used.

3.7 NETWORKS WITH NONLINEAR ARC COSTS

Convex and Concave Cost Models

Although the discussion above has been restricted to the assumption of linearity in all network attributes, nonlinear considerations can be introduced in some cases, albeit with a concomitant increase in modeling and solution complexity.

In general, the cost of flow in an arc may be a linear, convex, or concave function of flow. We discuss in this section how some *nonlinear arc costs* can be represented by a piecewise linear approximation. The resultant costs can be easily modeled with a linear network if the function is convex; unfortunately, if the function is concave, this statement is not true.

Consider Figure 3.13, which shows a piecewise linear convex arc cost function and its network representation where the three pieces of the curve are represented by three arcs. Since $h(\mathbf{f})$ is convex, $h_1 < h_2 < h_3$.

Therefore, as flow is increased from node i to node j, flow will first pass through arc 1. When the capacity of this arc is reached, flow begins to pass through arc 2. Arc 3 has nonzero flow if and only if arc 2 is full. Because the objective is to minimize cost, if two or more arcs are parallel the one with the lowest cost is used first. If the piecewise linear function were an approximation for an underlying continuous function, the approximation could be improved by breaking the interval into more pieces and hence defining more arcs. The cost of the increased accuracy is, again, a more complex network.

Consider Figure 3.14, which shows a piecewise linear concave cost function. This kind of function often arises in practice to represent economies of scale. Unfortunately, it cannot be modeled as a minimum cost linear network because

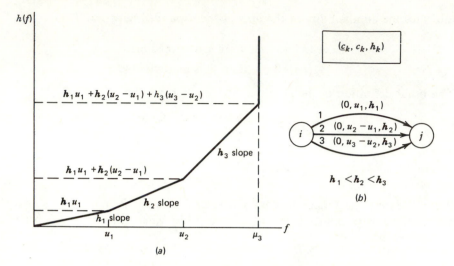

Figure 3.13

Convex Arc Cost Function (*a*) Piecewise Linear Convex Arc Cost Function (*b*) Network Model

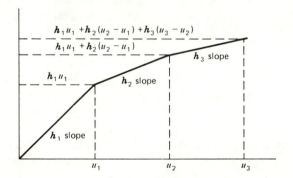

Figure 3.14

Concave Arc Cost Function (There Is No Equivalent Network Model)

$h_1 > h_2 > h_3$. If the parallel representation of Figure 3.13 were attempted, the solution method would direct that the first units of flow pass through arc 3 rather than arc 1. Although the linear algorithms are often used as a part of the solution procedure for concave cost problems, they are embedded in more time-consuming integer programming algorithms.

Because a benefit may be treated as a negative cost and the negative of a concave function is convex, concave benefit functions can be modeled but convex benefits cannot. The concave benefit case is important because it illustrates the economic concept of decreasing marginal returns.

Applications of Nonlinear Arc Costs—An Overview

As noted above, convex costs or concave benefits can be modeled using piecewise linear functions and solved with the linear algorithms. Chapter 11, however, attacks the convex cost problem in a way that does not require the storage inefficiencies introduced by multiple parallel arcs.

A very important generalization of network models is the incorporation of certain kinds of risk situations into the model. Certainly, in multiperiod planning applications there are many aspects of a situation that cannot be determined with certainty. The use of the "expected value" for a parameter with a known probability distribution in a deterministic model often yields solutions that are not reasonable. In Chapter 11, we show that under certain conditions situations involving risk will result in convex expected cost curves. Thus, the procedures for solving convex cost network problems will be useful in such cases.

Also related to the convex cost network problem are the "Kirchoff laws" network systems familiar to practitioners in the classical engineering disciplines. For example, the current–voltage solutions for dc electrical networks, the flow–pressure solutions for pipeline networks, and the heat flow–temperature solutions of thermodynamics all have equivalent convex cost network representations. The establishment of the links between physical and economic network problems has important implications for the modeling and solution procedures for both classes of problems.

There are many situations of practical importance that have the characteristic of a concave cost curve. For example, suppose there is an ordering cost for the activity of procuring raw materials. This situation would be reflected by a cost function for an arc

$$h(f) = \begin{cases} 0 & \text{if } f = 0 \\ h_F + h_v f & \text{if } f > 0 \end{cases}$$

where h_F is a fixed setup or ordering cost independent of flow and h_v is a variable cost per unit of flow. This cost function is illustrated as curve F in Figure 3.2b and it is concave. One familiar with inventory theory will recognize the setup cost as an important consideration of inventory management problems, which balance the holding cost against the setup cost. A linear or convex cost model can in no way reflect this setup cost. If one could solve efficiently a network model with concave cost, he would be able to solve a number of important problems in inventory and production management.

As another example of the concave cost model, consider a transportation problem with demands specified but supply points as decision variables. The graphical model could be represented as in Figure 3.15. The supply arcs will each have a concave cost curve with a fixed cost to represent the costs of obtaining a site and a curve showing a decreasing first derivative to represent decreasing marginal cost with total amount shipped from the site. This class of

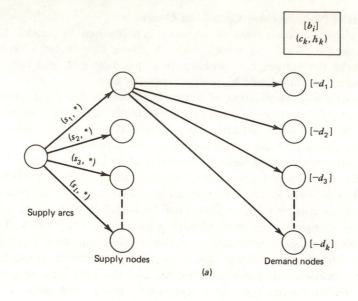

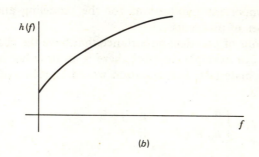

Figure 3.15
Location–Allocation Problem

problems has been called the location–allocation or warehouse location problem. Concave costs could also be associated with the arcs passing from supply to demand nodes to represent the cost of building and maintaining a transportation link.

There is no shortage of situations where concave costs would be a convenient modeling tool. Unfortunately, although algorithms do exist to solve these problems, they are very inefficient when compared to those for linear or convex cost functions. Where the latter may solve problems of thousands of arcs in seconds, the concave cost problems can be solved for fewer than one hundred arcs in minutes. Chapter 12 discusses branch and bound solution techniques, which use the linear solution algorithms as subroutines.

3.8 SOLUTION ALGORITHMS FOR THE PURE MINIMUM COST FLOW PROBLEM

A wide variety of solution algorithms have been introduced to solve network problems of the various types. In general, they are finite iterative procedures designed to obtain a solution that satisfies the conditions given at the end of Section 3.4. The principal differences between the algorithms is the order in which the conditions are satisfied. We consider four classes of algorithms: primal feasible, dual node infeasible, dual arc infeasible, and primal–dual. They are described in general in the sections to follow. Each of the algorithms generally takes on a different form as it is applied to the various problem classes, but the steps noted below are consistently followed. It is *this observation* that provides the unifying link between the large number of algorithms that have appeared in the literature.

It should be noted that the conditions for optimality do not require that the solution be basic; however, maintaining a basis will often result in savings in computational cost. A dichotomy may be made for network flow algorithms on whether a basic solution is maintained.

a. Primal Approach. The primal approach iteratively derives $\mathbf{f}$ and π such that $\mathbf{f}$ is primal feasible while attempting to achieve complementary slackness. Define e_{ck}, a measure of the violation of complementary slackness for arc k, as

$$e_{ck} = \text{Max}\left[d_k f_k, (f_k - c_k) d_k \right]$$

For feasible flow, e_{ck} will be positive only if complementary slackness is violated. Let $E_c = \sum_{k=1}^{m} e_{ck}$. When E_c is driven to zero, $\mathbf{f}$ is optimal.

ALGORITHM

1. Find $\mathbf{f}$ and π that satisfy primal feasibility.
2. Find an arc such that $e_{ck} > 0$. If there are none, stop. Otherwise go to step 3.
3. Find a new $\mathbf{f}$ and π that reduce e_{ck} for the arc while maintaining primal feasibility. Go to step 2.

For some network problems finding a primal feasible flow will be trivial; for others it will be the most time-consuming portion of the algorithm. A second consideration related to this is the "closeness" to the optimum of the first feasible solution. If step 1 finds a solution close to the optimum, the effort required for steps 2 and 3 will be less. Unfortunately, one generally pays a larger computational cost to obtain a close first solution. Thus, the subject of the trade-off between the costs of step 1 as opposed to the cost of steps 2 and 3 is of interest.

There are a number of computational options associated with step 2. For instance, do we select the first arc we find that violates the optimality condition, or do we look for the arc that violates the condition most, or is there some other option? The difficulty of this step depends again on the specific flow problem under consideration. Some algorithms we will present do not explicitly use the node potentials and, hence, the optimality conditions are not obviously imposed. We will show, however, that the conditions are checked *implicitly* for these algorithms.

The third step in the procedure must prescribe a means of changing flows or node potentials in a manner that will move the solution toward optimality. An important characteristic of this step is whether convergence to the optimum in a finite number of iterations is assured.

b. Dual Node Infeasible Algorithm. The dual node infeasible algorithm maintains complementary slackness and satisfies all primal feasibility requirements except conservation of flow. Let b_i' be the external flow requirement at node i that would satisfy conservation of internal flows at node i, under current arc flow assignments. Let $e_{Ni} = |b_i - b_i'|$ be a measure of the infeasibility for node i and $E_N = \sum_{i=1}^{N-1} e_{Ni}$. In this approach, we achieve optimality by iteratively changing $\mathbf{f}$ and π in such a way to ultimately force E_N equal to zero.

ALGORITHM

1. Find π and $\mathbf{f} \geqslant 0$ that satisfy complementary slackness and the arc capacity restrictions. Let b_i' be the node flows required to obtain conservation of flow.
2. Find a node such that $e_{Ni} > 0$. If there are none, stop. Otherwise go to step 3.
3. Find a new $\mathbf{f}$ and π that reduces the infeasibility of the node while maintaining complementary slackness and arc feasibility. Go to step 2.

As with the primal algorithm, the computational cost depends on the methods by which each of the steps is accomplished. The difficulty of step 1 depends on the type of network problem under consideration and on how "good" an initial solution is sought. In step 2, a decision rule must be specified to determine which infeasible node to select when there is more than one. Finally, the manner in which step 3 is carried out must assure finite termination of the algorithm.

c. Dual Arc Infeasible Algorithm. During each iteration of the dual arc infeasible approach, complementary slackness and conservation-of-flow conditions are satisfied; however, one or more arcs will have flows above their capacities or below zero. Let $e_{Ak} = \text{Max}[-f_k, 0, f_k - c_k]$ be a measure of infeasibility of arc k. Let $E_A = \sum_{k=1}^{m} e_{Ak}$ measure the infeasibility of the network. The dual arc infeasible algorithm achieves optimality by iteratively modifying $\mathbf{f}$ and π to force $E_A = 0$.

ALGORITHM

1. Find an initial f and π that satisfies complementary slackness and conservation of flow.
2. Find an arc k for which $e_{Ak} > 0$. If there are none, stop.
3. Modify f and π in order to reduce e_{Ak} while maintaining conservation of flow and complementary slackness. Go to step 2.

It is possible to modify any primal infeasible f to obtain either a node infeasible flow or an arc infeasible flow. Therefore, only one dual algorithm is required. However, since certain applications may favor either approach both algorithms have been described.

 d. Primal-Dual Approach. The primal-dual approach operates iteratively on solutions f and π that satisfy neither primal feasibility nor complementary slackness.

ALGORITHM

1. Start with an arbitrary f and π.
2. Find a node such that $e_{Ni} > 0$ or arc such that $e_{Ak} > 0$ or $e_{ck} > 0$. If there are none, stop. Otherwise go to step 3.
3. Find a new f and π which reduces e_{Ni} (or e_{Ak} or e_{ck}) for the node (or arc). Go to step 2.

The primary advantage of this kind of algorithm is the simplicity of step 1 contrasted to the relative difficulty of step 1 for the other approaches. This simplicity is offset by the relative difficulty present in step 3.

Because the application of each of these four basic approaches is so computationally oriented, it is appropriate to consider some of the basic computational building blocks and important underlying concepts before applying them to the various types of network flow problems. Chapter 4 is presented with this goal in mind and is essential to the complete understanding of the remainder of this book.

3.9 LIMITATIONS OF NETWORK FLOW PROGRAMMING

Although many interesting problems can be formulated as network programming models, we must remind the reader that this is a special class of models that has limitations. The modeling tools available to the analyst are the network structure, the arc parameters, and the node parameters. If the real situation cannot be modeled with these tools with acceptable accuracy, then network programming is not applicable. Some more general class of models must be applied. It is just as important to understand the limitations of a modeling tool as it is to recognize its possibilities.

Listed below are situations that arise in practice that are usually impossible to solve using the network programming techniques considered in this book.

1. The multicommodity flow problem in which more than one type of flow uses the arcs of the network.
2. Problems that have constraints that link flow on different arcs in addition to those describing conservation of flow.
3. The problem whose objective function includes a nonseparable function of arc flows. An objective of

$$\text{Min.} f_1 f_2 + f_3$$

presents formidable difficulties.
4. The problem in which arc and node parameters vary as a function of flow.
5. The problem in which network structure varies as a function of flow.

For special cases, a clever formulation might overcome some of these difficulties but the observance of any one of these characteristics in a problem situation should signal that network flow programming may not be directly applicable. One should note, however, that the network model will still be a valuable conceptual tool even if these characteristics are present, and it is quite likely that a network flow programming algorithm might be useful when embedded in a more complex solution procedure.

3.10 HISTORICAL PERSPECTIVE

The relationship between network flow programming and linear programming was recognized early in the study of both subjects. Early solution procedures were specializations of the simplex procedures: Dantzig (1951), Charnes and Cooper (1954), Ford and Fulkerson (1956), Dantzig and Fulkerson (1956), Charnes and Cooper (1961), and Dantzig (1963).

The representation of a basis as a tree is described in Dantzig (1963), and Johnson (1966) provides an efficient means of representing a basis tree. Modern investigators have had great success in tying the basic procedures to efficient computational schemes: Charnes et al (1973), Glover and Klingman (1975), and Jensen and Bhaumik (1977).

The integrality of the solutions for pure network flow problems is discussed in Dantzig (1963) and Ford and Fulkerson (1962). The constraint matrix for the pure network model has the characteristic of unimodularity. This characteristic assures integer solutions.

Several nonbasic algorithms have also been proposed for the pure minimum cost flow problem. The flow augmentation method of Busacker and Gowen (1961) is a dual feasible procedure. The negative cycle method of Klein (1967) is a primal procedure. The out-of-kilter method of Fulkerson (1961b) is a primal-dual procedure. References to extensions of these procedures will be provided in later chapters.

EXERCISES

1. Write the following lists for the above network.

$$N, M, O, T; M_{Oi} \text{ and } M_{Ti} \text{ for all } i$$

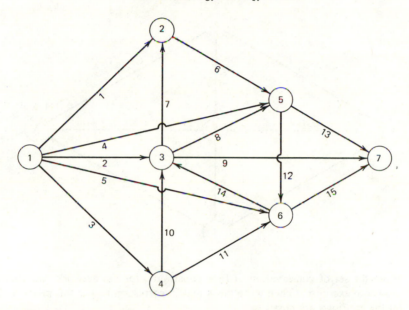

2. Consider the network given below. Is conservation of flow violated at any node? Are the arc capacity constraints satisfied? What is the total cost of the flow present in the network? External flows not shown are zero.

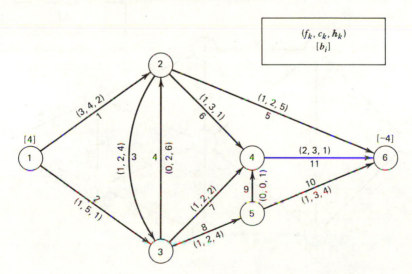

3. The network shown below has both fixed and slack external flows. Draw the equivalent network using arc parameters and a slack node to model the slack external flows.

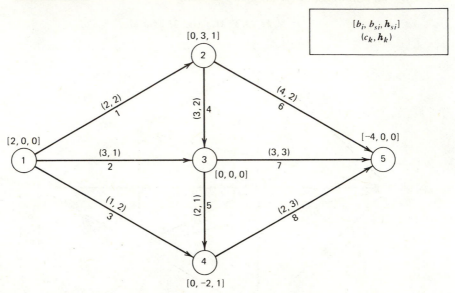

4. Write the set of conservation of flow constraints for the network you obtained in answer to exercise 3. Then write the **A** matrix corresponding to this problem if gains on the arc flows are possible.

5. Transform the network given below to an equivalent network where all arc lower bounds are zero.

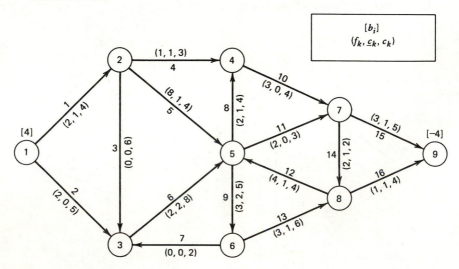

6. Give two examples of physical situations where capacity and cost parameters might be more conveniently associated with a node rather than an arc.

7. Draw the network corresponding to the algebraic network model given below.

$$\text{Min. } Z = 5f_1 + 6f_2 + 7f_3 + 4f_4 + 3f_5 + 2f_6 + 10f_7 + 8f_8 - 6f_9 - 20f_{10}$$

s.t.

$$
\begin{array}{rcl}
f_1 - f_2 + f_3 + f_4 & = & 3 \\
-f_1 + f_2 \quad\quad +f_6 +f_7 \quad\quad\quad -f_{10} & = & 0 \\
f_5 \quad -f_7 +f_8 -f_9 & = & 0 \\
-f_8 +f_9 +f_{10} & = & 0 \\
-f_4 -f_5 -f_6 & = & -3
\end{array}
$$

$$2 < f_1 < 5, 0 < f_2 < 4, 2 < f_3 < 4, 0 < f_4 < 2$$
$$1 < f_5 < 6, 0 < f_6 < 4, 0 < f_7 < 12, 1 < f_8 < 3$$
$$0 < f_9 < 2, 0 < f_{10} < 2$$

8. For the network given below, draw the following subnetworks and identify them by name where appropriate.

    (a) $M_s = \{2,6,10,12\}$, $N_s = \{1,3,6,8,10\}$
    (b) $M_s = \{1,5\}$, $N_s = \{1,2,5\}$
    (c) $N_s = \{1,2,3,5,6\}$, $M_s = \{1,2,3,5,6,9\}$
    (d) $N_s = \{5,6,3,2\}$, $M_s = \{4,5,6,9\}$
    (e) $M_s = \{8,9,10\}$, $N_s = \{5,6,8\}$
    (f) $N_s = \{1,2,3,4,5,6,7,8,9,10\}$, $M_s = \{1,2,3,6,7,9,10,11,12,13\}$
    (g) $N_s = \{1,2,3,5,6\}$, $M_s = \{2,4,6,9\}$
    (h) $N_s = \{1,2,3,4,5,6,7,8,9,10\}$, $M_s = \{1,2,6,7,9,10,11,13\}$
    (i) $N_s = \{1,2,3,5,6,8,9,10\}$, $M_s = \{1,2,13,9,8\}$

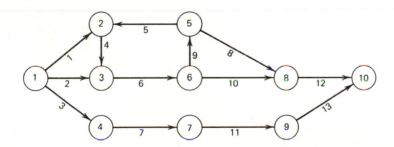

9. Draw the expanded and marginal networks for the network given in exercise 2.

10. A particular arc has a capacity of 10 units of flow and the arc's cost function is $h_k(f_k) = 2f_k + 3f_k^2$. Using no more than three arcs with strictly linear cost functionals, construct a linear approximation to this nonlinear arc cost.

11. Another arc has a capacity of 8 units of flow, and the profit for flow over the arc is inversely proportional to the flow. Construct a linear approximation to this arc if the problem is to minimize total costs.

12. Show that the matrix of the conservation-of-flow constraints for the network of Figure 3.6
    (a) Has Rank $(m-1)$ when f_8 is not included
    (b) Has Rank (m) when f_8 is included.

13. Show that the bases given in Figures 3.7 and 3.8 all have det $(\mathbf{B}) = +1$ or -1.

14. Show that, given $\mathbf{b}$ and $\mathbf{c}$ are integer, the relationship $\mathbf{B}\mathbf{f_B} = \mathbf{b} - \mathbf{R}\mathbf{f_R}$ joined to the fact that all $f_k \in \mathbf{f_R}$ are set to value zero or c_k is sufficient to conclude that $\mathbf{f_B}$ is integer.

15. In what order does the primal simplex algorithm satisfy the optimality conditions?

    (a) Primal feasibility.
    (b) Dual feasibility.
    (c) Complementary slackness.

16. Answer exercise 15 for the dual simplex algorithm.

17. Which algorithm of those given in Section 3.8 would be most appropriate to apply to the basic solutions given in Figure 3.7?

18. Verify that the optimality conditions are satisfied by the solution given in Figure 3.8.

19. Show that the complementary slackness conditions given in Theorem 1 and the result of Theorem 2 lead directly to the complementary slackness conditions of Theorem 3.

CHAPTER 4

NETWORK MANIPULATION ALGORITHMS

This chapter begins our consideration of the computational aspects associated with solution of network flow problems. We believe that a modular approach in our presentation of the computational considerations of network flow programming makes possible a clearer explanation of the methods used and this results in an enhanced understanding of the procedures encountered. Here a number of small algorithms used extensively in the following chapters to store, retrieve, and manipulate network and subnetwork representations will be introduced.

4.1 COMPUTATIONAL COST

When considering alternative algorithms one must consider two aspects of computational cost: time and space. The time cost is the cost of the time that is consumed by an algorithm on a digital computer. This can be described by actual computation time on a specific computer or by counts of operations required to carry out all or parts of procedures. Both approaches have desirable properties. The latter approach is desirable because it is computer independent. However, the former approach has the weight of empirical evidence in its favor.

Operation counts will be given by an order of magnitude as a function of n and m, the number of nodes and arcs, respectively. For example, $O(n+m)$ indicates the time cost is proportional to $n+m$; $O(n^2)$ indicates that time cost is proportional to n^2. Thus $O(g(n,m))$ indicates that time cost equals $a \cdot g(n,m)$ where a is a constant and $g(n,m)$ is a function of n and m. Usually the constant term will not be specified, and the time cost will generally be a worst-case estimate for the network problem under consideration.

Unfortunately, in some cases operation counts alone can be misleading. The simplex algorithm, upon which many network algorithms are based, would have to be considered an inferior approach if only operation counts were considered. However, this widely used algorithm rarely actually uses more than a small proportion of the operations that might be ascribed to it by worst-case operation counts.

Space cost is associated with the amount of computer storage required to represent the data necessary to carry out a procedure. All storage used is assumed to be direct access memory. Although in many instances it would be possible to reduce space cost by "packing" the data, this option will not be used. Conversely, the memory required for the storage of the computer program will not contribute to space cost. Space cost, usually in units of computer "words," will be specified as $b \cdot g(n,m)$ where b is a constant and $g(n,m)$ is again a function of n and m. It will be possible to estimate the constant term for space cost in most cases.

Time and space costs are generally complementary variables in the design of computational procedures. A decrease in one will usually be accompanied by an increase in the other. The most important quantity, time or space, depends on the total resources available. Since time is usually much less constrained than memory space, the ultimate limit for the capabilities of the algorithms described here will usually be in the space requirement. We therefore recognize the primary importance of space cost and eliminate those representations that slavishly use space in return for time. On the other hand, certain expenses in space cost are very effective at reducing time cost and these will be used. The result of our approach will be neither the least costly algorithms in time or space but rather a balance that will provide an acceptable compromise in most applications.

4.2 NETWORK REPRESENTATIONS

The network representation is the set of constants, lists, and matrices used to represent the structure of a network and its parameters for the computational algorithm. The representation used is particularly important to the time cost of an algorithm because a procedure may be much more efficient with one representation than with another. Most algorithms presented in the literature do not explicitly specify the representation used; thus the computational cost of these algorithms is very difficult to determine.

One useful dichotomy is the *arc-oriented representation* versus the *node-oriented representation*. With the arc-oriented representation, the cost of identifying the origin and terminal node of a given arc is small. With the node-oriented representation, the cost of obtaining the parameters of an arc given its origin and terminal nodes is small.

The simplest arc-oriented representation is the *arc list*. Example arc lists are $\mathbf{O} = [o_k]$ and $\mathbf{T} = [t_k]$. Given the arc index k, the origin and terminal nodes of the arc are simply the kth entries in $\mathbf{O}$ and $\mathbf{T}$, respectively. Arc parameters are similarly represented in lists such as $\mathbf{f} = [f_k]$, and $\mathbf{c} = [c_k]$. The space cost of this representation is $(p+2)m$ words, where p is the number of arc parameters (exclusive of the origin and terminal nodes), and m is the number of arcs. Figure 4.1 shows the arc list representation of an example problem.

Arc (k)	1	2	3	4	5	6	7
o_k	2	5	1	1	3	2	5
t_k	4	2	3	2	4	3	1
f_k	2	1	3	1	3	0	1
c_k	2	1	3	3	5	1	2
h_k	-1	-1	1	5	3	2	1

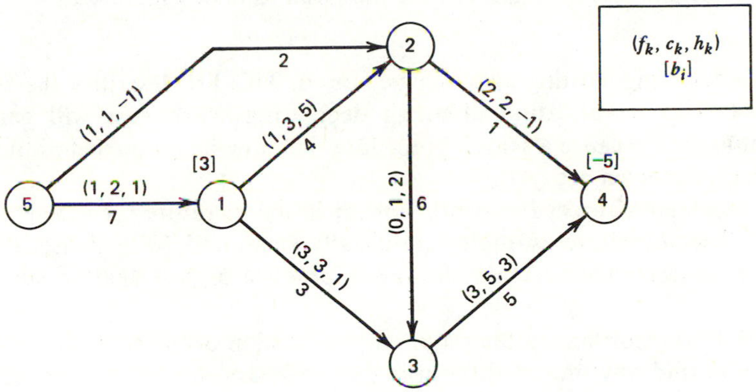

Figure 4.1
Example Arc List Representation

The *simplest node-oriented representation* is the *origin–terminal matrix*. Variables and parameters are stored in these matrices such that the entry in row i, column j corresponds to arc (i,j). Thus the representations include such matrices as $\mathbf{F} = [f(i,j)]$ and $\mathbf{C} = [c(i,j)]$ as shown in Figure 4.2. This representation requires $n^2 \cdot p$ words and has the best time and space cost for networks with near 100% density. The representation is both time and space costly for networks with considerably less than full density, where many matrix entries are wasted for arcs that do not exist in the network. For most practical applications, the density of the network will not be great and the large data requirement for even small networks that actually have high density argues for this proposition. With full density and 100 nodes, there will be 10,000 arcs, each requiring a specification of parameters as data. In practice, one would expect the number of arcs to be considerably reduced by heuristic considerations to obtain relatively low density networks.

It is possible to condense the space required for a node-oriented representation by storing only entries corresponding to existing arcs.

Thus, for each row, one can store a list of columns that represent existing arcs. Such a list would describe the set M_{Oi} for node i. Similarly for each column a list

		Terminal				
		1	2	3	4	5
	1	--	1	3	--	--
	2	--	--	0	2	--
Origin	3	--	--	--	3	--
	4	--	--	--	--	--
	5	1	1	--	--	--

Figure 4.2
Origin–Terminal Matrix for the Arc Flows of Figure 4.1.

of rows representing existing arcs can be stored. This list describes the set M_{Ti} for node i. This modification although decreasing space cost will generally increase time cost because a search procedure will now be required to obtain the parameters for a given arc (i,j).

Origin–terminal matrices frequently appear in the literature for transportation problems. Since literature examples are usually small and 100% dense, this is a reasonable representation. Again, however, practice argues against such high density.

Network flow algorithms often require identification of the sets M_{Oi} or M_{Ti}. It is possible to find any one of these using an ordinary arc list by performing a complete pass through **O** or **T**. Since the operation must be performed many times in the solution of a large problem, it is beneficial to provide some node-oriented information in the network representation. This increase in computational efficiency, as usual, requires some decrease in storage efficiency.

The first modification is to store the arc list in order of *increasing origin node*. Figure 4.3 shows the results of *renumbering* the arcs of Figure 4.1 according to this stipulation. An additional n-length list is required. This list, $\mathbf{P}_O = [p_O(i)]$ (for origin pointer), contains the index of the lowest numbered arc originating from node i; if no arc originates from node i, $p_O(i) = p_O(i+1)$.

Arc (k)	1	2	3	4	5	6	7
o_k	1	1	2	2	3	5	5
t_k	2	3	3	4	4	2	1
f_k	1	3	0	2	3	1	1
c_k	3	3	1	2	5	1	2
h_k	5	1	2	-1	3	-1	1

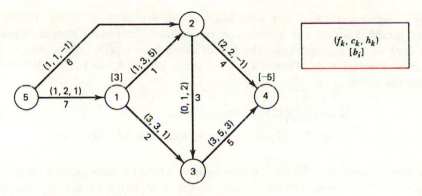

Figure 4.3
Arc List of Figure 4.1 Ordered by Origin Node

Thus the list of arcs is ordered such that

$$
\begin{aligned}
&\text{if } o(k_1) < o(k_2) &&\text{then } k_1 < k_2 \\
&\text{if } o(k_1) = o(k_2) &&\text{then order } k_1 \text{ and } k_2 \text{ arbitrarily}
\end{aligned}
\tag{1}
$$

Furthermore,

$$
\begin{aligned}
p_O(1) &= 1 \\
p_O(i) &= \{ k \mid o(k) \geqslant i, o(k-1) < i \}\ 1 < i \leqslant n
\end{aligned}
\tag{2}
$$

and

$$
p_O(n+1) = m+1
$$

The set of arcs originating at node i, M_{Oi}, is

$$
M_{Oi} = \{ k \mid p_O(i) \leqslant k < p_O(i+1) \}
\tag{3}
$$

By this definition M_{Oi} is empty if $p_O(i) = p_O(i+1)$. Figure 4.4 illustrates the origin pointer list for the network of Figure 4.3.

Node	1	2	3	4	5	6
p_O	1	3	5	6	6	8

Figure 4.4
Origin Pointer for Arc List of Figure 4.3

Rapid determination of the sets M_{Ti} can be provided, if necessary, by a further modification of the network representation. This modification requires additional m- and n-length lists. The first of these, $\mathbf{L}_T = \{l_T(i)\}$ orders the arcs by *increasing terminal node*. Thus if k'_w is the index of arc k_w in $\mathbf{L}_T$, the following defines the ordering

$$\text{if } t(k_w) < t(k_y) \quad \text{then } k'_w < k'_y$$
$$\text{if } t(k_w) = t(k_y) \quad \text{order } k_w \text{ and } k_y \text{ arbitrarily} \tag{4}$$

Since this list contains all arcs, it is of length m. The aforementioned n-length list is $\mathbf{P}_T = \{p_T(i)\}$ whose element points to the first entry in the $\mathbf{L}_T$ list that terminates at node i. Thus

$$p_T(1) = 1$$
$$p_T(i) = \{k' \mid t(l_T(k')) \geqslant i, t(l_T(k'-1)) < i\} \tag{5}$$

for

$$i \leqslant n$$

and

$$p_T(n+1) = m+1$$

The set

$$M_{Ti} = \{l_T(k') \mid p_T(i) \leqslant k' < p_T(i+1)\} \tag{6}$$

Figure 4.5 illustrates the terminal auxiliary list and terminal pointer, consistent with Figure 4.3.

k'	1	2	3	4	5	6	7
$\mathbf{L}_T$	7	1	6	2	3	4	5
Node	1	2	3	4	5	6	
$\mathbf{P}_T$	1	2	4	6	8	8	

Figure 4.5
Auxiliary Terminal Node-Oriented List with Terminal Pointer

4.3 METHOD OF ALGORITHM DESCRIPTION

All algorithms presented in this book will be stated at two levels of abstraction: a verbal statement of the procedure and a flowchart. Each level has certain advantages that aid in overcoming the disadvantages of the other level. The verbal statement of the algorithm is closest to human perception and, therefore, may be best at describing the individual steps in a procedure. However, it does not clearly display complex logic transfers that are often necessary. Also the verbal statement completely neglects the relation between data storage structure and the computational procedure. The flowchart excels in visually displaying logical transfers and is also quite useful in illustrating the relation between data storage and procedure.

The presentation adopted in this book provides the verbal statement and the flowchart statement in parallel. This results in an easily readable description of each procedure.

Because our flowchart format differs somewhat from the norm, some additional explanation is appropriate.

Our decision to use a different flowchart format requires only a small initial investment of effort by the reader. This small effort is justified because our flowchart format is both more compact and easier to comprehend than the traditional format.

A *subroutine* is a collection of program operations identified by a name within a rectangle with rounded ends, as shown below:

$$(\text{ NAME })$$

When this symbol appears outside a box, as pictured above, the flowchart that follows the symbol defines the logic of the subroutine. When the symbol appears within a box, as pictured below, the steps of the subroutine are to be performed as part of another subroutine or the master program. This is equivalent to the CALL statement in FORTRAN.

$$\boxed{(\text{ NAME })}$$

A "box" is a rectangular shape. Every program statement and subroutine reference must appear in a box. Boxes may be any dimension required for convenient representation. Statements arranged within a box are to be performed one line at a time, from left to right, and the lines are to be performed from top to bottom. Boxes may be included in boxes, as below:

NAME

The flow of logic proceeds from the top down.

A box may be labeled with a NAME above the upper left-hand corner. This name is used for reference in written descriptions of the program. It also may identify the box for transfer from another part of the program. Two boxes may be connected by a vertical line:

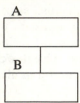

This line indicates that, when the bottom line of box A is reached, control passes to box B.

A *variable* is a letter or set of letters representing a memory location within the computer. The variable usually represents some quantitative characteristic of the model or quantity obtained by computation. A variable may have subscripts. A(K) represents a member of a list or vector. K is the general index that may take on the values $1, 2, 3, \ldots N$, where N is the length of the list. A(I, J) represents a member of a matrix with the indices I and J indicating the location of the member in the matrix. A(I, J, K) is a member of a three-dimensional matrix.

The *Assign Statement* is symbolized

$$A := B$$

This means that the value of variable B is assigned to variable A. A must be a variable; B may be a variable or expression.

The *If statement* always appears in a box and has the form

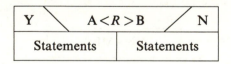

where the symbol R represents a logical relation. The "if" statement asks the truth of the relationship $A < R > B$. If the relationship is true, the statements in the box below Y are performed. If the relationship is not true, the statements in the box below N are performed. The "Go To" statement appears as below:

$$\text{NAME} \quad \text{or} \quad \rightarrow \text{NAME}$$

Either representation transfers control to the box labeled NAME.

The word "READ" followed by a list of variable names indicates that the program is to obtain values for the variables from an outside source. Generally, the identity of the source or the format of the data is not indicated. If an array is specified without indices, the entire array is read.

The word "PRINT" followed by a list of variable names indicates that the program is to print the values of the variables on some output device. Generally, the identity of the device or format of the data is not indicated.

When the STOP statement is encountered, the entire operation of the program is to stop. When the RETURN statement is encountered, control is to leave the subroutine in which the statement appears and return to the calling subroutine. The next statement to be executed follows the statement in which the subroutine call appeared.

The *"DO" loop* is a mechanism for causing a set of statements to be performed a repeated number of times. It is represented by the form

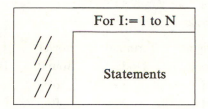

The statements in the inner box are performed for each integer value of the variable I as I is incremented from an initial value as 1 to the value N in steps of 1. Thus I and N are integer program variables. The paired diagonal lines indicate the extent of the "DO" loop. After the statements have been performed for each value of I, control is passed through the bottom line of the box containing the loop.

The flowcharts do not indicate the means of *communication* between subroutines usually provided by COMMON and ARGUMENTS in FORTRAN programs. The *arguments* shown with a subroutine name indicate the principal inputs and outputs of the subroutine.

4.4   READING AND STORING THE NETWORK

Suppose the analyst has reduced a system to a network model. The nodes have been numbered consecutively, each node's fixed external flow has been determined, and the bound and cost for the slack external flows have been determined. We recall that the direction for the slack flow is indicated by the sign of the bound: positive for an incoming slack flow and negative for outgoing slack flow.

The node data is prepared with a data item for each node that is read in the sequence: node number, fixed external flow, bound on slack external flow, cost for slack external flow. A blank data item follows the data items for all nodes.

Following the node data, the arc data is read for one arc at a time in the sequence: origin node, terminal node, lower bound, upper bound, cost. A blank data item follows the description of the arcs of the network.

The data read is that required for the pure linear minimum cost flow problem. The total content of the data set will be different for other problem types but the logic of the procedure remains correct. Figure 4.6 illustrates the data for the example problem of Figure 4.1.

Node data:	Node	External Flow	Slack Bound	Slack Cost
	1	3	2	1
	2	0	1	−1
	3	0	0	0
	4	−5	0	0
	Blank			

Arc Data:	Origin Node	Terminal Node	Lower Bound	Upper Bound	Cost
	1	2	0	3	5
	1	3	0	3	1
	2	3	0	1	2
	2	4	0	2	−1
	3	4	0	5	3
	Blank				

Figure 4.6
Data for Example Problem

In this section of the chapter, we present algorithms READ, ORIGS, ORIG, TERMS, and TERM. These five basic algorithms perform the following functions:

READ—reads the network data and uses the node data and a slack node to modify the network to have only fixed external flows.
ORIGS—orders all arcs in order of increasing origin node, derives the node length origin pointer list P_O, and performs lower bound transformations if required.
ORIG—finds M_{Oi} for a stipulated node i.
TERMS—derives both the auxiliary arc length list L_T ordered by ascending terminal nodes and a node length terminal pointer list P_T.
TERM—finds M_{Ti} for a stipulated node i.

In essence, these five algorithms simply implement the concepts contained in equations (1) through (6). The detailed presentation of the above algorithms appears on the following pages.

READ ALGORITHM

Purpose: To read and store node and arc data for the minimum cost flow problem.

1. (INITIAL) Initialize number of arcs to zero. Read number of nodes, create slack node, set all fixed external flows to zero.
2. (NODE) Read a node data item. If the item is blank, go to step 3. Else put fixed flow in storage. If the slack external flow is zero, repeat step 2. Otherwise, create a slack arc and store the arc data in the correct positions in the arc lists.
3. (ARC) Read an arc data item. If the item is blank, the complete data set is read. Go to step 4. Else, store the arc data in the correct positions in the arc lists. Repeat step 3.

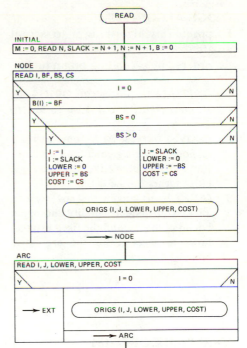

4. (EXT) Place each arc's data in the correct position on the terminal list.

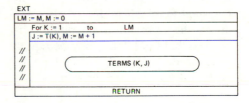

ORIGS ALGORITHM

Purpose: This algorithm is to accept an arc data item and store it in an arc list ordered by ascending origin node.

1. (INITIAL) If first call to ORIGS, set all node pointers to 1. Else go to step 2.
2. (MOVE) Increase M by one. Increase all node pointers greater than I by one. Move all arcs above the new entry one index higher on the list.
3. (ARC) Insert arc in the last position allotted to node I. Modify the upper bound of the arc and the fixed external flows to account for the arc lower bound.

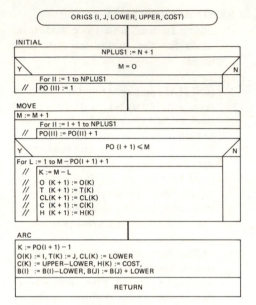

TERMS ALGORITHM

Purpose: This algorithm creates LT, a list of the arc indices in order of increasing terminal node. It also produces the terminal node pointer list (PT) so that the LT list may be rapidly referenced. It is called one time for each of the arcs in the network.

1. (INITIAL) Set node pointers to one the first time through.
2. (MOVE) Increase the pointers for all nodes greater than J by one. Move all references for arcs with terminal node greater than J one index higher in the list.
3. (ARC) Insert the new entry in the list in the last position allotted to node J.

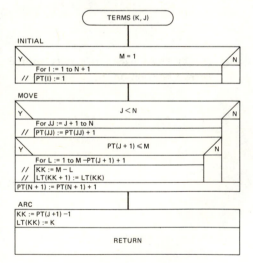

ORIG ALGORITHM

Purpose: To determine the list of arcs (and their terminal nodes) originating at node I.

Find pointers to start and end of list of arcs originating at node I.

If there are no such arcs, return. Otherwise, put arcs originating at I in the arc list. Find the terminal node of each arc and put it in the node list.

TERM ALGORITHM

Purpose: To determine the list of arcs (and their origin nodes) terminating at node I.

Find pointers to start and end of arcs terminating at node I. If there are no such arcs, return. Otherwise, find name of arcs in the auxiliary arc list. Put each arc in the arc list. Find the origin node of each arc and put it on the node list.

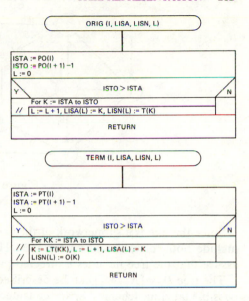

4.5 TREE REPRESENTATION

Section 3.4 established that the arcs in a basic solution to a pure network flow problem form a spanning tree. We will consistently represent this tree as a directed spanning tree, in the expanded network, rooted at the slack node. Since several of the algorithms use a basis, it is important that the basis tree be represented in storage in such a way that it is easily searched and manipulated. Several kinds of operations must be performed on the tree. Among these are:

1. Find the unique path from node i to node j in the tree.
2. Find the set of nodes in that portion of the tree rooted at some node i.
3. Delete one arc from the tree, and add another.

To illustrate the tree representation, we introduce a new example network in Figure 4.7. Two spanning trees for this network are shown in Figure 4.8. The root node appears at the bottom of the figure and is said to be at depth 0. The tree arcs originating at the root emanate up from the root node and their terminal nodes are all placed on the same horizontal line. These nodes are said to be at depth 1 of the tree. The tree arcs originating from nodes at depth 1 can be drawn up from these nodes to terminate at a node on depth 2. This process continues until eventually there will be some depth at which no tree arcs originate.

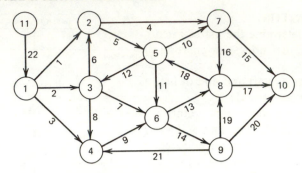

Figure 4.7
Example Network

Since the tree is a subnetwork of the expanded network of Section 2.10, it may include both forward and mirror arcs. The directions of the arcs are chosen to assure that the tree includes a directed path from the root to all other nodes.

The tree $D_T = [N_T, M_T]$ can be described completely by using any one of many possible labeling schemes. Space considerations force us to limit our attention to two specific schemes. These were selected because they are each representative of one of the two more popular and well known methods of representing a tree.

Let us first consider a triple labeling scheme. For this representation, each node is provided three labels, each indicating either an arc or a node of the tree. These labels for node i are called the back pointer, $p_B(i)$, the forward pointer, $p_F(i)$, and the right pointer, $p_R(i)$. The back pointer is the unique *arc* that terminates at node i. The forward pointer is the left-most *node* of the tree that is a terminal node of an arc that originates at node i. The right pointer is the *node* of the tree that appears directly to the right of node i; that is, $p_R(i)$ points to the node "right adjacent" to node i and on the same level as node i. Figure 4.8 gives the associated pointer representation for each of the spanning trees contained therein; $p_D(i)$ is the depth label for each node and $p_P(i)$ is the "preorder traversal" label for each node. p_P will be fully discussed when the second method of tree representation is presented.

It is characteristic of a tree that each node is the terminal of at most one arc. However, each node may originate more than one arc. Thus, for a given D_T, the assignment of back pointers is unique, but the assignment of forward and right pointers is not necessarily unique.

Tree Path-Finding Algorithm

Given any pair of nodes i and j, there exists a unique path from i to j defined by the arcs of the tree. A path will contain a junction node ℓ and may contain either a *reverse path* or a *forward path* or both.

The junction node is the last common node on the two unique paths from the root to nodes i and j, respectively. If i lies on the path from the root to j, then

Node	1	2	3	4	5	6	7	8	9	10	11
P_B	22	6	-8	3	-12	9	4	19	-21	17	0
P_F	4	7	5	9	0	0	0	10	8	0	1
P_R	0	0	6	0	2	0	0	0	3	0	0
P_D	1	4	3	2	4	3	5	4	3	5	0
P_P	4	7	5	9	2	11	6	10	8	3	1

Node	1	2	3	4	5	6	7	8	9	10	11
P_B	22	6	2	3	18	7	10	13	14	20	0
P_F	3	0	6	0	7	9	0	5	10	0	1
P_R	0	0	4	0	0	2	0	0	8	0	0
P_D	1	3	2	2	5	3	6	4	4	5	0
P_P	3	4	6	11	7	9	2	5	10	8	1

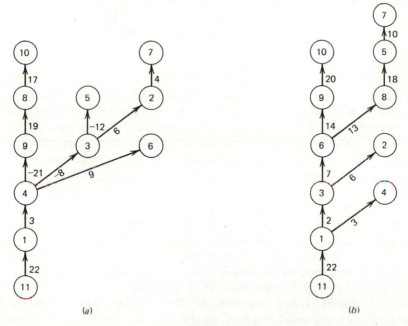

(a)   (b)

Figure 4.8
Example Spanning Trees from Figure 4.7.

the junction is node i and there is no reverse path. If j lies on the path from the root to node i, then j is the junction and there is no forward path. Otherwise, the reverse path proceeds from node i to node ℓ traversing arcs in the reverse direction to their orientation in the tree, while the forward path proceeds from node ℓ to node j following the arcs in the same direction as their tree orientation.

Let M_{rj} and M_{ri} be the set of arcs on the path from the root to j and i, respectively. The set of arcs common to the two paths is $M_{r\ell}$. Thus the arcs in the forward path are

$$M_F = M_{rj} - M_{r\ell} \tag{7}$$

The arcs in the reverse path are

$$M_R = M_{ri} - M_{r\ell} \tag{8}$$

If $M_{\bar{R}}$ is defined as the set of mirror arcs of M_R, then we may write the set of arcs in the path from i to j as

$$M_{ij} = M_F \cup M_{\bar{R}} \tag{9}$$

A conceptually simple, but computationally naive, method of determining M_{ij} would be to first use the back pointers from node j to find M_{rj}. Next, the back pointers from node i could be traced until the first node common to M_{rj}, node ℓ, is encountered. Equations (7), (8), and (9) above could then be directly applied. In small networks, the user would not pay a large penalty in using this method. However, in large networks, $M_{r\ell}$ will usually be much larger than both M_F and M_R. Consequently, the essentially unnecessary tabulation and subsequent deletion of the elements of $M_{r\ell}$ would comprise almost all of the computation effort.

Fortunately, this problem can be overcome by associating an additional label $p_D(i)$ with each node. Then $p_D(i)$ will contain the depth of node i in the tree. If the depths of nodes i and j are known, we first identify which of the pair of nodes, if either, is deepest in the tree. Suppose, without loss of generality, that node j is deepest. Next, we backtrack from node j visiting one node at a time until the node currently "visited" has the same depth as node i.

From that point, we first determine if the node currently visited on both backpaths is the same node. If this is not the case, we backtrack simultaneously on both backpaths until the junction node is found. Algorithm TPATH implements the above conceptual approach as part of our subroutine library.

TPATH-ALGORITHM

Purpose: To find the path in the tree from node (IS) to node (IT). The outputs are the list of arcs on the path (LISA), the list of nodes (LISN), the number of arcs in the path (IC), the junction (JUNC) between reverse and forward parts of the path. The arcs and nodes of the path are listed in reverse order. If there is no path from IS to IT, NP is returned as 1.

1. (INITIAL) Initialize variables and compute the difference in the depths of nodes IS and IT.
2. (DECIDE) Decide if additional backtracking is necessary to achieve equal depths for the two nodes currently visited. If not, go to COMPARE. Otherwise, determine which node has greater depth and stipulate one backtrack iteration from that node.
3. (ITBACK) Perform one backtrack iteration on the backpath from node IT. Record the arc and node encountered. If the currently visited node has no backpointer, no path exists between IS and IT.
4. (ISBACK) Perform one backtrack iteration on the backpath from node IS. Record the node and arc encountered. If the currently visited node has no back pointer, no path exists between IS and IT.
5. (COMPARE) Since the currently visited nodes are at the same depth, they may be the same node. Otherwise, go to ITBACK.
6. (COMBINE) Combine the nodes (arcs) recorded on the two backpaths into a single contiguous list.

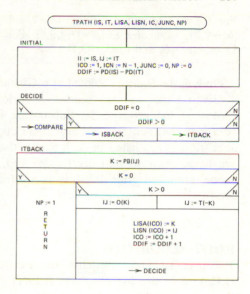

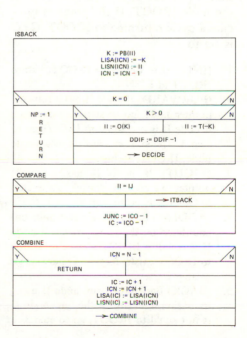

For Figure 4.8*a* with IS=8 and IT=2, subroutine TPATH will return IC=4, LISA=
[6, −8,21, −19], and LISN=[3,4,9,8]. For IS=2 and IT=4, the results are IC=2,
LISA=[8, −6], and LISN=[3,2].

Finding the Subtree Rooted at a Given Node Using Triple Labels

Given a tree and any one of its nodes, there is a unique subtree of the tree
rooted at the given node. Subroutine ROOT finds the arcs and nodes that
comprise that subtree. Because some applications of ROOT will require it, the
arcs (nodes) are listed in such a way that all the predecessors of an arc (node)
are listed prior to that arc (node) and, further, all successors of a given arc
(node) are listed prior to listing an arc (node) that is not a successor of the given
arc (node).

ROOT ALGORITHM

Purpose: To find the list of arcs
(LISA) and the list of nodes (LISN)
that are in the directed tree rooted at
the node IROOT. If the pointers indicate a cycle returning to IROOT, CYC
is set to 1.

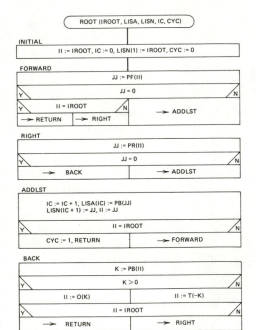

1. (INITIAL) Let II = IROOT; store
   IROOT in LISN.
2. (FORWARD) If node II has a forward
   pointer node, go to step 4. Otherwise,
   if node II is not equal to IROOT go to
   step 3. The subtree consists of a single
   node, IROOT, Return.
3. (RIGHT) If node II has no right
   pointer, we must backtrack; go to step
   5. Otherwise, go to step 4.
4. (ADDLST) Store the arc and node encountered on the latest search. If node
   II is equal to IROOT a cycle returning
   to IROOT exists. Return. Otherwise go
   to step 2.
5. (BACK) Backtrack from node II using
   its backpointer arc. If the new node II
   is not equal to IROOT, go to step 3.

Suppose we have the tree of Figure 4.8a and the given root, node 4. The procedure searches from left to right through all paths originating at the root. The left-most path is searched first by tracing through forward pointers. This search discovers the arcs $(-21, 19, 17)$ and nodes $(4, 9, 8, 10)$. When this path is complete, the node farthest from the root on this path with a nonzero right pointer is found and the path is traced through this node and through subsequent forward pointers. For our present example, this adds the arcs $(-8, -12)$ and nodes $(3, 5)$. The process continues until the complete subtree is identified, which for our present example is

$$M_{T4} = (-21, 19, 17, -8, -12, 6, 4, 9)$$

$$N_{T4} = (4, 9, 8, 10, 3, 5, 2, 7, 6)$$

Deleting an Arc from the Basis Tree Using Triple Labels

The deletion of arc $k_L(i_L, j_L)$ from the basis tree is accomplished by modifying the pointer representation of the tree. There are two ways an arc may appear:

As a "forward pointer arc"; that is, for some pair of nodes

1. $p_B(j_L) = k_L$ and $p_F(i_L) = j_L$

or as a "right pointer arc"; that is, for some pair of nodes

2. $p_B(j_L) = k_L$ and $p_R(\ell) = j_L$

In Figure 4.8a, arcs -21 and -8 are forward and right pointer arcs, respectively.

In either case, node j_L becomes the root of a subtree when $k_L(i_L, j_L)$ is deleted from the original tree. Therefore, after the arc deletion is complete, node j_L can have no back or right pointer node. Then $p_R(j_L)$ cannot be set immediately to zero since one other thing must be done to update the triple label structure of the original tree for each of the two cases.

1. $p_F(i_L) := p_R(j_L)$
2. $p_R(\ell) := p_R(j_L)$

Following the performance of the appropriate step above, we need only implement

$$p_B(j_L) := 0   \text{and}   p_R(j_L) := 0$$

In case 2 a simple search through right pointer nodes beginning at $p_F(i_L)$ is performed to identify node ℓ (such that $p_R(\ell) = j_L$).

Figures 4.9a and 4.9b illustrate the two cases for arcs deleted from the tree of Figure 4.8a. These examples illustrate that the deletion of an arc from a tree forms two subtrees, each having all the properties of a tree.

Node	1	2	3	4	5	6	7	8	9	10	11
P_B	22	6	-8	3	-12	9	4	19	0	17	0
P_F	4	7	5	3	0	0	0	10	8	0	1
P_R	0	0	6	0	2	0	0	0	0	0	0
P_D	1	4	3	2	4	3	5	4	3	5	0
P_P	4	7	5	3	2	11	6	10	8	9	1

Node	1	2	3	4	5	6	7	8	9	10	11
P_B	22	6	0	3	-12	9	4	19	-21	17	0
P_F	4	7	5	9	0	0	0	10	8	0	1
P_R	0	0	0	0	2	0	0	0	6	0	0
P_D	1	3	2	2	5	3	6	4	3	5	0
P_P	4	7	5	9	2	11	3	10	8	6	1

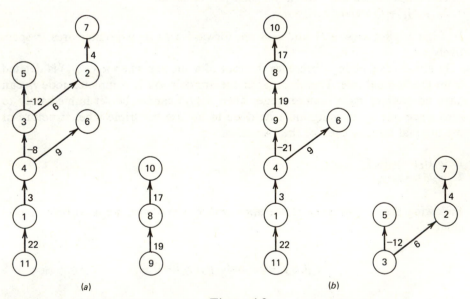

Figure 4.9
Examples of Arc Deletions from Tree of Figure 4.8a (a) Arc $-21(4,9)$
Deleted (b) Arc $-8(4,3)$ Deleted

Subroutine DELTRE, described below, performs these operations as part of our network manipulation library.

DELTRE ALGORITHM

Purpose: To delete an arc from the pointer representation of the tree.

1. (FORWARD) Arc $k_L(i_L, j_L)$ is to be deleted from the tree. If $p_F(i_L) = j_L$, set $p_F(i_L) := p_R(j_L)$ and go to step 3. If $p_F(i_L) \neq j_L$, go to step 2.
2. (RIGHT) Find the node ℓ for which $p_R(\ell) = j_L$. Set $p_R(\ell) = p_R(j_L)$ and go to step 3.
3. (DELETE) Make node j_L a root node; that is, $p_B(j_L) := 0$, $p_R(j_L) := 0$.

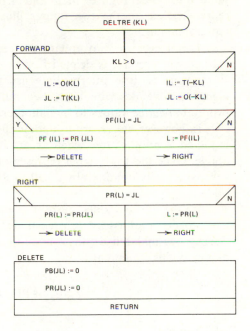

Adding an Arc to a Forest Using Triple Labels

A forest may consist of two or more directed trees, $D_{T1}, D_{T2}, \ldots$. When an arc $k_E(i_E, j_E)$ is added to the forest such that node i_E lies in one tree and node j_E is the root of another, two trees are linked to obtain a single tree and the number of trees in the forest is reduced by one. Two modifications to the pointer representation are necessary to accomplish the addition of an arc. Again two cases must be considered for the first modification.

1. If $p_F(i_E) = 0$, then $p_F(i_E) := j_E$
2. If $p_F(i_E) = \ell$ (i.e., $p_F(i_E) \neq 0$), let $p_R(j_E) := \ell$ and $p_F(i_E) := j_E$

The second modification is:

$$p_B(j_E) := k_E$$

This is possible since node j_E is the root of a tree; hence, the back pointer for node j_E is zero prior to the assignment operation. Figure 4.10 illustrates the tree obtained by adding an arc to the forest of Figure 4.9b. Note that the progression of Figures 4.8a, 4.9b, and 4.10 describes the deletion and the addition of the same arc. Since Figures 4.8a and 4.10 represent the same tree, the pointer representations in the two figures illustrate the nonunique character of these representations.

It is not necessary to update the depths of the nodes in the newly created subtree when an arc is deleted from a tree. There are two reasons for this fact. First, there is little intermediate use made of such subtrees in the network algorithms we describe later. Second, the node depths remain at their appropriate *relative* values to one another.

It is, however, necessary to update the depths of the nodes in the subtree that is joined to another tree when an arc is added to a forest of trees as described above. Once the triple label representation is updated, the list of nodes in the subtree is easily obtained by a call to ROOT with node j_E as the stipulated root node.

Since the subtree nodes' current depths have the correct relative values, all we need to do is adjust all subtree node depths by a uniform amount, PDADJ. Clearly, the updated depth of node j_E must be one greater than $p_D(i_E)$. Mathematically this statement is

$$p_D(j_E) + \text{PDADJ} = p_D(i_E) + 1$$

which implies

$$\text{PDADJ} = p_D(i_E) - p_D(j_E) + 1$$

Adding PDADJ to the depths of the subtree nodes completes the operations required.

Subroutine ADDTRE performs the operations described above for our network library.

Node	1	2	3	4	5	6	7	8	9	10	11
P_B	22	6	8	3	-12	9	4	19	-21	17	0
P_F	4	7	5	3	0	0	0	10	8	0	1
P_R	0	0	9	0	2	0	0	0	6	0	0
P_D	1	4	3	2	4	3	5	4	3	5	0
P_P	4	7	5	3	2	11	9	10	8	6	1

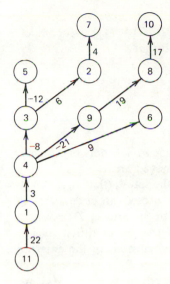

Figure 4.10

Arc $-8(4,3)$ Added to the Forest of Figure 4.9 b

ADDTRE ALGORITHM

Purpose: To add arc $k(i,j)$ to a forest of trees. Nodes i and j must be in different trees and node j must be the root of a tree.

1. (FORWARD) If the forward pointer node i_E is zero, go to step 2. Otherwise, let the right pointer of j_E equal the forward pointer of i_E and go to step 2.
2. (BACK) Let the back pointer of j_E equal k_E and let the forward pointer of i_E equal j_E.
3. (DEPTH) Update depths of nodes in subtree rooted at node j_E.

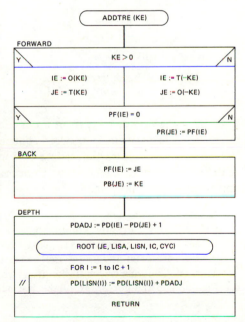

BASIS CHANGE ALGORITHM

When an arc $k_L(i_L, j_L)$ is removed from the basis tree $D_T = [N, M_T]$ two components are formed, $D_1 = [N_1, M_1]$ and $D_2 = [N_2, M_2]$, such that

$$j_L \in N_2 \qquad i_L \in N_1$$
$$N_1 \cup N_2 = N$$
$$M_1 \cup M_2 = M_T - k_L$$

The network D_1 is a directed tree rooted at slack node n. The network D_2 is a directed tree rooted at the terminal node j_L of the removed arc k_L.

The entering arc $k_E(i_E, j_E)$ is chosen to terminate in the set N_2 (that is, $j_E \in N_2$). The new basis tree $D_T' = [N, M_T']$ is a directed tree rooted at node n. It is constructed by combining D_1, the arc k_E, and D_2, with the arcs in D_2 possibly redirected to make D_T' a directed tree. Let $D_2' = [N_2, M_2']$ be the modified version of D_2. D_2' must form a directed tree rooted at the terminal node of the entering arc, node j_E. The only change required to obtain D_2' from D_2 is to reverse arcs on the path from node j_L to node j_E in D_2. If this path is $D_P = [N_P, M_P]$, we desire to replace it with $D_{\bar{P}}' = [N_P, M_{\bar{P}}]$. The change is accomplished by the following.

$$M_2' = (M_2 - M_P) \cup M_{\bar{P}}$$

For the new basis:

$$M_T' = M_1 \cup k_E \cup M_2'$$

Perhaps the easiest way, from a conceptual viewpoint, to effect a tree change operation is to perform an ordered sequence of arc deletions and additions. This sequence should assure that a tree or forest of trees is maintained at all times. Although there are several ways to accomplish this, one particularly appealing approach is the following.

1. Delete the leaving arc k_L.
2. Delete the arc currently terminating at j_E and add the entering arc k_E.
3. Proceed down the reverse path going from j_E to j_L, first deleting the next arc on the reverse path, then adding the mirror arc of the arc that was deleted in the previous iteration. Continue this until the arc originating at j_L is deleted.
4. Finally, add the mirror arc of the last deleted arc and the tree change operation is complete.

This sequence of deletions and additions always assures that any arc being added is terminating at the root of a subtree (a requirement for the use of algorithm ADDTRE).

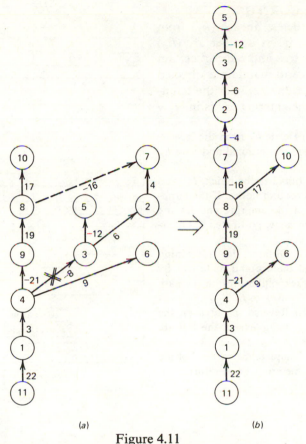

(a)   (b)

Figure 4.11
Basis Change Operation (a) Original Basis (b) Basis Formed by Deleting
$-8(4,3)$ and Adding $-16(8,7)$

Figure 4.11 illustrates the results of a specific tree change operation. Here arc $-8(4,3)$ is deleted from the tree of Figure 4.8a and arc $-16(8,7)$ is added.

This tree change could be interpreted as the result of deleting arc -8, deleting arc 4, adding arc -16, deleting arc 6, adding arc -4, and finally adding arc -6. Note that in each case the arc being added terminated at the root of a subtree.

Subroutine TRECHG, with the help of DELTRE, and ADDTRE, may be used to perform the change of basis manipulations described above. The algorithmic details are presented below.

TRECHG ALGORITHM

Purpose: To delete an arc (k_L) from the basis tree, insert another arc (k_E) into the basis tree and redirect certain arcs in the tree to maintain a directed tree. The algorithm assumes the terminal node of the entering arc is in N_2.

1. (DELETE) Delete k_L from the basis.
2. (FIND) Find the terminal nodes for k_E and k_L.
3. (CHECK) Initialize the index, IC, and the list of arcs and nodes. If the terminal nodes of the entering and leaving arcs are the same, go to step 6. Otherwise go to step 4.
4. (OBTAIN) Obtain the remaining members of the arc and node lists by using the backpointer arcs on the path in the tree from j_L to j_E.
5. (REVERSE) Reverse the arcs on the path from j_L to j_E, except the last arc on the path.
6. (FINISH) Complete the reversal of the last arc on the arc list and return.

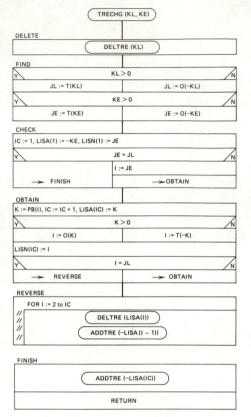

4.6 EQUIVALENT ALGORITHMS USING A PREORDER TRAVERSAL LIST

Let us consider another way to represent the list of nodes in the order that they are returned by algorithm ROOT.

Suppose ROOT had been called for the tree of Figure 4.8a and node 11 was the specified root node, IROOT. The list of nodes would be returned as:

$i =$	1	2	3	4	5	6	7	8	9	10	11
LISN(i)	11	1	4	9	8	10	3	5	2	7	6

Notice, that by following the order in LISN, node 11 "points" to node 1, node 1 points to node 4,..., node 7 points to node 6. Further if one envisions a closed circular list, node 6 may be viewed as pointing to node 11, the slack node.

The preorder traversal list P_P simply stores this natural pointer representation such that $p_P(i)$ is set equal to the node number that node i "points at." For Figure 4.8a, P_P is

i	1	2	3	4	5	6	7	8	9	10	11
P_P	4	7	5	9	2	11	6	10	8	3	1

Among its other advantages, which will be discussed below, the list P_P effectively replaces the lists P_F and P_R. This reduces the number of node length lists required to represent the network by one list.

Finding the Subtree Rooted at a Given Node Using a Preorder List

The preorder list P_P may be used to great advantage in performing the function of algorithm ROOT. With P_P at hand, it is very simple to obtain any node's set of successor nodes and arcs, that is, the nodes and arcs that are members of the subtree rooted at that node. We simply begin at the stipulated root, say node r, and record it in the first position of LISN. The next node entered in the list is $s = p_P(r)$ and the first entry into the arc list is $p_B(s) = p_B(p_P(r))$, the back pointer arc of the *immediate preorder traversal successor* of r. The second node, and all subsequent preorder traversal successors, are found through the recursive relation embodied in the list P_P; that is, $p_P(s) = p_P(p_P(r))$. It follows directly that the next arc to be placed on the arc list is $p_B(p_P(p_P(r)))$.

This process is continued until we arrive at a node r' such that $p_D(r') \leqslant p_D(r)$. Since only nodes with depths strictly greater than the depth of r can be a successor of r, we have visited and recorded all nodes and arcs in the subtree rooted at r. Only one thing remains to be checked: If $r' = r$ and $p_B(r) \neq 0$, a cycle returning to r exists.

Suppose we desired the list of nodes in the subtree rooted at node 6 in Figure 4.8b. Using the P_P and P_D lists given in the figure, we would find that

$$LISN = [6, 9, 10, 8, 5, 7]$$

The preorder traversal successor of node 7 is node 2, but $p_D(2) = p_D(6) = 3$ and therefore node 2 cannot be a successor of node 6.

Algorithm ROOTP implements the concepts presented above and is an alternative to algorithm ROOT. Note that the coding of algorithm ROOTP is given the same name, ROOT, as the coding of algorithm ROOT was given earlier. This is done in order that the set of equivalent algorithms using a preorder traversal list may be substituted interchangeably for its set of triple label counterparts.

ROOTP ALGORITHM

Purpose: (Same as ROOT) To find the list of arcs (LISA) and the list of nodes (LISN) that are in the directed tree rooted at the node IROOT. If a cycle that returns to IROOT exists, CYC is set to 1.

1. (INITIAL) Initialize list index, cycle flag, and depth of stipulated root. Store IROOT as first member of node list and access second node in subtree.
2. (GETLSTS) If the depth of the currently visited node is greater than the depth of the root node, store next elements of node and arc lists. Access next successor node.
3. (CKCYC) If the terminal node of the search is the root node, a cycle is present; set CYC=1. Otherwise, return with arc and node lists of subtree rooted at IROOT.

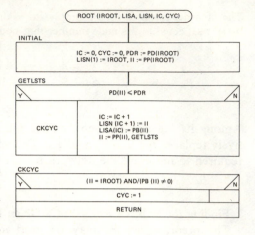

Deleting an Arc from the Basis Tree Using a Preorder Traversal List

Deletion of an arc from a tree that is represented by P_P and P_B is also straight forward. Node j_L can be either a forward or right pointer node. Conceptually it makes no difference which it is; computationally it makes little difference either. Suppose $k_L(i_L, j_L)$ is leaving. Define j_L' to be the last preorder successor node in the tree rooted at j_L. This is easily found by incrementing through P_P from j_L until the depth of a node $p_P(j_L') \leqslant$ depth of j_L, that is, $p_D(p_P(j_L')) \leqslant p_D(j_L)$. Also let us define q to be the node such that $p_P(q) = j_L$, which may be found by incrementing P_P from i_L. Once j_L' and q are identified, the following three operations are all that are required to update P_P and P_D.

$$p_P(q) := p_P(j_L')$$

$$p_P(j_L') := j_L$$

$$p_B(j_L) := 0$$

P_D is not updated for the same reasons that were given in the discussion of algorithm DELTRE. Algorithm DELTREP, given below, implements the above concepts.

DELTREP ALGORITHM

Purpose: To delete a specified arc $k_L(i_L, j_L)$ from a tree represented by P_P and P_B lists.

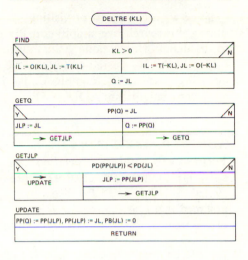

1. (FIND) Obtain the origin node i_L and terminal node j_L of the leaving arc k_L. Initialize the search for node q, the preorder traversal predecessor of node j_L.
2. (GETQ) Increment P_P until node q is found; that is, until $p_P(q) = j_L$. Initialize the search for node j'_L, the last preorder successor node in the tree rooted at node j_L.
3. (GETJLP) Increment P_P until node j'_L is found.
4. (UPDATE) Update the P_P and P_B lists.

For examples of the use of DELTREP, we refer the reader to Figures 4.8 and 4.9, which also contain the P_P lists for the example networks they contain.

Adding an Arc to a Forest Using a Preorder Traversal List

When adding an arc $k_E(i_E, j_E)$ to a forest of trees represented by a preorder traversal list, we will again assume that node j_E is the root of one of the trees. In updating the P_P list, we will stipulate that node j_E will become the new preorder traversal successor of node i_E and that i_E's old preorder traversal successor will become the new successor of node j'_E, the last preorder successor node in the subtree rooted at node j_E. Node j'_E is found by incrementing P_P starting at node j_E until $p_P(j'_E) = j_E$.

Therefore, after node j'_E is identified, performing the operations below *in the order given* will update the P_P and P_B lists resulting from the addition of arc $k_E(i_E, j_E)$.

$$p_P(j'_E) := p_P(i_E)$$

$$p_P(i_E) := j_E$$

$$p_B(j_E) := k_E$$

Updating the node depths is particularly simple when the P_P list is used since it can be implemented simultaneously with the search for node j'_E (thus avoiding a call to subroutine ROOT).

Since we are assured of visiting each node in the subtree rooted at node j_E during this search, we simply add PDADJ (computed exactly as in ADDTRE) to the depth of each node visited during the search.

The reader is referred to Figures 4.8, 4.9, and 4.10 for three examples of this updating process. The computational details of the techniques described above are given in algorithm ADDTREP.

ADDTREP ALGORITHM

Purpose: To add a specified arc $k_E(i_E, j_E)$ to a forest of trees represented by P_P and P_B lists. It is assumed that j_E is the root node of one of the trees.

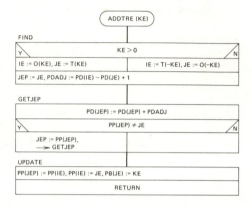

1. (FIND) Obtain the origin node i_E and the terminal node j_E of the entering arc k_E. Compute the depth adjustment factor PDADJ and initialize the search for j'_E.
2. (GETJEP) Update the depth of the subtree node currently visited. If this node is j'_E, go to step 3. Otherwise increment P_P and go to step 2.
3. (UPDATE) Update the P_P and P_B lists and return.

From the above descriptions of ROOTP, DELTREP, and ADDTREP and the earlier discussions of ROOT, DELTRE, and ADDTRE, it is easily seen that the algorithms TRECHG and TPATH are independent of which set of three algorithms is used. The only restriction is that they be used *as sets* and not mixed together.

Tree Initialization

One additional algorithm associated with trees is useful to have available. Algorithm TREINT is used to initialize the pointer representation of a tree. All that is required is stipulation of the arcs in the tree through the back pointer list P_B. Furthermore, TREINT is independent of the type of representation chosen.

After initialization of the pointers P_P, P_F, P_R, and P_D, TREINT simply makes one call to ADDTRE for each member arc of P_B. Initialization of P_F, P_R, and P_D simply requires that each element be set to zero.

Initialization of P_P requires that each $p_P(i)$ be set equal to i. This follows directly from the definition of the circular P_P list for any tree and the fact that an isolated node is only a special kind of tree.

TREINT ALGORITHM

Purpose: To construct a pointer representation of a tree given knowledge of the back pointer arcs which make up the tree.

1. Initialize the lists P_P, P_F, P_R, and P_D.
2. Call ADDTRE for each backpointer arc and RETURN.

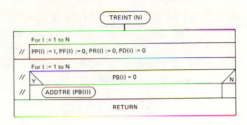

4.7   FLOW MANIPULATION ALGORITHMS

Many of the algorithms to be described require the flow to be changed on a path or cycle of the original network. The path (or cycle) is described as a set of positive and negative arc indices. For example, a cycle in Figure 4.12 is

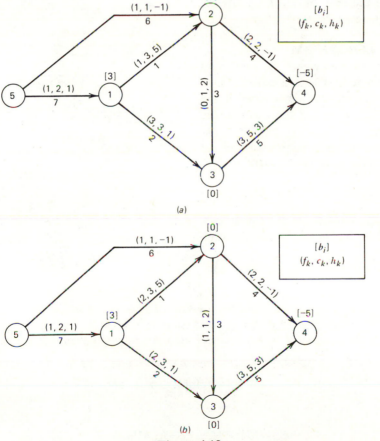

Figure 4.12

Example of Flow Change on a Cycle (a) Flows before the Change
(b) Flows after the Change on Cycle, $M_p = (1, 3, -2)$, by $\Delta = 1$

described by the arc set $M_P = (1, 3, -2)$. Changing the flow on this sequence of arcs by an amount Δ corresponds to increasing the flows on the arcs with positive indices by Δ and decreasing the flow on arcs indicated with a negative index by Δ. Thus, if f_k and f_k' are the flows on arc k before and after the flow change, respectively, then

$$f_k' = f_k + \Delta \qquad \text{for } k \in M_P \quad \text{and} \quad k > 0$$

$$f_{-k}' = f_{-k} - \Delta \qquad \text{for } k \in M_P \quad \text{and} \quad k < 0.$$

This operation is accomplished in the FLOCHG Algorithm.

FLOCHG ALGORITHM

Purpose: To change the flows on a path (LISA).

For each arc on the path, if $k > 0$, increase the flow on arc k by MF. If $k < 0$, decrease the flow on arc $-k$ by MF.

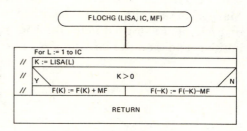

MFLO ALGORITHM

Purpose: To determine the maximum flow change (MF) on a path (LISA). The algorithm also determines if the maximal flow causes the flow on the limiting arc k_L to go to zero or to arc capacity and stores the location of the arc k_L in the path list in variable ILC.

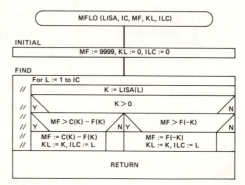

1. (INITIAL) Set MF = R (large number).
2. (FIND) Go through the list of arcs. Find arc k_L which obtains the minimum

$$\text{MF} = \text{Min.} \left(\underset{k>0}{\text{Min}} \, c_k - f_k, \, \underset{k>0}{\text{Min}} \, f_{-k} \right).$$

An operation that often precedes the flow change operation is the determination of the maximum flow change in a sequence of arcs. The maximum flow change is the maximum value of Δ such that the new *flows are feasible* or

$$f_k' = f_k + \Delta \leqslant c_k \qquad \text{for } k \in M_P \quad \text{and} \quad k > 0$$

$$f_{-k}' = f_{-k} - \Delta \geqslant 0 \qquad \text{for } k \in M_P \quad \text{and} \quad k < 0$$

Thus

$$\Delta_m = \text{Min.} \left[\underset{k>0}{\text{Min.}} \, c_k - f_k, \, \underset{k<0}{\text{Min.}} \, f_{-k} \right]$$

This operation is accomplished in the MFLO Algorithm.

The application of MFLO to the cycle $M_P = [1, 3, -2]$ of Figure 4.12a yields $\Delta_m = 1$. The application of FLOCHG yields the network flow of Figure 4.12b.

In this chapter, the most important basic structural representations of networks have been discussed and a number of algorithms designed to manipulate these representations have been presented. The remaining chapters will be dedicated to applying this information to the solution of the network flow programming problems described in Chapter 1.

4.8   HISTORICAL PERSPECTIVE

The triple label method was introduced by Johnson (1966) to represent trees for network flow programming applications. Glover, Karney, and Klingman (1972) and Srinivasan and Thompson (1972, 1973) describe the application of the technique to transportation problems. Langley, Kennington, and Shetty (1974) use the triple label method for the capacitated transportation problem. Bennington (1973) uses the method for a shortest path application. Glover, Karney, and Klingman (1974) solve the pure minimum cost flow problem using the triple label method (called the augmented predecessor index method). Maurras (1972), Glover, Klingman, and Stutz (1973), and Jensen and Bhaumik (1977) use the method for the generalized minimum cost flow problem.

The preorder traversal list was first used for network flow applications by Glover, Klingman, and Stutz (1974). In this paper, it is called the *augmented threaded index method*. Bradley, Brown, and Graves (1977) describe a detailed primal simplex procedure for the minimal cost flow problem. Glover, Klingman, and Stutz (1974) report a 10% reduction in computation time using the preorder traversal list for basis representation rather than the triple label method.

The flowchart format introduced in this chapter and used throughout the book is due to Chapin (1974).

EXERCISES

All exercises below refer to Figure E.4.

1. Construct an origin–terminal matrix for the arc capacities c_k.
2. Assign arc numbers to the arcs such that an origin ascending arc list (i.e., a list in order of increasing origin node) is formed.
3. Formulate the lists P_O, P_T, L_T.
4. For each node, formulate M_{Oi} and M_{Ti}.
5. Suppose you have the spanning tree formed by the arcs: $(10, 1)$, $(1, 4)$, $(4, 3)$, $(4, 6)$, $(6, 9)$, $(6, 8)$, $(3, 5)$, $(5, 2)$, $(5, 7)$

   (a) Draw this tree in the format given in Section 4.5.
   (b) Generate the P_B, P_F, and P_R lists.
   (c) Generate the P_P list.
   (d) Generate the P_D list.

6. For the tree of exercise 5, assume arcs $(10, 1)$ and $(4, 6)$ are deleted.

    (a)   Draw the resulting forest with each tree in the format given in Section 4.5.

    (b)   Obtain the resulting P_B, P_F, P_R, and P_P lists.

7. For the tree of exercise 5, assume the following TRECHG operations are performed in the order given below.

    (a)   Delete arc $(6, 8)$ and add arc $(4, 8)$.

    (b)   Delete arc $(4, 3)$ and add arc $(9, 5)$.

    (c)   Delete arc $(4, 6)$ and add arc $(8, 9)$.

    For each of steps a, b, and c: Draw the resulting tree and obtain the resulting P_F, P_B, P_R, P_P, and P_D lists.

8. The modular approach incorporated into TRECHG through the use of ADDTRE and DELTRE (ADDTREP and DELTREP) involves some redundancy in operations.

    a)   Identify where redundant steps are taken when an arc is deleted and its mirror arc is added during the TRECHG process.

    b)   Formulate a conceptual approach that eliminates those redundancies and updates the pointer representation in one "pass" through the reverse path from j_E to j_L. Hint: see Bradley et al. (1977) for a preorder traversal update procedure and see Glover et al. (1972) for a triple label update procedure.

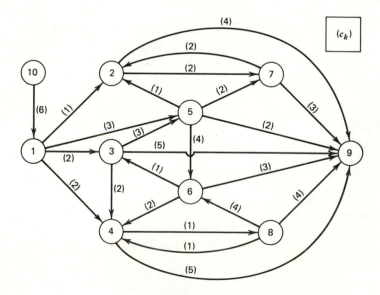

Figure E.4
Example Network

CHAPTER 5
THE SHORTEST PATH PROBLEM

5.1 INTRODUCTION

If each arc has a cost parameter that defines the *length* of the arc, the length of a path may be defined as the sum of the lengths of the arcs on the path. The problem addressed in this chapter is to find the shortest path from a specified source node s to either a specific sink node t or to all other nodes of the network. In the latter case, the problem is often called the *Shortest Path Tree Problem*. Three cases must be considered.

1. All arc lengths nonnegative.
2. Some arc lengths negative but no negative length cycles.
3. One or more negative length cycles present.

In the latter case, it will not be possible to solve the shortest path problem but the algorithms will lead to the discovery and identification of negative length cycles.

In this chapter, we will describe solutions for the problem of finding the shortest path when only forward arcs are admissible. This restriction is quite appropriate when one is interested only in solving shortest path problems. However, as we describe in detail in Chapter 7, the shortest path algorithm applied to the entire marginal network can be of great use when embedded in a minimum cost flow algorithm.

The algorithms to be described will find either the shortest path from node s to a single node t or to all nodes. There is very little difference between the two problems in computational difficulty because finding the shortest path to a single node usually requires that the shortest paths to all other nodes also be determined. We identify the case in which the shortest path to all nodes is to be determined by defining $t = 0$.

5.2 REPRESENTATION AS A MINIMUM COST FLOW PROBLEM

To represent the shortest path tree problem in the general minimum cost flow format, we identify the length parameter as the arc cost, set $b_s = n - 1$, and set $b_i = -1$ for all other nodes. Although the presence of arc capacities is somewhat redundant, the addition of an arc capacity of n units of flow to each arc makes a clearer and more consistent presentation possible.

With this representation, the shortest path problem on the network $D = [N, M]$ can be stated as the primal linear program.

$$\text{Min. } \sum_{k \in M} h_k f_k \tag{1}$$

$$\text{s.t. } \sum_{k \in M_{Oi}} f_k - \sum_{k \in M_{Ti}} f_k = -1 \qquad i \neq s \tag{2}$$

$$\sum_{k \in M_{Os}} f_k - \sum_{k \in M_{Ts}} f_k = n - 1 \tag{3}$$

$$f_k \geqslant 0 \qquad k \in M$$

$$f_k \leqslant n \qquad k \in M$$

The total unimodularity of the conservation-of-flow equations assures that all optimal flows will be integer amounts. The dual of this problem becomes

$$\text{Min. } - \sum_{i \in N - s} \pi_i + (n - 1)\pi_s + n \sum \delta_k \tag{4}$$

$$\text{s.t. } \pi_i - \pi_j + \delta_k \geqslant -h_k \qquad \text{for } k(i,j) \in M \tag{5}$$

$$\pi_i \text{ unrestricted for all } i$$

$$\delta_k \geqslant 0 \qquad k \in M$$

Once again Theorem 3 allows us to write the optimality conditions for this dual pair of problems as:

1. Primal feasibility.

   (a) Conservation of flow.
   (b) $0 \leqslant f_k \leqslant n$.

2. Restricted dual feasibility.
   $\delta_k = \text{Max. } [0, -d_k] \ (d_k = \pi_i - \pi_j + h_k)$.
3. Complementary slackness.

   (a) If $0 < f_k < n$, $d_k = 0$.
   (b) If $d_k > 0$, $f_k = 0$.
   (c) If $d_k < 0$, $f_k = n$.

Notice, however, that f_k can never exceed $n-1$ units of flow for any arc. This implies that the d_k must always be greater than or equal to zero for all k, which in turn implies that each δ_k must equal zero.

These observations allow the following simpler optimality conditions to be written.

1. Primal feasibility.

    (a) Conservation of flow.
    (b) $f_k \geqslant 0$.

2. Restricted dual feasibility.

    $$\delta_k = 0 \qquad \text{all } k.$$

3. Complementary slackness.

    (a) If $f_k > 0$, $d_k = 0$.   (6)
    (b) If $d_k > 0$, $f_k = 0$.   (7)
    (c) $d_k \geqslant 0$.   (8)

Substituting $\delta_k \equiv 0$ into equations (4) and (5) above and rewriting the dual in the form below makes a physical interpretation simpler.

$$\text{Max.} \sum_{i \in N-s} \pi_i - (n-1)\pi_s = \text{Max.} \sum_{i \in N-s} (\pi_i - \pi_s) \qquad (4a)$$
$$\text{s.t. } \pi_j - \pi_i \leqslant h_k \qquad \text{for } k(i,j) \in M \qquad (5a)$$
$$\pi_i \text{ unrestricted}$$

It is convenient, and informative, to view each dual variable π_j as representing the distance from some arbitrary origin point to its associated node j. Since we are interested in the distance from node s to all nodes $i \in N-s$, we assign that origin point to be node s itself; that is, $\pi_s = 0$.

This yields

$$\text{Max.} \sum_{i \in N-s} \pi_i$$
$$\text{s.t. } \pi_j - \pi_i \leqslant h_k \qquad \text{for } k(i,j) \in M$$
$$\pi_i \text{ unrestricted} \qquad i \in N-s$$
$$\pi_s = 0$$

Let us consider the network of Figure 5.1 with $s = 1$. After dropping the redundant arc

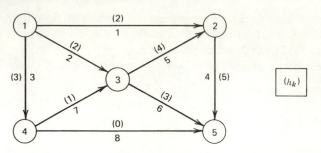

Figure 5.1
Example Network

flow capacity restrictions, we find that the primal and dual formulations are:

Primal:

Min. $2f_1 + 2f_2 + 3f_3 + 5f_4 + 4f_5 + 3f_6 + f_7 + 0f_8$

s.t.
$$
\begin{aligned}
f_1 + f_2 + f_3 &= 4\\
-f_1 \qquad\qquad +f_4 -f_5 &= -1\\
-f_2 \qquad\qquad +f_5 +f_6 -f_7 &= -1\\
-f_3 \qquad\qquad +f_7 +f_8 &= -1\\
-f_4 \qquad -f_6 \qquad -f_8 &= -1\\
\text{all } f_k &\geqslant 0
\end{aligned}
$$

Dual:

Max. $\pi_2 + \pi_3 + \pi_4 + \pi_5$

s.t.
$$
\begin{aligned}
\pi_2 &\leqslant 2\\
\pi_3 &\leqslant 2\\
\pi_4 &\leqslant 3\\
-\pi_2 \qquad +\pi_5 &\leqslant 5\\
\pi_2 - \pi_3 &\leqslant 4\\
-\pi_3 \quad +\pi_5 &\leqslant 3\\
\pi_3 - \pi_4 &\leqslant 1\\
-\pi_4 + \pi_5 &\leqslant 0
\end{aligned}
$$

π_i unrestricted,
$i = 2, 3, 4, 5$

We can now state that, while the primal problem seeks to minimize the total distance traveled by $n-1$ units of flow while requiring that each of the $(n-1)$ nodes, $i \in N - s$, receives exactly one unit of flow, the dual attempts to maximize the total distance from node s to all other nodes while requiring that the difference between the distances from node s to nodes i and j be no greater than $h(i,j)$ if $k(i,j)$ exists.

As we shall see, the algorithms to follow are strongly based on this interpretation of the dual.

A number of statements can be made concerning the solutions to these problems.

1. A basic solution to the primal problem is defined by $n-1$ arcs, which form a directed spanning tree $D_T=[N, M_T]$ where $M_T \subset M$.
2. Arcs not in the basis tree are called nonbasic arcs and have zero flow.
3. For the shortest path tree problem, the flows in the network serve only to define the current tree that is represented by the back pointer $p_B(i)$ for each node. Thus the algorithms used do not explicitly keep track of the flows; rather, they construct the representation of the tree.

5.3 ALL ADMISSIBLE ARC COSTS POSITIVE

When all admissible arc costs are positive, the inequalities of equations (5) or (5a) of this chapter are optimally satisfied by an efficient "greedy" algorithm usually credited to Dijkstra. Starting with the value of $\pi_s = 0$, the algorithm finds the shortest path from the source node to one additional node in each subsequent iteration. The procedure requires at most $n-1$ iterations. Each iteration requires $O(n)$ computations; hence, the time cost for the entire procedure is $O(n^2)$. The details of the procedure are shown in the DSHORT algorithm.

DSHORT ALGORITHM

Purpose: To derive the shortest path from node s to node t when all admissible arc costs are positive. If $t=0$, the algorithm derives the shortest path tree.

1. (INITIAL) Set $\pi_i = R$ (R some large number) for all nodes except s. Set all back pointers equal to zero. Define the node set $S = \{s\}$. Let $i := s$. Let $\pi_s := 0$.
2. (FORWARD) For each admissible forward arc originating at node i, $k(i,j)$, such that $j \notin S$, calculate the length of the path to node j through node i, $\pi_i + h_k$. If this value is less than π_j, replace π_j by $\pi_i + h_k$. Replace the back pointer for j with i.

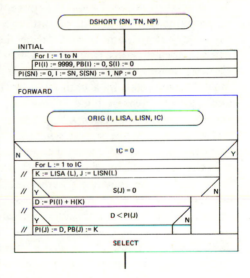

3. (SELECT) Find $D = \text{Min}_{j \in N-S}[\pi_j]$. Let i_E be the node for which the minimum is obtained. IFIN is set to zero if any node is not yet a member of S.

4. (ADD) If $D < R$ include i_E in S. If $i_E = t$, stop with the shortest path from s to t. If $i_E \neq t$ let $i := i_E$ and go to step 2. If $D = R$ and all the nodes are in the set S, stop with the shortest path tree. If all the nodes are not in S, there is no path from s to t, and set $NP = 1$.

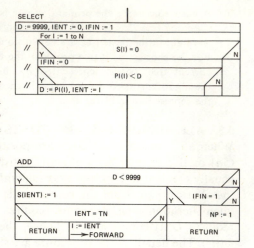

The algorithm is based upon the concept of the next nearest node. Since initially no paths have been found, we initialize the values of the π_j, $j = 1,\ldots,n (j \neq s)$ to some very large number and define the node set S to consist of those nodes whose shortest paths from node s have been found. Initially S has only one element, node s. Thus, the first iteration starts at node s. The next nearest node must be connected by a path of order 1; that is, only one arc separates node s and the next nearest node. Therefore, at the first iteration, we only need to search through M_{Os}. Each $k(s,j) \in M_{Os}$ has a terminal node whose node potential is assigned to be $\pi_j = \pi_s + h_k = h_k$. Taking the minimal value of π_j, say π_δ, from those computed above, the next nearest node is node δ. After adding node δ to the set S we proceed to the following iterations.

At any of the remaining iterations, $\ell = 2,\ldots,n-1$, the next nearest node is connected to node s by a path of, *at most*, order ℓ. The next nearest node is found by first searching through $M_{O\delta}$ and computing $\pi_i + h_k$ for $k(i,j) \in M_{O\delta}$ and $j \notin S$. If $\pi_i + h_k < \pi_j$ then π_j is updated to equal $\pi_i + h_k$. After the search of $M_{O\delta}$ is completed, the minimal value of π_j for all $j \notin S$ identifies the next nearest node found at the ℓ'th iteration. That node becomes the ℓ'th member of the set S and δ is updated equal the index of that node.

The procedure stops when node t is encountered or when all nodes have been included in S; that is, $S = N$.

Figure 5.2 illustrates Dijkstra's algorithm as it might be accomplished with hand calculations. Figure 5.2a shows the example network. Figure 5.2b shows the partial tree after the fourth iteration when $S = \{1,3,4,2,6\}$. Sequential values of π_i are shown adjacent to the nodes. The crossed out values have been replaced by lower values. The dotted lines indicate tentative tree arcs; that is, arcs through which a shorter path than previously found was possible. Those arcs that are crossed out have been replaced in a

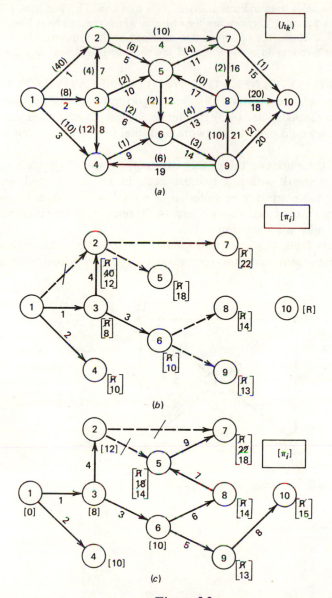

Figure 5.2

Example Application of Dijkstra's Shortest Path Algorithm ($s = 1, t = 0$)

later iteration. The solid lines indicate shortest path tree arcs. The numbers adjacent to the arcs in Figures 5.2b and 5.2c indicate the order in which the arcs have been added to the tree. Figure 5.2c shows the optimum tree obtained at the ninth and final iteration. The value of π_i at termination is the length of the shortest path from node 1 to node i.

5.4 SOME ARC COSTS NEGATIVE WITH NO NEGATIVE CYCLES

When some arc costs are negative, Dijkstra's algorithm does not necessarily lead to the optimum. Suppose arc (4, 3) in Figure 5.1 has its cost changed so that $h(4, 3) = -2$. This would yield the network and dual formulation given in Figure 5.3.

We note that there are two things that can happen when Dijkstra's algorithm is applied to a network with negative arc cost. Either the optimal solution is found or one or more arcs are in violation of equation (8) above. In the above example, the latter case occurs with arc (4, 3) remaining in violation at the termination of Dijkstra's algorithm.

In light of these facts, it is natural to follow Dijkstra's algorithm with a primal algorithm that iteratively adds and removes arcs from the primal basis until the

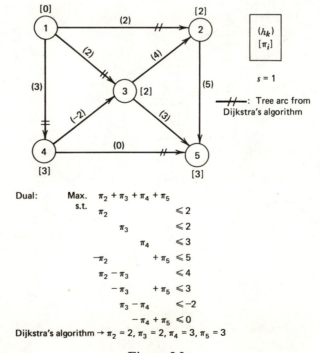

Dual: Max. $\pi_2 + \pi_3 + \pi_4 + \pi_5$

s.t.

$$\pi_2 \qquad\qquad\qquad \leqslant 2$$
$$\pi_3 \qquad\qquad \leqslant 2$$
$$\pi_4 \qquad \leqslant 3$$
$$-\pi_2 \qquad\quad + \pi_5 \leqslant 5$$
$$\pi_2 - \pi_3 \qquad\quad \leqslant 4$$
$$-\pi_3 \quad + \pi_5 \leqslant 3$$
$$\pi_3 - \pi_4 \qquad \leqslant -2$$
$$-\pi_4 + \pi_5 \leqslant 0$$

Dijkstra's algorithm $\rightarrow \pi_2 = 2, \pi_3 = 2, \pi_4 = 3, \pi_5 = 3$

Figure 5.3

Application of Dijkstra's Algorithm to a Network with a Negative Cost

node potentials, the π_i, $i \in N$, are such that all arcs satisfy equation (8). Algorithm PSHORT performs that function for the purposes of this chapter.

Naturally, some method must be provided to connect DSHORT and PSHORT. Subroutine SHORT, to be discussed later, performs that function by combining DSHORT and PSHORT with an intervening step that forms the desired pointer representation of DSHORT's feasible basis tree.

PSHORT ALGORITHM

Purpose: Starting with a feasible basis tree and node potentials that violate the dual feasibility conditions for some nonbasic arcs, this algorithm iteratively modifies the basis tree and node potentials to obtain an optimum solution.

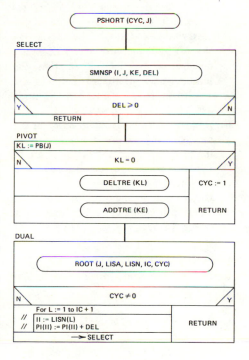

1. (SELECT) Find a nonbasic arc $k_E(i,j)$ for which $\pi_i + h_{k_E} < \pi_j$. If there is no such arc, stop with the optimum solution. Otherwise, let $d_E = \pi_i + h_{k_E} - \pi_j$.
2. (PIVOT) Let $k_L(i',j)$ be the basic arc that terminates at node j. Delete this arc from the basis tree. Add arc k_E to the basis tree.
3. (DUAL) Find the nodes in the tree rooted at node j. If a negative cycle is found in this step, stop. There is no solution to the problem. Otherwise add d_E to the potentials for all nodes in the set.

The primal algorithm must begin with a basic feasible solution for the primal problem and dual variables that satisfy complementary slackness for the basic arcs. Thus, for the shortest path tree problem, any directed spanning tree will suffice for an initial basis. The dual variables are set so that for each basic arc $k(i,j)$:

$$\pi_i + h_k = \pi_j \tag{9}$$

These goals are accomplished if the results of Dijkstra's algorithm are used as the starting solution. There are other ways to obtain an initial basis but they will not be described here. However, given an initial tree, algorithm STARTD will determine the associated initial node potentials.

STARTD ALGORITHM

Purpose: Given a basic tree, this algorithm sets the node potentials so that $\pi_i + h_k = \pi_j$ for each basic arc $k(i,j)$. Only forward arcs are allowed.

1. (TREE) List the nodes and arcs of the tree in an order so that each node (arc) appears in the list after all its predecessor nodes (arcs). The node list (LISN) and the arc list (LISA) are constructed by ROOT so that the terminal node of the ith entry in the arc list is the $i+1$ entry in the node list. Let $\ell = 1$. Set $\pi_s = 0$.
2. (DUAL) Let k be the ℓth entry in the arc list. Let j be the $(\ell + 1)$th entry in the node list. Find the origin of arc k. Let $\pi_j := \pi_i + h_k$. (π_i has already been set since node i is precedent to node j). If the end of the arc list has been reached, stop. Otherwise, increase ℓ by one and repeat step 2.

SMNSP ALGORITHM

Purpose: To find the arc with the most negative value of $\pi_i + h_k - \pi_j$.

1. Let $D := 9999$ and $K_E := 0$.
2. For each arc $k(i,j)$, calculate $d_k = \pi_i + h_k - \pi_j$. If $d_k \geqslant 0$, go on to next arc. If $d_k < 0$ and $d_k < D$, let $K_E = k$ and $D := d_k$ and go on to next arc.
3. After completing arc scan, if a $d_k < 0$ has been found, let $I := O(KE)$ and $J := T(KE)$ and return.

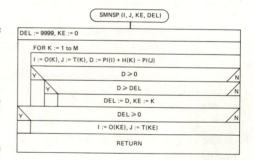

Algorithm PSHORT, the primal shortest path algorithm, has three parts: SELECT finds an arc that violates equation (8) for some nonbasic arc. PIVOT adds that arc to the basis and deletes another, and DUAL modifies the node potentials for the new basis. There are many ways to implement SELECT. For now, we will use a subroutine (SMNSP algorithm) which selects the arc with the most negative value of $d_k = \pi_i + h_k - \pi_j$. This is equivalent to the usual simplex algorithm's basis entry criterion. Other implementations are discussed in Chapter 7. The basis change operation (PIVOT) is particularly simple for the shortest path problem. If arc $k(i,j)$ is to enter, then the tree arc that currently

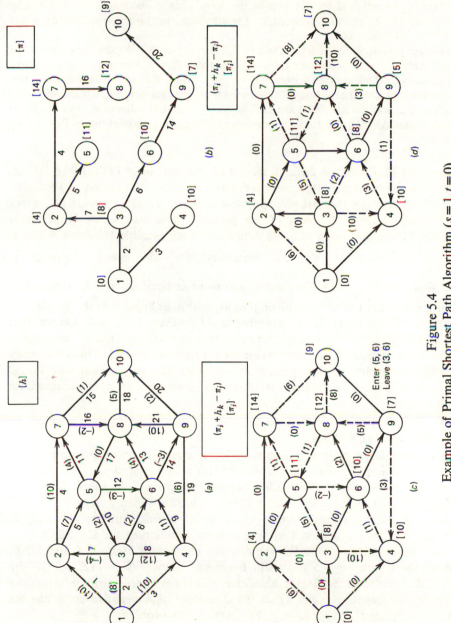

Figure 5.4.

Example of Primal Shortest Path Algorithm ($s = 1, t = 0$)

terminates at node j must leave. When arc $k(i,j)$ is inserted in the tree, the node potentials must be modified so that the complementary slackness condition of equation (6) is satisfied for all arcs in the tree. This is accomplished by adding $d_k = \pi_i + h_k - \pi_j$ to the node potentials for all nodes in the tree rooted at node j.

The algorithm is illustrated for the example in Figure 5.4. Figure 5.4a shows the original network with several negative length arcs. Figure 5.4b shows a basis tree determined with Dijkstra's algorithm. Figure 5.4c shows this tree with nonbasic arcs shown as dotted arcs. The values of d_k are shown on the arcs. Since $d_k < 0$ for some arcs the solution is not optimal. Since arc 12(5,6) has the smallest value of d_k it is chosen to enter the basis. Arc 6(3,6) is forced to leave the basis and the optimal tree in Figure 5.4d is obtained.

Since SELECT must go through the list of arcs and since PIVOT and DUAL operate on the basis tree, the computational cost for each iteration of PSHORT is $O(n+m)$. For the shortest path problem degeneracy never occurs and every basis change reduces the cost of the solution. The number of iterations is bounded by the maximum possible number of basic feasible solutions, which is $\dfrac{(n+m)!}{n!m!}$. Therefore, an order statement for the entire algorithm is $O\left(n^2 + (n+m)\dfrac{(n+m)!}{n!m!}\right)$. This order statement is large since it is identical to the order statement of the primal simplex algorithm as applied to the problem of equations (1) through (3) of this chapter. However, it is well known that simplex-based algorithms rarely take more than a small fraction of the number of iterations presented in their associated order statement. Therefore since Dijkstra's algorithm usually provides an excellent starting solution, one should expect a relatively small number of primal iterations in typical algorithm applications.

5.5   NEGATIVE CYCLES

If the network has a cycle with a negative total length, the shortest path problem becomes unbounded. When the primal problem is unbounded, the dual problem has no feasible solution. When this happens, the primal algorithm of the last section will not terminate with the optimum solution. When a negative cycle is present, the pointer representation, which under normal circumstances represents the basis tree, will show the cycle at some iteration of the algorithm.

This is illustrated in Figure 5.5 where, in the network of Figure 5.4, arc 12(5,6) is given the length $h = -5$ to obtain the negative cycle $M_c = \{12, 13, 17\}$. The cycle is discovered in the ROOT algorithm and the primal algorithm terminates indicating the negative cycle. Thus, the discovery of a negative cycle has the same computational cost as solving the shortest path problem.

As we mentioned earlier, when some arc lengths are negative, we combine the Dijkstra and primal algorithms to obtain Algorithm SHORT. The algorithm terminates when either a negative cycle is encountered or the optimal solution is found.

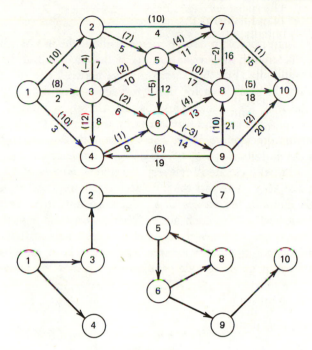

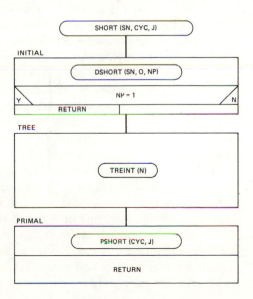

Figure 5.5
Problem with a Negative Cycle

SHORT ALGORITHM

Purpose: To find the shortest path tree in a network that may have negative arc lengths. The algorithm terminates when the shortest path tree is determined or a cycle is discovered.

1. (INITIAL) Use Dijkstra's algorithm to find a basic feasible solution for the shortest path tree. If there is no spanning tree, indicate by NP=1 and return.

2. (TREE) Create the pointer representation of the basis tree from DSHORT.

3. (PRIMAL) Use a primal algorithm to test the initial solution for optimality and make any changes necessary to obtain the shortest path tree. If a cycle is discovered in this process, indicate this with CYC equal to 1 and J some node on the cycle.

SHORT (SN, CYC, J)

INITIAL

DSHORT (SN, 0, NP)

NP = 1

Y   RETURN   N

TREE

TREINT (N)

PRIMAL

PSHORT (CYC, J)

RETURN

5.6   A NONBASIC ALGORITHM

The shortest path tree problem can be solved without the use of a basis tree. The algorithm uses the condition of equation (8) of this chapter but does not require basis tree manipulations. For this algorithm, the network can be represented by a simple arc list. The space cost for this approach is smaller but the time cost is generally greater than for the basic approaches already suggested.

The algorithm initially sets the shortest path lengths (π_i) to a large number for all nodes other than the source node. The source node is assigned the value zero ($\pi_s = 0$). Then the condition $\pi_i + h_k \geqslant \pi_j$ is checked for each arc $k(i,j)$. Obviously, the condition is initially violated for every arc that originates at the source node. Whenever an arc $k(i,j)$ is found that violates this condition, π_j is set equal to $\pi_i + h_k$. In this manner, the values of π_j decrease until they are equal to the lengths of the shortest path to each node. When a complete pass through the list of arcs is accomplished with no changes, $d_k \geqslant 0$ must be satisfied for every arc.

The process of the algorithm assures that there is at least one arc $k(i,j)$ terminating at each nonsource node j (assuming the network is connected) such that

$$\pi_i + h_k = \pi_j$$

A selection of one such arc for each node (except the source) yields a directed spanning tree rooted at the source. This tree can be used to find a feasible primal solution. Since complementary slackness is satisfied for primal and dual solutions, this tree must be the shortest path tree.

Note that although the argument for optimality uses the existence of a primal basis at termination, the algorithm itself does not maintain a basis representation and only operates on the dual variables. Thus we call this algorithm a *nonbasic algorithm*. This algorithm is similar to the primal algorithm PSHORT in that both move from a dual solution that violates equation (8) to an optimal solution. The primal algorithm, however, maintains and uses a primal basis tree throughout the solution process.

The space cost for this algorithm is $2n + 2m$ (π and P_B lists and origin and terminal lists). The time cost is $O(nm)$. This is justified by noting that every iteration passes through the arc list once using $O(m)$ computations. Each iteration adds at least one new arc to the shortest path tree. Since the tree has $n - 1$ arcs, the number of iterations is $O(n)$. Thus $O(nm)$ is required for the complete procedure. If the algorithm has not found an optimal solution in $n - 1$ iterations, it terminates and notes that a negative cycle is present in the network. The details of the procedure are described in NBSHORT.

NBSHORT ALGORITHM

Purpose: To find the length of the shortest path from the source node to all others through arcs. The algorithm uses no basis representation.

1. (INITIAL) Let $\pi_i := 9999$, for each node i. Let $\pi_s := 0$ where node s is the source node.
2. (ITER) For each arc $k(i,j)$, if $\pi_i + h_k < \pi_j$, replace π_j by $\pi_i + h_k$ and record back pointer. Otherwise take no action.
3. (TEST) If no node potentials were changed in step 2, stop with the optimum solution. Otherwise return to step 2.

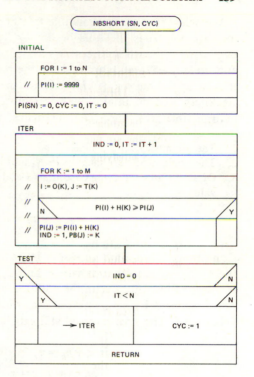

5.7 A DUAL SHORTEST PATH ALGORITHM

Suppose an optimal shortest path tree has been obtained for some network, but one or more arcs of the optimal tree are now made inadmissible. The solution now satisfies the complementary slackness conditions but is no longer feasible. Thus a solution algorithm that proceeds from the current solution to a new optimal solution can be classed as a *dual arc infeasible* algorithm. The problem situation might occur in a variety of applied situations. The algorithm developed below is also useful for minimal cost flow problems to be discussed in Chapter 7.

Consider the initial tree $D_T = [N, M_T]$. Because this is a shortest path tree, the complementary slackness conditions hold.

$$\pi_i + h_k = \pi_j \qquad \text{for each tree arc } k(i,j) \in M_T$$
$$\pi_i + h_k \geqslant \pi_j \qquad \text{for each nonbasic arc } k(i,j)$$

Now assume one or more tree arcs are made inadmissible. We do not specify here the criteria for admissibility since that will vary with the application. To form a feasible tree, the inadmissible arcs must be removed and replaced with admissible arcs, so that the resulting tree satisfies complementary slackness and is dual feasible.

We select first an inadmissible tree arc that has no inadmissible predecessors to leave the basis. Let this be arc $k_L(i_L,j_L)$. Removal of arc k_L partitions the basic tree into two subtrees $D_1=[N_1,M_1]$ and $D_2=[N_2,M_2]$. Here node $s\in N_1$ and node $j_L\in N_2$ since arc k_L is on the path from s to j_L. We want to find an arc to add to the basis $k_E(i_E,j_E)$, so that the new basis will satisfy complementary slackness and maintain dual feasibility, while perhaps still including some inadmissible arcs. Three characteristics must be true for the entering arc.

1. The arc must be admissible.
2. The origin of the arc must be in N_1 and its terminal must be in N_2.
3. Of all the arcs satisfying conditions 1 and 2, the entering arc must have the minimum value of

$$d_k=\pi_i+h_k-\pi_j$$

Conditions 1 and 2 are necessary in order that the new arc form a basis tree. If there are no arcs that satisfy 1 and 2, no spanning tree exists. Condition 3 requires justification. With the addition of arc k_E and the deletion of k_L, the new tree is formed by reversing the arcs on the path from j_L to j_E as described in Section 4.5. The arcs reversed still satisfy complementary slackness since

$$\text{If } \pi_i+h_k=\pi_j \quad \text{for arc } k(i,j)$$
$$\text{then } \pi_j+h_{-k}=\pi_i \quad \text{for arc } -k(j,i)$$
$$\text{since } h_{-k}=-h_k$$

Let π_i' indicate the node potentials on the new tree. If we define

$$\pi_i'=\pi_i \quad \text{for } i\in N_1$$
$$\pi_i'=\pi_i+d_{k_E} \quad \text{for } i\in N_2$$

then the complementary slackness conditions hold for all tree arcs including arc k_E. For all arcs satisfying conditions 1 and 2

$$d_k=\pi_i'+h_k-\pi_j'\geqslant 0$$

by the rule for choosing the entering arc. Thus the new tree does satisfy complementary slackness and dual feasibility.

If no inadmissible arcs are resident, the optimal solution is present. However, the tree may still include inadmissible arcs and thus be infeasible. In particular, some of the arcs on the path from j_E to j_L may be inadmissible since the mirror of an admissible arc is not necessarily admissible. The process continues by choosing a new inadmissible arc. The removal of this arc forms the two trees D_1 and D_2 and the process repeats as previously described. Since the set N_1

increases by at least one node (node j_E) at each iteration, the algorithm eventually yields a tree that contains only admissible arcs. Thus an upper bound to the number of iterations is $n-1$. This procedure is accomplished by the DUALSP algorithm. As will be seen in the discussion below, the procedure depends on the order in which arcs and nodes are listed by ROOT.

DUALSP ALGORITHM

Purpose: To obtain a new optimal shortest path tree when one or more arcs of the shortest path tree are made inadmissible. The algorithm begins with a shortest path spanning tree rooted at the source node with node potentials satisfying the dual feasibility and complementary slackness conditions.

1. (INITIAL) List the tree arcs in precedence order.
2. (LEAVE) From the inadmissible tree arcs select one that has no inadmissible predecessors to leave the basis. If there are no inadmissible arcs, stop with an optimum tree. If an inadmissible arc is found, let $k_L(i_L, j_L)$ be the leaving arc and go to step 3.
3. (TREE) Find the set of nodes in the tree rooted at node j_L. Let this be the set $N_2(j \in N_2$ if $S(j) = 1)$.
4. (ENTER) Find the admissible arc that originates in N_1 and terminates in N_2 to enter the basis tree with the smallest value of:

$$d_k = \pi_i + h_k - \pi_j.$$

Let the arc with the smallest value enter the basis. Let this arc be $k_E(i_E, j_E)$. If there is no entering arc indicate infeasibility.

5. (PIVOT) Change the basis tree by deleting arc k_L, reversing the arcs in the path from j_L to j_E, and adding arc k_E. Find the list of arcs and nodes in the tree rooted at j_E. Add d_{k_E} to the potentials of the tree rooted at j_E. Go to step 1.

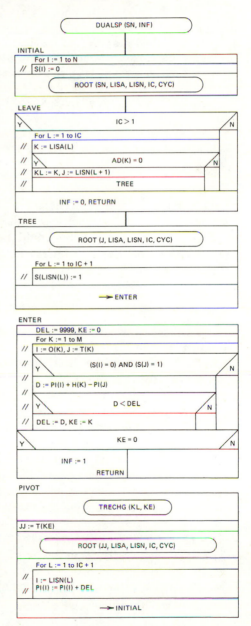

Figure 5.6 illustrates this process. The original network appears in Figure 5.6a. Only forward arcs are admissible (mirror arcs are inadmissible). The alternate optimum shortest path tree to that derived in Figure 5.4 is shown in Figure 5.6b. We now declare arc 12(5,6) inadmissible and delete it from the original network. This forms the two trees defined by the node sets

$$N_1 = \{1,2,3,4,5,7\} \quad \text{and} \quad N_2 = \{6,8,9,10\}$$

The arc that originates in N_1 and terminates in N_2 with the smallest value of d_k is 16(7,8). Allowing this arc to enter the basis forms the tree in Figure 5.4c with node potentials as shown. This tree includes the inadmissible arc $-13(8,6)$. Deleting this arc forms the sets

$$N_1 = \{1,2,3,4,5,7,8\} \quad \text{and} \quad N_2 = \{6,9,10\}$$

The arc to enter the basis is 6(3,6) with $d_k = 2$. The resulting tree in Figure 5.6d includes only admissible arcs and the procedure is complete.

The time cost for this algorithm is $O(mn + n^2)$ since for each iteration the entire list of arcs must be searched to find the entering arc and several operations require node list searches. There are at most $n-1$ iterations. This bound will usually not be reached since it is unlikely the $n-1$ iterations will be needed.

The function of DUALSP could be performed using Dijkstra's algorithm (DSHORT). Consider an arbitrary assignment of π values to nodes and a new definition for arc lengths according to the expression

$$h'_k = \pi_i + h_k - \pi_j \qquad \text{for } k(i,j)$$

It has been shown [Bazaraa and Langley (1974)] that solving the shortest path problem with the new lengths h'_k will yield the same solution as with the original lengths h_k. Furthermore, if the π values define a dual feasible solution (i.e., $\pi_i + h_k \geqslant \pi_j$) then the values of h'_k are positive. Thus, even if some h_k are negative, the modified lengths are all positive, and Dijkstra's algorithm may be used.

Recall that DUALSP begins with a dual feasible but primal infeasible solution. Thus the approach above would allow the use of Dijkstra's algorithm to find a new primal feasible shortest path tree. Since the computational cost of this algorithm is $O(n^2)$ it may be computationally superior to DUALSP. However, in the authors' experience DUALSP requires few iterations to obtain a feasible tree if only a single arc is made inadmissible. Computational tests have not been made to determine the superior approach.

5.8   HISTORICAL PERSPECTIVE

The algorithm described in Section 5.2 for the positive distance problem is a variant of those described in Dijkstra (1959) and Whitting and Hillier (1960). Modifications to the labeling procedure to improve efficiency are suggested in Yen (1971) and Williams (1963).

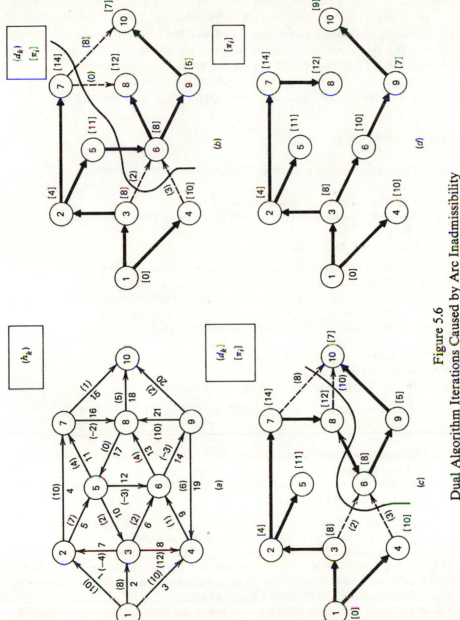

Figure 5.6
Dual Algorithm Iterations Caused by Arc Inadmissibility

143

The nonbasic algorithm of Section 5.5 is a derivative of Bellman (1958) and Bellman and Dreyfus (1962). Various procedures to speed up convergence are described in a survey by Dreyfus (1969). Nemhauser (1972) describes how a proper selection of dual variables can transform the negative distance problem to one that can be solved by a positive distance algorithm. Bazaraa and Langley (1974) describe a procedure for obtaining the proper dual variables.

Algorithms to find the shortest paths between all pairs of nodes are given by Hu (1969), Yen (1971), Hoffman and Winograd (1972), and Spira (1973).

APPLIED EXERCISES

1. Using Dijkstra's algorithm (algorithm DSHORT) and hand calculations

    (a) Obtain the shortest path tree for Figure 5.1.
    (b) Perform the iterations that are necessary to move from Figure 5.2a to Figure 5.2b.
    (c) Obtain the results of Figure 5.3.

2. Using algorithm PSHORT and hand calculations, obtain the shortest path tree for Figure 5.3. Begin with the tree yielded from Dijkstra's algorithm.
3. Using algorithm SHORT and hand calculations, obtain evidence of the negative cycle given in Figure 5.5.
4. Using algorithm NBSHORT and hand calculations, obtain

    (a) The shortest path tree for Figure 5.3.
    (b) The shortest path tree for Figure 5.4.
    (c) Evidence that the negative cycle exists in Figure 5.5.

5. Suppose the tree formed by $M_T = [(1,2), (1,3), (2,5), (2,7), (5,6), (6,9), (9,4), (9,8), (8,10)]$ is given for the network of Figure 5.2.

    (a) Use hand calculations and the methods of algorithm STARTD to generate the initial dual solution.
    (b) Use PSHORT to move from this initial feasible basis to the optimal solution. (This takes seven iterations.)

6. Suppose that arc $(3,2)$ is declared inadmissible in Figure 5.6b instead of arc $(5,6)$. Use algorithm DUALSP to achieve the new optimal solution.
7. At the end of Section 5.7, a technique is explained that allows the application of Dijkstra's algorithm DSHORT to be applied to a network with negative costs if a dual feasible solution is available for the network. Use this method on the network o Figure 5.6 after arc 12(5,6) is declared inadmissible.
8. Use one or more methods of this chapter to obtain the optimal solution to each of the problems presented in Section 2.6.
9. A silent alarm has just been activated at McGillicutty's Liquor Store on the corner of

Baldwin and Fourth Avenues. Three 2-man patrol cars have answered the call. Given below is a police map of the area of the city where the liquor store and the three police cars are situated. Notice that each street segment between intersections has the approximate time in seconds that a car answering an emergency call should take to pass over it. As is typical in many metropolitan downtown areas, all streets are one-way streets as indicated on the map by the arrows. Determine the shortest amount of time Mr. McGillicutty should have to wait for police assistance. How long will it be before all three cars arrive on the scene?

	1st	2nd	3rd	4th	5th	6th	7th	8th	9th	
	10	11	10	8	7	5	7	8 [3]		Baker
Agnes (6)	8^8	10^9	8^8	13^6	15^6	12^5	11^5	10^7		Agnes
(8)	12^9	10^9	8^{11}	12^{14}	14^{14}	11^9	9^{12}	7^{10}		Baldwin
(10)	13^{11}	14^9	15^{12}	15^{12}	15^9	12^{10}	13^9	12^{13}		Cathoway
(10)	12^{10}	10^{11}	12^{12}	14^{11}	13^{12}	13^{10}	10^9	10^{10}		Demster
(8)	10^9	8^{10}	11^8	12^7	12^9	10^{12}	8^{10}	7^7		Epley
(6) [2]	9^8	7^8	6^9	7^9	8^7	9^8	7^7	7^6 [1]		Forkney

THEORETICAL EXERCISES

1. Show that the primal and dual problems, without the capacity restrictions on the arcs, lead to optimality conditions that are identical to those given in equations (6), (7), and (8) of this chapter.
2. Show that, in algorithm PSHORT when arc $k(i,j)$ is to enter the tree, the tree arc that currently terminates at node j must leave. Explain your reasons fully.
3. In step DUAL of PSHORT, you are instructed to add d_E to the node potentials for all nodes in the subtree rooted at the terminal node of arc k_E. Give a physical interpretation of why this is done.
4. Construct a proof of the statement that the number of iterations required by NBSHORT is $O(n)$.
5. It is possible to modify the optimal network when one or more arcs in the optimal solution are declared inadmissible so that PSHORT can be used instead of DUALSP. Develop such a modification scheme and clearly explain your rationale.

CHAPTER 6
THE MAXIMUM FLOW
PROBLEM

6.1 PROBLEM STATEMENT

Given a directed network $D = [N, M]$ with each arc having a capacity parameter, we consider in this chapter the problem of finding the maximum total network flow from one node, the *source*, to another node, the *sink*. Figure 6.1 illustrates an example problem.

The algorithm introduced in this chapter has historically been used to obtain the maximum flow (thus the title of the chapter) but the same algorithm can be used to find a flow that achieves some required source-to-sink value v_r, if that required value does not exceed the maximum flow. Let node s be the source and node t be the sink, and let v be the *external flow* into node s and out of node t. The linear programming statement of the problem is

$$\text{Max.} v \tag{1}$$

$$\text{s.t.} \sum_{k \in M_{Oi}} f_k - \sum_{k \in M_{Ti}} f_k = 0 \qquad i \in N - \{s, t\} \tag{2a}$$

$$\sum_{k \in M_{Os}} f_k - \sum_{k \in M_{Ts}} f_k - v = 0 \tag{2b}$$

$$\sum_{k \in M_{Ot}} f_k - \sum_{k \in M_{Tt}} f_k + v = 0 \tag{2c}$$

$$0 \leqslant v \leqslant v_r \tag{3}$$

$$0 \leqslant f_k \leqslant c_k \qquad k \in M \tag{4}$$

The dual of this problem is

$$\text{Min.} \sum_{k=1}^{m} c_k \delta_k + v_r \delta_{m+1} \tag{5}$$

147

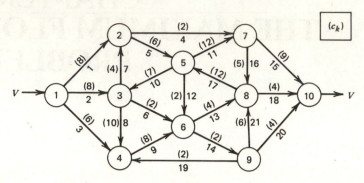

Figure 6.1

Example Maximum Flow Problem $(s = 1, t = 10)$

s.t. $\pi_i - \pi_j + \delta_k \geqslant 0$ for $k(i, j) \in M$ (6)

$\pi_t - \pi_s + \delta_{m+1} \geqslant 1$ (7)

π_i unrestricted   for all i

$\delta_k \geqslant 0$ $k = 1, 2, \ldots, m+1$ (8)

6.2 PHYSICAL INTERPRETATION OF THE DUAL

For the moment, let us remove equation (3) above from the constraints by allowing v_r to be a *large* number (implying an unattainable flow). Further, let us consider the conservation-of-flow equation for node s to be the redundant equation, which also removes it from the constraint system. Applying these stipulations to the network of Figure 6.2, we give the associated primal and dual linear programming problems in Figure 6.3. It is clear that the primal problem seeks, in a physical interpretation, to maximize the total flow from node $s = 1$ to node $t = 5$ while not exceeding the arc capacities and while assuring conservation of flow.

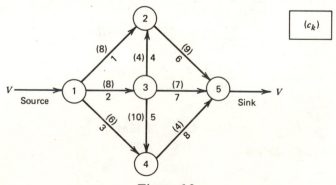

Figure 6.2

Example for Physical Interpretation of Dual

Primal:

Max. v
s.t.

$$-f_1 \qquad\quad -f_4 \qquad +f_6 \qquad\qquad = 0$$
$$-f_2 \quad +f_4+f_5 \qquad +f_7 \qquad = 0$$
$$-f_3 \qquad -f_5 \qquad\qquad +f_8 = 0$$
$$-f_6-f_7-f_8+v = 0$$

$$f_1 \qquad\qquad\qquad\qquad\qquad \leqslant 8$$
$$f_2 \qquad\qquad\qquad\qquad \leqslant 8$$
$$f_3 \qquad\qquad\qquad \leqslant 6$$
$$f_4 \qquad\qquad \leqslant 4$$
$$f_5 \qquad\qquad \leqslant 10$$
$$f_6 \qquad \leqslant 9$$
$$f_7 \qquad \leqslant 7$$
$$f_8 \leqslant 4$$

$$f_k \geqslant 0 \qquad k \in M$$
$$v \geqslant 0$$

Dual:

Min. $\qquad\qquad 8\delta_1+8\delta_2+6\delta_3+4\delta_4+10\delta_5+9\delta_6+7\delta_7+4\delta_8$

s.t.
$$-\pi_2 \qquad\qquad\qquad\quad +\delta_1 \qquad\qquad\qquad\qquad\qquad \geqslant 0$$
$$-\pi_3 \qquad\qquad\qquad\quad +\delta_2 \qquad\qquad\qquad\qquad \geqslant 0$$
$$-\pi_4 \qquad\qquad\qquad\quad +\delta_3 \qquad\qquad\qquad \geqslant 0$$
$$-\pi_2+\pi_3 \qquad\qquad\qquad +\delta_4 \qquad\qquad\qquad \geqslant 0$$
$$\pi_3-\pi_4 \qquad\qquad\qquad +\delta_5 \qquad\qquad \geqslant 0$$
$$\pi_2 \qquad -\pi_5 \qquad\qquad\qquad +\delta_6 \qquad \geqslant 0$$
$$\pi_3 \qquad -\pi_5 \qquad\qquad\qquad\qquad +\delta_7 \qquad \geqslant 0$$
$$\pi_4-\pi_5 \qquad\qquad\qquad\qquad\qquad +\delta_8 \geqslant 0$$
$$\pi_5 \qquad\qquad\qquad\qquad\qquad\qquad\qquad \geqslant 1$$

$$\delta_k \geqslant 0 \qquad \text{for all } k$$
$$\pi_i \text{ unrestricted} \qquad \text{for all } i$$

Figure 6.3
Primal Dual Formulation

It is helpful, from a conceptual viewpoint, to view the δ_k as "identifying variables." If, in the optimal solution, $\delta_k > 0$, arc k is a member of a not necessarily unique "bottleneck" set of arcs that limit the maximal flow in the network. Taking this viewpoint, we may state that the dual problem seeks to select the set of arcs of minimal total combined capacity while satisfying the dual constraints. The dual constraints, in turn, only allow sets of arcs whose removal from the network would eliminate all paths from node s to node t.

A set of arcs with this characteristic is called a *cut*. The cut with the minimum total capacity is the *minimal cut*. We will see that the value of the maximum flow is equal to the capacity of the minimal cut. The algorithms to be described in this chapter solve both the maximum flow and minimal cut problems simultaneously. This is not surprising because of the duality relation between the two problems.

6.3   THEORETICAL RESULTS

A number of interesting results are available for the solutions to the primal and dual problems [Ford and Fulkerson (1962); Dantzig (1963); Bazaraa and Jarvis (1977)]. We consider two cases for the optimal solution to the primal: when $v < v_r$ and when $v = v_r$.

For both cases, the following statements are true.

1. If the capacity c_k for each arc and the bound v_r are integer, a basic solution to the primal problem is integer (all flows have integer values).
2. Regardless of the integrality of c_k and v_r, a basic solution to the dual problem is integer (all π_i and δ_k have integer values).
3. There is an optimal dual solution for which all node potentials are either 0 or 1 ($\pi_i = 0$ or 1).
4. From the complementary slackness conditions, we obtain the conditions for optimality. For each arc $k(i,j) \in M$,

$$\text{If } \pi_i - \pi_j > 0 \qquad \text{then } f_k = 0.$$
$$\text{If } \pi_i - \pi_j < 0 \qquad \text{then } f_k = c_k.$$
$$\text{If } \pi_i - \pi_j = 0 \qquad \text{then } 0 \leqslant f_k \leqslant c_k.$$

When an optimal solution has $v = v_r$, we have obtained the upper bound on flow and the following statements are true.

5. There is an optimal dual solution for which all node potentials are zero.
6. An optimal basic solution forms a directed spanning tree in the expanded network rooted at the source node. Figure 6.4 illustrates a basic solution for Figure 6.1 when $v_r = 9$. The flows and capacities are shown not only for basic arcs but also for nonbasic arcs which have flow at capacity. All other arcs have zero flow and are not shown. For basic arcs, the small black arrowhead indicates the direction of the arc in the original network. The large arrowhead indicates the direction of the arc in the tree. Where these directions are the same, only the large arrow is shown and the basic arc is a forward arc. Where the arrows are oppositely directed, the basic arc is a mirror arc. The back pointers for the nodes identify the tree arcs.

When an optimal solution has $v < v_r$, we have obtained the maximum flow from source to sink and the following statements are true.

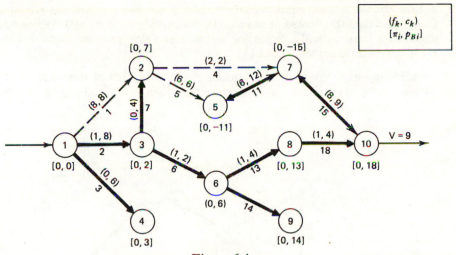

(f_k, c_k)
$[\pi_i, p_{Bi}]$

Figure 6.4

Basic Optimal Solution when $v_r = 9$

7. The nodes can be divided into two sets N_1 and N_2 with

$$s \in N_1, \quad t \in N_2, \quad N_1 \cup N_2 = N, \quad \text{and} \quad N_1 \cap N_2 = \varnothing$$

such that

$$\pi_i = \begin{cases} 0 & \text{if } i \in N_1 \\ 1 & \text{if } i \in N_2 \end{cases} \quad \text{and} \quad \delta_k(i,j) = \begin{cases} 1 & i \in N_1, j \in N_2 \\ 0 & \text{otherwise} \end{cases}$$

for an optimum dual solution. The arcs that originate in N_1 and terminate in N_2 form the minimal cut.

8. In the optimal primal solution, the flow is equal to the capacity for arcs in the minimal cut

$$f_k(i,j) = c_k(i,j) \quad \text{for } i \in N_1, j \in N_2$$

and, for arcs passing from N_2 to N_1, the flow is zero.

$$f_k(i,j) = 0 \quad \text{for } i \in N_2, j \in N_1$$

The value of maximum flow is equal to the capacity of the minimal cut.

$$\sum_{\substack{i \in N_1 \\ j \in N_2}} c_k(i,j) = v$$

9. A basic optimal solution consists of the variable v and two subtrees, one rooted at node s and the other rooted at node t. [Dantzig (1963) pp. 398–403]. Figure 6.5a illustrates a basic optimal solution for the maximum flow problem. Note that the minimum cut is the arc set $\{4, 5, 13, 14\}$.

The reader may have noted that in equations (2b) and (2c) of this chapter the variable v "looks like" a flow variable associated with a directed arc having the

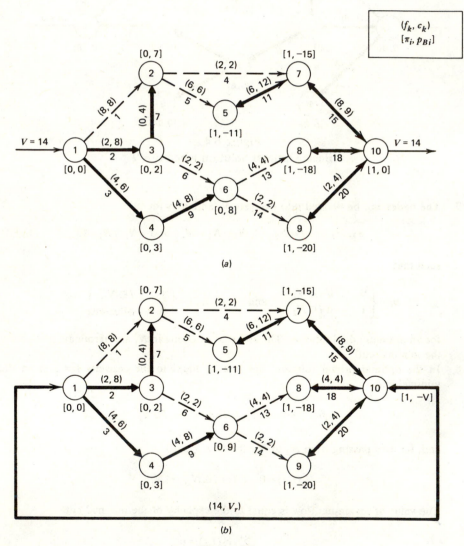

Figure 6.5
Basic Optimal Solution for $v_r > 14$

origin node t and terminal node s. If this viewpoint is chosen, then the optimal basic solution may once again be viewed as a spanning tree rooted at the source node. Figure 6.5b illustrates the basic optimal solution of Figure 6.5a when v is viewed as an arc flow variable.

6.4 BASIC AND NONBASIC ALGORITHMS

Figures 6.4 and 6.5 illustrate basic solutions to the specified flow and maximum flow problems, respectively. It should be noted, however, that there are often alternative optimum basic solutions and nonbasic optimum solutions to the maximum flow problem. Condition 8 holds for all arcs in a minimal cut; however, flows on other arcs are frequently less restricted. For example, a modification of the flows of Figure 6.5 such that $f_6 = 1$, $f_8 = 1$, and $f_9 = 5$ yields a nonbasic optimal solution. Figure 6.6 illustrates this solution where arcs with zero flows are not shown. A nonbasic solution is characterized by having a cycle of arcs whose flows are strictly between their bounds ($0 < f_k < c_k$). For Figure 6.6, arcs 6, 8, and 9 or arcs 2, 3, and 8 define such cycles.

Throughout this book, we will consider both basic and nonbasic algorithms. The nonbasic algorithms usually are simple iterative procedures. The basic algorithms maintain a basis tree and require more complex computational procedures to update and use the information stored in the tree. Which procedure is more efficient is not always clear. For smaller problems, the overhead required for maintaining the basis probably exceeds the time savings provided by the additional information. For large problems, it appears that the opposite is true. Recent advances in basis tree storage and manipulation procedures such as used in this book make the basic procedures much more competitive. Most current research and application work involves basic procedures.

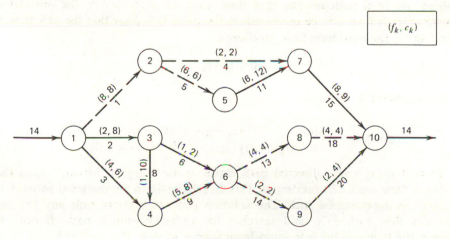

Figure 6.6
Nonbasic Optimal Solution

6.5 FLOW AUGMENTING ALGORITHMS

Consider the marginal network $D^*=[N,M^*]$ with the admissibility function defined as

$$A_d(k)=1 \quad \begin{cases} \text{if } k>0 \quad \text{and} \quad f_k<c_k \\ \text{if } k<0 \quad \text{and} \quad f_{-k}>0 \end{cases}$$

$$A_d(k)=0 \quad \begin{cases} \text{if } k>0 \quad \text{and} \quad f_k=c_k \\ \text{if } k<0 \quad \text{and} \quad f_{-k}=0 \end{cases} \qquad (9)$$

A verbal interpretation of this admissibility function is:

1. A forward arc is admissible as long as the flow on it is less than the arc capacity.
2. A mirror arc is admissible as long as flow on the forward arc is greater than zero.

Let $D_P=[N_P,M_P]$ be a directed path from s to t in D^*. Then M_P may contain both admissible forward arcs and admissible mirror arcs. We can define new flows on the expanded network as follows.

$$\begin{aligned} f'_k&=f_k & &\text{if } k\notin M_P \quad \text{and} \quad -k\notin M_P \\ f'_k&=f_k+\Delta & &\text{for } k\in M_P, k>0 \\ f'_{-k}&=f_{-k}-\Delta & &\text{for } k\in M_P, k<0 \\ v&=v+\Delta \end{aligned} \qquad (10)$$

where f_k is the flow in arc k before the flow change, f'_k is the flow after the flow change, and Δ is the amount of flow change.

Notice that the mirror arcs are only a conceptual construct used in the representation of a particular type of path from the source to the sink. Each mirror arc in a path implies that flow must be *decreased* in the associated forward arc if flow is to be increased on the path. It is clear that the new flow is feasible for the maximum flow problem if

$$0\leqslant f'_k\leqslant c_k$$

This is assured if

$$\Delta=\text{Min.} \quad \begin{Bmatrix} \text{Min.} & (c_k-f_k), & \text{Min.} & f_{-k} \\ k\in\{M_P|k>0\} & & k\in\{M_P|k<0\} \end{Bmatrix}$$

Thus if $v<v_r$ and a directed path exists in the marginal network, then the network flow cannot be maximum. This directed path in the marginal network is called the *augmentation path*. The solution procedure starts with any feasible network flow with $v<v_r$ and searches for an augmentation path. If none is found, the flow v is the maximum from source to sink. If a path is found, then the flow is augmented according to equations (10) above. A new augmentation

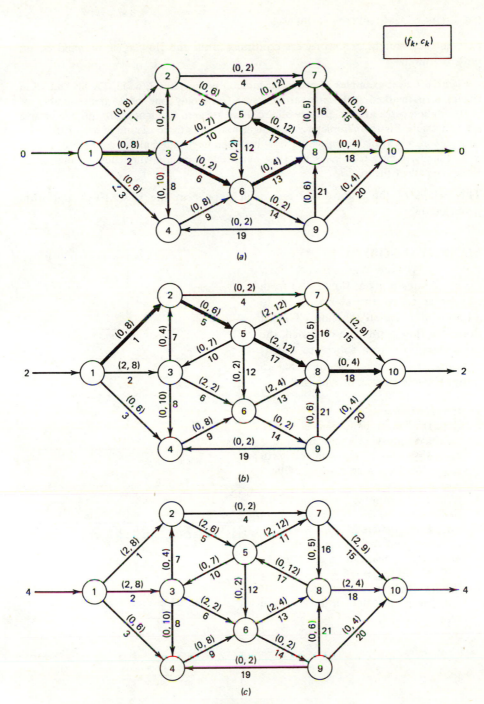

(f_k, c_k)

Figure 6.7
Examples of Flow Augmentation Paths

path is then sought. This process continues until the flow v_r is reached or no path is found.

Figure 6.7 gives examples of two augmentation paths. Arcs 2, 6, 13, 17, 11, and 15 in Figure 6.7a (marked with large arrowheads) form only one of a great number of augmentation paths available. Since all arc flows are zero in Figure 6.7a, no mirror arcs are admissible. Figure 6.7b pictures the augmentation path consisting of arcs 1, 5, −17, and 18, which is available only after flow has been augmented along the path of Figure 6.7b. Finally, Figure 6.7c gives the flow pattern that results when flow is augmented along the path of Figure 6.7b.

The details of the algorithmic approach appear in the MAXFLO algorithm given below.

MAXFLO ALGORITHM

Purpose: To find the network flow which obtains a given flow (VR) from one node called the source (SN) to another node called the sink (TN). If the given flow cannot be obtained, the maximum flow from source to sink is obtained. This process assumes that an initial feasible flow is provided.

1. (INITIAL) Let $v := 0$.
2. (PATH) Find a path from source to sink that consists of admissible arcs. If no path is found, stop with the maximum flow from source to sink. Otherwise go to step 3.
3. (ARCLST) Form the list of arcs on the path from node *SN* to node *TN* using node back pointers.
4. (FLOW) Find the maximum flow increase (DEL) that can be made on the path. If DEL exceeds $VR - V$, increase the flow by this amount and stop. If not, increase the flow in the path and return to step 2.

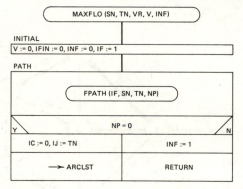

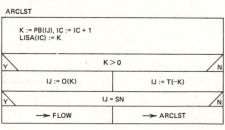

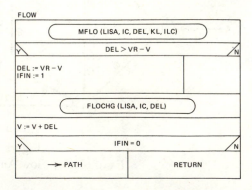

The principal steps in this algorithm are:

1. Find an augmenting path.
2. Determine the maximum flow augmentation possible.
3. Augment the flow by that amount.

Whether the solution is basic or nonbasic depends on the path-finding procedure used. The LABEL1 procedure is used when a basic solution is not required and the LABEL2 procedure is used to obtain a basic solution.

LABEL1 ALGORITHM

Purpose: To find a path from node SN to node TN using only admissible forward or mirror arcs.

1. (INITIAL) Set all labels (s_i) and back pointers (p_{Bi}) to zero. Include node SN in the set S—i.e., $S(SN) = 1$. Let $i :=$ SN; let label counter $I_c := 1$; let iteration counter $I_t := 1$.
2. (FORWARD) Find the set of arcs originating at node i. If arc $k(i,j)$ is in the set and is admissible and j is not labeled, label node j. Let $I_t := I_t + 1$; let $s_j := I_t$ and $p_{Bj} := k$.
3. (MIRROR) Find the set of arcs terminating at node i. If arc $k(j,i)$ is in the set, $-k$ is admissible, and j is not labeled, label node j. Let $I_t := I_t + 1$; let $s_j := I_t$ and $p_{Bj} := -k$. If the node TN is labeled, a path has been found; stop. If not, go to step 4.
4. (SELECT) Let $I_c := I_c + 1$. Select the node with label $S_j = I_c$. If one is found, let this node be i and go to step 2. If there is no such node, no path exists from SN to TN; stop.

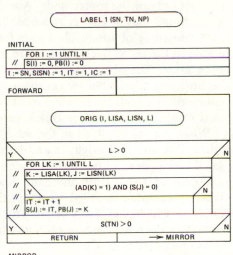

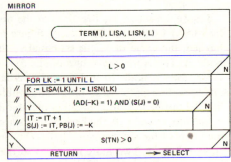

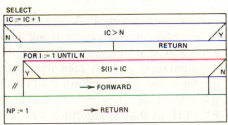

FPATH1 ALGORITHM (NONBASIC)

Purpose: To call LABEL1 for the nonbasic maximum flow algorithm.

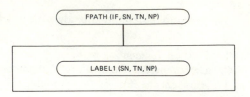

Nonbasic Augmenting Path Algorithm

The LABEL1 algorithm, called by the FPATH1 algorithm, is similar to Dijkstra's shortest path algorithm. A node set S is constructed that initially consists only of the source node. A node is added to the set when an admissible arc is found that originates at a node in S and terminates at a node not in S. A path consisting of admissible arcs is found and the process stops when node t is added to the set S. Back pointers are used to keep track of the path arc entering each node. The algorithm is usually more efficient than Dijkstra's algorithm because any admissible arc can be used to add a new node to the set S and no comparison operation is required. The time cost for the labeling algorithm is thus $O(n^2)$, the same as for Dijkstra's algorithm.

When a node is added to the set S, it is said to be *labeled*. When all of the admissible arcs that originate and terminate at a labeled node i are scanned to see if new nodes can be labeled, node i is said to be *scanned*. In order to provide a bound of $O(n^3)$ [Edmonds and Karp (1972)] on the number of times the labeling algorithm is called, we adopt the procedure of scanning the nodes in the order in which they are labeled. This is equivalent to finding the augmenting path with the fewest arcs in each iteration. Thus, the time cost for the maximum flow procedure is $O(n^5)$.

Figure 6.8 shows the iterations of the nonbasic algorithm as they would be performed by MAXFLO for the network of Figure 6.1 with $v_r = 10$. Notice that, when a node i is labeled, the value of $S(i)$ is set equal to the current number of labeled nodes (counting node i). Therefore, the values of the $S(i)$ tell us in what order the nodes were labeled. Once node t, node 10 in Figure 6.8, is labeled, we are assured of the existence of an augmentation path made up of admissible arcs. For this reason, node 9 does not get labeled during the initial iteration pictured in Figure 6.8a.

Starting at node 10, the back pointers are easily used to identify arcs 15, 4, and 1 as the members of the augmentation path. The capacity of arc 4 limits the amount of additional flow to 2 units. Two more iterations suffice to obtain the required flow of 10 units.

Basic Augmenting Path Algorithm

In order to maintain a basic solution, we introduce a flow augmentation approach that utilizes a variant of the DUALSP algorithm to find a flow augmentation path at each iteration after the first iteration. This approach is implemented in algorithm FPATH2. Notice that, to facilitate interchangeability, the coding of algorithm FPATH2 is given the same name, SUBROUTINE FPATH, as that of algorithm FPATH1.

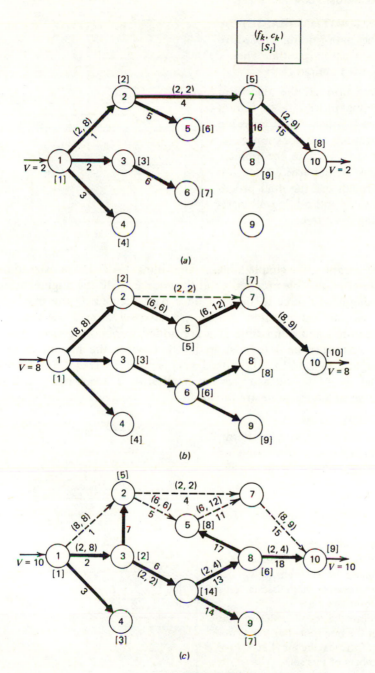

(f_k, c_k)
$[S_i]$

(a)

(b)

(c)

Figure 6.8
Example of the Nonbasic Maximum Flow Algorithm with $s = 1$, $t = 10$, and $v_r = 10$

159

FPATH2 ALGORITHM (BASIC)

Purpose: The path-finding procedure for the shortest path algorithm when a basic solution is required.

The first time through the algorithm use the labeling procedure to find a spanning tree consisting of admissible arcs. If a spanning tree does not exist, stop. Otherwise calculate the pointer representation of the tree. After the first time through use the dual procedure for deleting and adding admissible arcs to the basis tree.

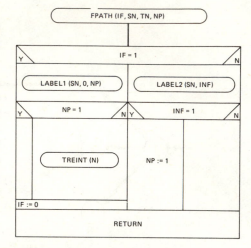

In the first iteration, the simple labeling algorithm (LABEL1) is used to find a spanning tree of admissible arcs rooted at the source. With the augmentation of the flow through the source to the sink on a path defined by the tree, one or more arcs become inadmissible. These are discovered and deleted in the LABEL2 algorithm and admissible arcs are added to obtain a new basis tree. The process continues until the required sink flow or the maximum flow is obtained. Algorithm LABEL2 is indeed very similar to DUALSP. DUALSP stipulates that the admissible arc with the smallest d_k is to enter the basis while LABEL2 allows any admissible arc to enter.

LABEL2 ALGORITHM

Purpose: To derive a new spanning tree of admissible arcs after one or more arcs in the tree are made inadmissible.

1. (INITIAL) Set initial conditions JJ:= SN.
2. (LEAVE) Initialize the set S to 0. List the nodes and tree arcs in precedence order. Find an inadmissible arc k_L that has no predecessors inadmissible. This arc is to leave the basis.
3. (TREE) Find the set of nodes in that portion of the tree rooted at the terminal node of the leaving arc. Let $S(I) = 1$ for members of this set.

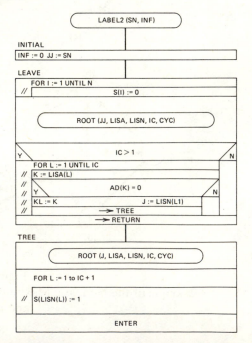

4. (ENTER) Go through the list of arcs. If forward arc $k(i,j)$ is admissible, $i \in \bar{S}$ and $j \in S$, let arc k enter the basis and go to step 5. If mirror arc $-k(j,i)$ is admissible, $j \in \bar{S}$ and $i \in S$, let arc $-k$ enter the basis and go to step 5. Otherwise go to the next arc. If there are no admissible arcs that can enter, stop; the maximum flow has been found.

5. (PIVOT) Let the entering arc be $k_E(i,j)$. Change the tree to include k_E and delete k_L. Find the terminal node of k_E and let $JJ = t(k_E)$. Go to step 2.

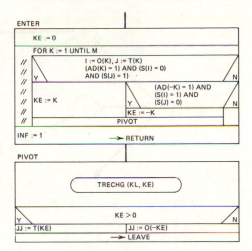

If two or more arcs are found to be inadmissible after a flow augmentation, the arc with the least depth in the tree is the first leaving arc chosen. Subsequent leaving arcs, if any, are also chosen by means of this criterion. This process continues until all arcs in the basis tree are admissible. From arguments very similar to those presented during the discussion of the DUALSP algorithm, we know that this process limits the number of degenerate iterations between two iterations where flow is augmented to $O(n)$. This process assures that no more than $n - 1$ degenerate paths will be found before a flow change is possible.

Figure 6.9 illustrates the application of the basic algorithm to obtain the maximum flow from node 1 to node 10. In Figures 6.9a through 6.9h solid lines indicate basis tree arcs, large arrowheads indicate the orientation of the basis arcs, dashed lines indicate nonbasic arcs with flow at the arc capacity, and nonbasic arcs with zero flow are not shown. Figure 6.9a gives the initial spanning tree obtained from LABEL1. Augmenting the flow by two units along arcs 1, 4, and 5 causes arc 4 to become inadmissible. Removing arc 4 and adding arc 11 results in the tree of Figure 6.9b, which has all its arcs admissible.

This makes it possible to increase the flow by six units along arcs 1, 5, 11, and 15. This augmentation drives both arc 1 and arc 5 to inadmissibility. Since arc 1 has less depth than arc 5, it is removed first and is replaced by arc 7. The resulting tree is given in Figure 6.9c. Because arc 5 is still inadmissible, no additional flow can be passed along arcs 2, 7, 5, 11, and 15. The *degenerate* change of basis that is performed by removing arc 5 and adding arc 13 results in the tree given in Figure 6.9d.

Unfortunately, arc 16 has zero flow. Therefore arc -16 is inadmissible and no additional flow may pass over arcs 2, 6, 13, -16, and 15. Removing arc -16 and adding arc 17 produces the tree given in Figure 6.9e. Since all its arcs are admissible, we may augment the flow by 1 unit along arcs 2, 6, 13, 17, 11, and 15. Removing arc 15 and adding arc 18 yields the tree of Figure 6.9f. Notice that, if v_r were set equal to 10 units, the algorithm would terminate with this tree.

Figure 6.9g results from increasing the flow by one unit along arcs 2, 6, 13, and 18, removing arc 6, and adding arc 9. Since all arcs are admissible, we may immediately increase the flow by two units along arcs 3, 9, 13, and 18. This increase makes arc 13

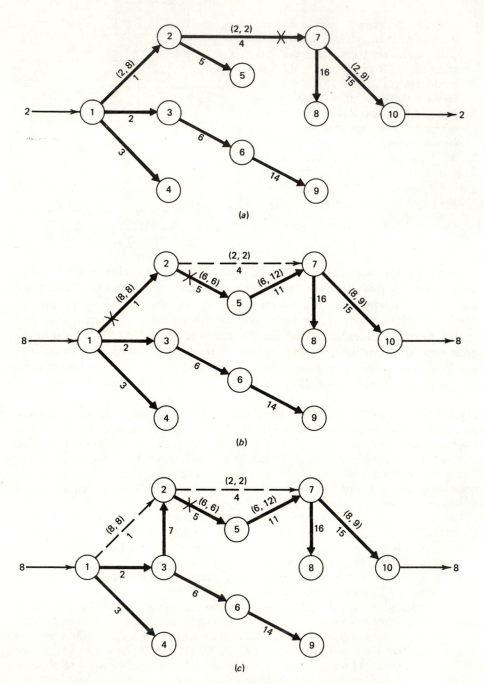

Figure 6.9
Example of Basic Maximum Flow Algorithm with $v_r = 99$

162

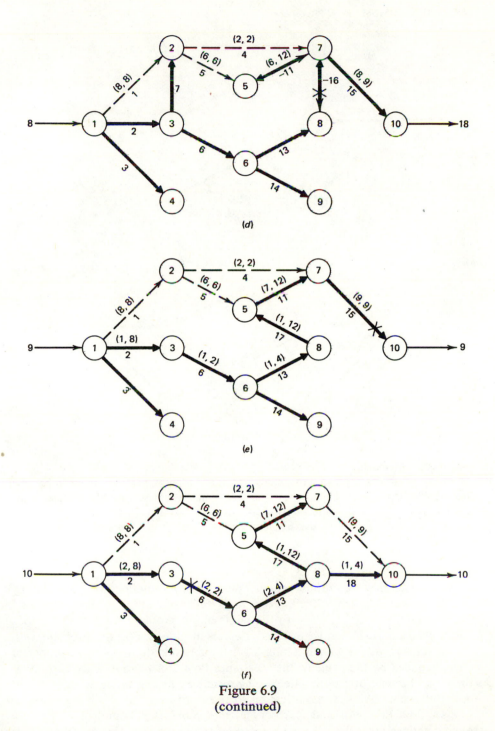

Figure 6.9
(continued)

163

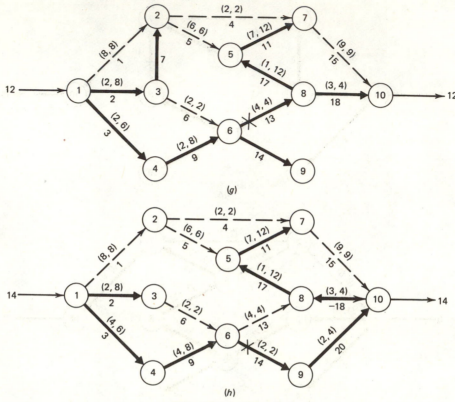

Fig. 6.9 (continued)

inadmissible. Removing arc 13 and adding arc 20 yields the tree of Figure 6.9*h*.

Increasing the flow by two units along arcs 3, 9, 14, and 20 causes arc 14 to become inadmissible. Upon removing arc 14, we discover that a new basis cannot be formed since no *admissible arc* exists to join the two disjoint node sets, that is, nodes 1,2,3,4,6, and nodes 5,7,8,9,10. Therefore the maximum flow from node 1 to node 10 has been achieved.

6.6 HISTORICAL PERSPECTIVE

Flow augmenting algorithms for the pure network model are provided by Fulkerson and Dantzig (1955), Ford and Fulkerson (1957a), and Ford and Fulkerson (1962). The relation to the minimal cut problem appears in Dantzig and Fulkerson (1956) and Ford and Fulkerson (1962). For the problem with integer capacities, the augmentation algorithms terminate in a finite number of steps bounded by the value of the maximum flow. For irrational arc capacities, the algorithm may not terminate or may terminate at the wrong solution (Ford and Fulkerson, 1962). A modification to the Ford and Fulkerson algorithm is suggested by Edmonds and Karp (1972) that provides a bound of $O(n^3)$ flow augmentations regardless of the character of the capacities. This modification is

included in the algorithm described in this chapter. We reserve consideration of
maximum flow in networks with gains until Chapter 10.

APPLIED EXERCISES

1. Suppose that $v_r = 99$ in Figure 6.8. Complete the iterations of the nonbasic flow
    augmenting algorithm.
2. Suppose the following initial spanning tree for Figure 6.1 were used instead of the
    tree yielded by LABEL1. Perform the basic flow augmentation iterations to achieve
    the maximal flow of 14 units.

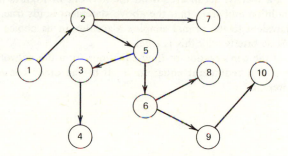

3. Solve each of the example problems of Section 2.7 in Chapter 2.
4. Given below is a network of one-way streets leading from the parking lot of a college
    football stadium. Each street has, as indicated, a capacity in cars per hour. Develop

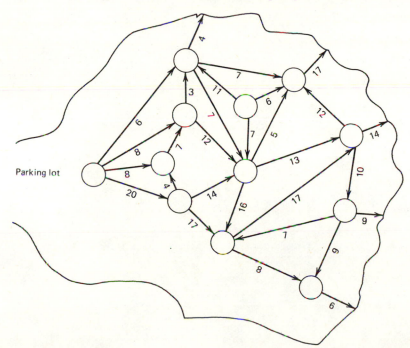

the optimal traffic flow pattern to disburse traffic after the big game. What would be your recommendations for modifying streets to provide quicker flow in the future?

THEORETICAL EXERCISES

1. Why does $v_r \to \infty$ imply removal of equation (3) of this chapter?
2. Any set of arcs that eliminates all paths from the source to the sink is dual feasible. Select a specific set of arcs from the network of Figures 6.2 and 6.3 that satisfies this condition and show that the above statement holds true.
3. Any set of arcs that does not eliminate all paths from source to the sink is dual infeasible. Select a specific set of arcs from the network of Figures 6.2 and 6.3 that satisfies this condition and show that the above statement holds true.
4. LABEL2 is equivalent to the primal simplex algorithm in its choice as the entering variable (arc). State briefly why this is true.
5. Select any one of the conservation-of-flow equations from the network of Figure 6.2 and show that it is a redundant equation if all of the other conservation-of-flow equations are present.

CHAPTER 7
PURE MINIMUM COST FLOW PROBLEMS

The goal of this chapter is to describe solution algorithms for the pure network minimum cost flow problem with linear costs. The linear programming and network flow models for this case were introduced in Chapter 3. We first consider primal algorithms for solution and later a dual node infeasible algorithm.

The primal algorithms have three basic steps.

1. Find an initial network flow that satisfies primal feasibility.
2. Determine if the network flow is optimal. If the flow is optimal, stop. Otherwise go to step 3.
3. Modify the network flow by changing the flows on a cycle. Return to step 2.

The algorithms that follow differ both in the way these three steps are carried out and in whether the algorithm maintains a basis.

The first step of obtaining an initial feasible solution is perhaps the most important to the total time cost. The "quality" or closeness to the optimum of the first feasible solution has a marked effect on the number of iterations required to obtain an optimal solution. There is always a trade-off between the quality of the initial solution and the computational cost of the procedure that obtains that solution. For transportation or assignment problems where every source is connected to every sink by an arc, there are a number of documented procedures for obtaining an initial solution. For the minimum cost flow problem with less than 100% arc density, the literature is not very helpful. We describe here two approaches.

1. The use of the maximum flow algorithm.
2. The use of artificial arcs.

7.1 THE MAXFLOW METHOD TO OBTAIN A PRIMAL FEASIBLE SOLUTION

When the maximal flow algorithm is used, we simply attempt to obtain a feasible solution, a solution that satisfies node-fixed external flows and does not violate the bounds on arc flow. The arc costs have no effect on the procedure and the obtained solution is not necessarily basic.

Starting with all flows zero, the procedure arbitrarily chooses a pair of nodes, one with unsatisfied positive external flow and one with unsatisfied negative external flow. Using the maximum flow algorithm, we select the node of the chosen pair whose external flow is the smallest in absolute value and attempt to satisfy that external flow. One of two results can occur—the maximum flow algorithm satisfies the fixed external flow for the selected node, or the maximum flow between the pair of nodes is established. Whatever the result, we choose another pair of nodes with fixed external flows not yet satisfied and repeat the process.

If there exist only nodes with negative unsatisfied external flow or if there exist only nodes with positive unsatisfied flow, we choose the slack node and one of the unsatisfied nodes for the maximum flow operation. If all the external flows cannot be satisfied in this way there is no feasible solution to the problem. This procedure is coded as the PHASE1 algorithm.

PHASE1 ALGORITHM

Purpose: To find a feasible network flow that satisfies node external flows.

1. (INITIAL) Set all flows equal to zero. Let b_{fi} be the net external flow unsatisfied by the network flow $b_{fi} = b_i$ initially. Let $b_{fn} = 0$.
2. (SOURCE) Find a node with positive unsatisfied external flow. Let this be the source node s. Let b_s be the unsatisfied flow for this node. If there are no such nodes, let $s = n$ (the slack node) and let $b_s = R$ (where R is some large number).
3. (SINK) Find a node with negative unsatisfied external flow which has not previously been considered with node s. Let this be the sink node t. Let b_t be the magnitude of the unsatisfied flow for this node. If there are no such nodes, let $t = n$ and let $b_t = R$.

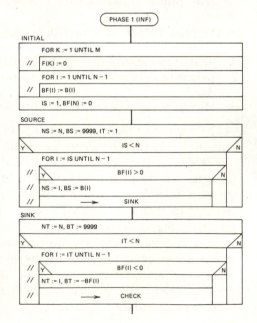

4. (CHECK) If both s and t are the slack node, stop; a feasible solution has been found. If not, go to step 5.

5. (FLOW) Use the maxflow algorithm to find the maximum flow from s to t bounded by $\mathrm{Min}(|b_s|, |b_t|)$. The algorithm establishes this flow. Reduce the magnitude of the unsatisfied external flows for nodes s and t by the amount of flow established between the nodes by the maxflow algorithm.

6. (CONTROL) If $s = n$ and $b_{ft} = 0$, all the positive external flows have been satisfied. Return to step 3 and seek another sink node with a negative unsatisfied external flow. If $s = n$ and $b_{ft} \neq 0$, there is no flow that can satisfy node t; stop; the problem is infeasible. If $s \neq n$ and $b_{fs} = 0$, the external flow for node s is satisfied; go to step 2 and find another source node. If $s \neq n$, $b_{fs} \neq 0$, and $t = n$, there is no way to satisfy the external flow of node s; stop; the problem is infeasible. If $s \neq n$, $b_{fs} \neq 0$, and $t \neq n$, return to step 3 to choose another sink.

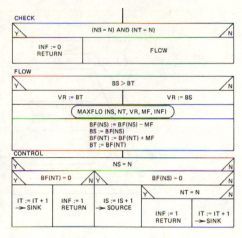

Figure 7.2 illustrates the procedure applied to the example problem of Figure 7.1, where algorithm FPATH1 is used to find the augmentation paths. FPATH2 cannot be used in this approach without modifying the network. (C.f. exercise 2 at the end of this chapter.) All arcs not shown have zero flow. Figure 7.2a shows the flows after three maximum flow iterations with $s = 1$ and t successively assuming the values $4, 5, 6$. Notice that each source node is visited only once during this process. If the algorithm ever leaves a source node, that means that its fixed external flow has been satisfied. For this reason, node 1 is the selected source node until its fixed external flow is satisfied at the end of iteration 3. Conversely, the number of visits to a sink node is bounded by the number of source nodes. Of course, once a sink node's fixed external flow is satisfied, no more visits occur.

Figure 7.2b shows the flows after three more maxflow iterations. In each of the first 6 maxflow iterations, a single arc and thus a single augmentation path is used. Figures 7.2c and 7.2d each show one iteration. The maximum flow iteration that yields Figure 7.2c requires three augmentation paths, each allowing one unit of flow. They are implemented in the following order, one at a time:

Arc 9.
Arcs 7, −1, 3.
Arcs 7, −4, 6.

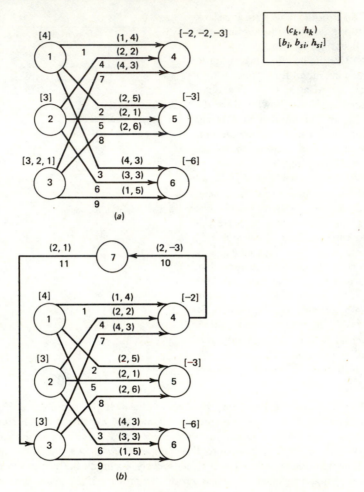

Figure 7.1
Example Minimum Cost Flow Problem (*a*) Original Network
(*b*) Network with Slack Node

A single augmentation path, arcs 11, 8, −2, and 3, carrying one unit of flow is used to satisfy the remaining fixed external flow at node 6. Augmenting flow along that path yields Figure 7.2*d*. Thus, eight applications of the maxflow algorithm are required to obtain the initial feasible solution.

7.2 THE ARTIFICIAL ARC METHOD TO OBTAIN A PRIMAL FEASIBLE SOLUTION

The artificial arc method can be used if one desires certainty in achieving an initial feasible solution that is also basic. The method supplies artificial arcs to

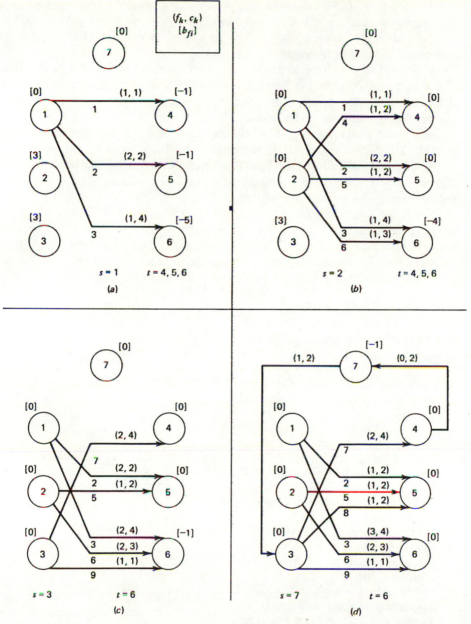

Figure 7.2
Finding a Feasible Solution Using the Maximum Flow Algorithm

satisfy node feasibility constraints but places a large positive cost on these arcs. Any primal procedure will eventually cause the flow on these arcs to become zero if a feasible solution exists. The approach is equivalent to the big-M method of linear programming.

An alternative would be to place a cost of 1 on these arcs and give the other arcs a zero cost. Minimizing network cost would yield a solution with zero cost if a feasible solution exists. Nonzero minimum cost indicates that the network has no feasible solution. This method is equivalent to the Phase I method of general linear programming. This alternative is not recommended because a problem completely unrelated to the original problem must be solved prior to attempting the solution of the original problem. Conversely, the big-M method allows progress toward the solution of the original problem while the artificial arcs are being driven from the solution.

The coding of this algorithm is given in algorithm ARTIFIC. The algorithm provides each nonslack node with an artificial arc connecting it to the slack node. If a node's fixed external flow is greater than or equal to zero, the artificial arc terminates at the slack node. For nodes with negative external flow, the arc originates at the slack node and terminates at the original node.

ARTIFIC ALGORITHM

Purpose: To form an initial all-artificial basis for a minimum cost flow network.

1. (INITIAL) Set all flows to zero and compute the number of nodes (excluding slack node).
2. (ARCS) Determine orientation and upper bounds on artificial arcs. Set lower bounds to zero, set arc costs to a large value, and formulate the new P_O list.
3. (NEWLSTS) Formulate the new L_T and P_T lists.
4. (FLOWS) Obtain the set of arcs originating at the slack node. If there are no such arcs, go to step 6. Otherwise, go to step 5.
5. (POSITIVE) For each arc in the set originating at the slack node, determine if it is an artificial arc. If not, go to the next arc. For each artificial rc k, set $f_k = c_k$, and stipulate that arc k is the back pointer of node t_k.

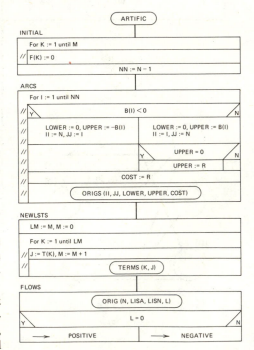

6. (NEGATIVE) Obtain the set of arcs terminating at the slack node. If there are no such arcs, go to step 7. Otherwise, for each arc in the terminating set, determine if it is an artificial arc. If not, go to the next arc. For each artificial arc, set $f_k = c_k$. If the resulting $f_k \geqslant R$, set $f_k = 0$. [This implies $b(o_k) = 0$.] Stipulate that arc $-k$ is the back pointer of node o_k.

7. (TREE) Form the initial basis tree and compute the associated node potentials.

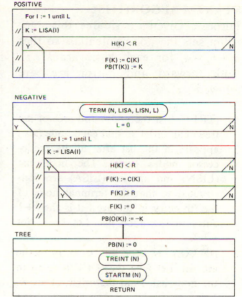

The flow on each artificial arc is set equal to the absolute value of each node's fixed external flow. The arc capacity is also set to the absolute value of the fixed external flow unless the fixed flow has value zero. In the latter case, the artificial arc capacity is set to a very large value, R.

Since new arcs have been added to the network, it is necessary to update the lists L_T, P_T, and P_O. This is accomplished by the indicated calls to subroutines ORIGS and TERMS. Following these operations, the initially all-artificial basis tree is formed by a call to TREINT. (The back pointer list P_B has been filled prior to this time with the indices of the artificial arcs.)

Finally, all that remains to be done is to compute the node potentials for the initial basis tree. This is performed through a call to algorithm STARTM. STARTM is very similar to STARTD. It is, however, more general since it can accommodate not only forward arcs but also mirror arcs. For further clarification, the statements that have been added to STARTD to obtain STARTM are shaded in the flowchart for STARTM.

STARTM ALGORITHM

Purpose: Given a basic tree, this algorithm sets the node potentials so that $\pi_i + h_k = \pi_j$ for each basic arc $k(i,j)$. Both forward and mirror arcs are allowed.

1. (TREE) List the nodes and arcs of the tree in an order so that each node (arc) appears in the list after all its predecessor nodes (arcs). The node list (LISN) and arc list (LISA) are constructed so that the terminal node of the ith entry in the arc list is the $i+1$ entry in the node list. Let $\ell = 1$. Set $\pi_s = 0$.

2. (DUAL) Let k be the ℓth entry in the arc list. Let j be the $(\ell + 1)$th entry in the node list. Find the origin of arc k. Let $\pi_j := \pi_i + h_k$. (π_i has already been set since node i is precedent to node j.) If the end of the arc list has been reached, stop. Otherwise, increase ℓ by one and repeat step 2.

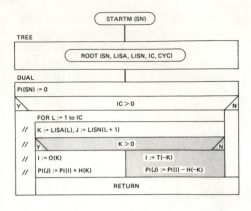

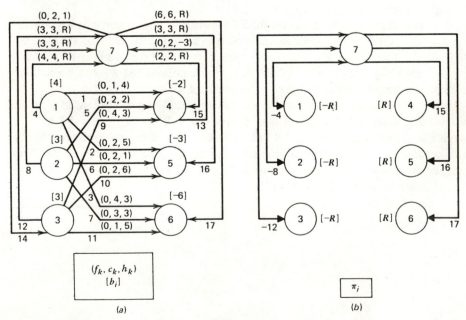

(a)

(b)

Figure 7.3
Results of Applying ARTIFIC to the Network of Figure 7.1

Recalling that a mirror arc's cost is the negative of its associated forward arc's cost, it is a simple matter to compute the proper value of π_j if node j has a mirror arc for its back pointer in the tree. Suppose that $\pi_5 = 3$, $h_{11} = 5$, and $p_B(7) = -11$. We desire, in general, that STARTM set the values of the node potentials such that $\pi_i + h_k = \pi_j$ [i.e., PI(I) + H(K) = PI(J)]. In this specific case I = 5, K = -11, and J = 7. Therefore, we require that PI(5) + H(-11) = PI(7). However, this is the same as requiring that PI(5) - H(11) = PI(7) and this is the expression used in STARTM.

Figure 7.3a shows the results of applying ARTIFIC to the network given in Figure 7.1. Figure 7.3b presents the fully artificial basis tree that ARTIFIC produced.

7.3 A NONBASIC PRIMAL ALGORITHM

The nonbasic primal algorithm appears in many introductory texts primarily because of its simple statement. The algorithm requires an initial feasible solution for flows. If there are no negative cycles in the marginal network defined by a solution, the flows are optimum. If there is a negative cycle, a set of flows may be found that will still be feasible but that will have a lower total cost. This new set of flows is easily found by increasing the flows in the negative cycle until an arc in the cycle is saturated. The total cost of the flows will decrease by an amount equal to the product of the increase in flow around the cycle times the value of the negative cycle. The process continues until no negative cycles remain. The algorithm is primal because it maintains primal feasible solutions throughout. The solutions are not necessarily basic. The steps of this procedure are implemented in the PRIMAL1 Algorithm.

PRIMAL1 ALGORITHM

Purpose: Starting from a primal feasible solution, find the optimum solution by iteratively discovering and saturating negative cost cycles in the marginal network.

1. (CYCLE) Find a cycle with negative cost in the marginal network. If none exists, stop with the optimum solution. If one is found, go to step 2.
2. (ARCLST) Find the list of arcs on the negative cycle using the node back pointers.
3. (CHANGE) Determine the maximum flow change in the cycle that maintains arc feasibility. Change the flow in the cycle arcs by this amount. Go to step 1.

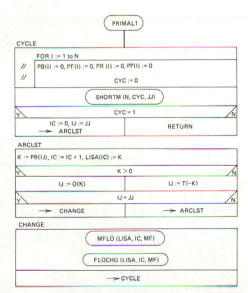

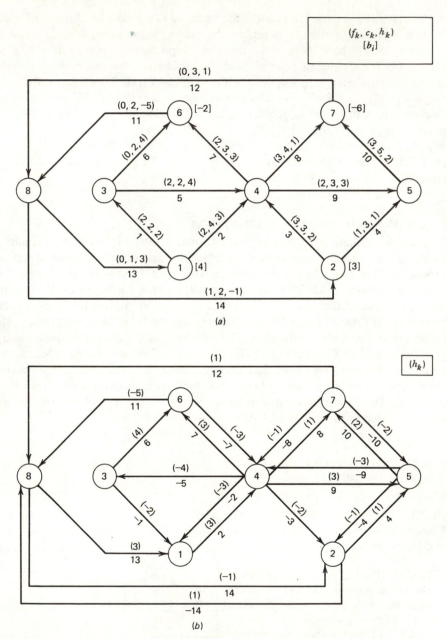

Figure 7.4
Example Marginal Network with Negative Cycles (*a*) Original Network
(*b*) Marginal Network (*c*) Resulting Flows

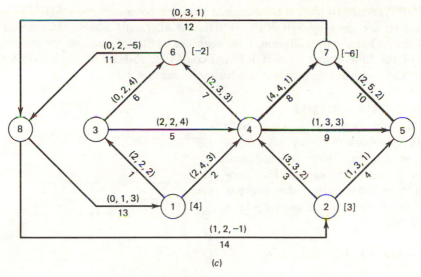

Figure 7.4
(continued)

To illustrate the method, consider the network of Figure 7.4*a* and its associated marginal network given in Figure 7.4*b*. Note that the given flows do constitute a feasible solution with a total cost of 43 units. However, the marginal network is not free of negative cycles. For example, a negative cycle is formed by arcs -10, -9, 8 and the cycle value is -4 (i.e., $h_{-10} + h_{-9} + h_8 = -2 - 3 + 1 = -4$).

The cycle in the marginal network only identifies, in an analytical manner, what could be observed directly from the original network. In Figure 7.4*a*, sending one unit of flow from node 4 to node 7 by means of arcs 9 and 10 will incur a total charge of 5 units of cost. Sending that same unit of flow from node 4 to node 7 by means of arc 8 incurs only 1 unit of cost. Hence, a savings of 4 units of cost per unit flow is present when the latter path of flow is selected. Note this savings is clearly given by the cycle value of -4.

The maximum increase around the cycle of arcs -10, -9, and 8 is 1 unit of flow, which saturates arc 8 driving it to its capacity of 4 units of flow. The resulting flows are given in Figure 7.4*c* where the total cost has been reduced to 39 units.

In summary, the approach described above requires two operations: find a negative cycle, and change the flow in the cycle by the maximum amount. The second of these is straightforward and is accomplished using network manipulation algorithms already introduced with $O(n)$ time cost for each iteration. The operation of finding negative cycles has a number of possible alternative implementations. We choose here to use algorithm SHORTM to identify the negative cycles. Just as STARTM was seen to be similar to STARTD, SHORTM is similar to SHORT. Once more, however, SHORTM considers all admissible arcs in the marginal network, that is, both forward and mirror arcs, during its search process. As indicated in the flowchart for SHORTM, instead of

DSHORT and PSHORT, their marginal network counterparts, DSHRTM and PSHRTM, are used by SHORTM. PSHRTM also calls SMNSPM instead of SMNSP. As before, the differences between the similar algorithms are shaded in each of the flow charts of SHORTM, DSHRTM, PSHRTM, and SMNSPM. SHORTM requires $O(n^3)$ operations for each iteration.

SHORTM ALGORITHM

Purpose: To find the shortest path tree in a network that may have negative arc lengths. Both forward and mirror arcs are considered. The algorithm terminates when the shortest path tree is determined or a cycle is discovered.

1. (INITIAL) Initialize DM, DZ, SPNTRE.
2. (SPAN) Use Dijkstra's algorithm to find a basic feasible solution for the shortest path tree. If there is no spanning tree, go to step 3. Else, go to step 4.
3. (REINIT) Add a large number of the π_i's of nodes not yet labeled as part of a tree. Update the values of DM and DZ and set flag indicating no spanning tree has yet been obtained. Go to step 2.
4. (TREE) Create a pointer representation of the spanning forest or spanning tree from DSHORT.
5. (PRIMAL) Use a primal algorithm to test the initial solution for optimality and make any changes necessary to obtain the shortest path tree. If a cycle is discovered in this process, indicate this with CYC equal to 1 and J some node on the cycle.

DSHRTM ALGORITHM

Purpose: To derive the shortest path from node s to node t when all admissible arc costs are positive. (Both forward and mirror arcs are allowed.) If $t = 0$, the algorithm derives the shortest path tree.

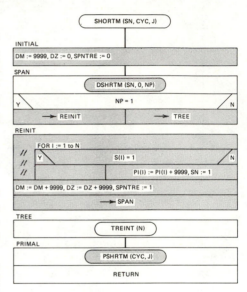

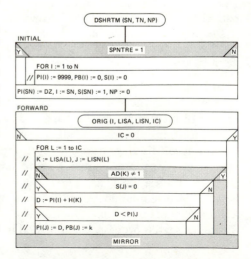

1. (INITIAL) Set $\pi_i = DM$ (DM some large number) for all nodes except s. Set all back pointers equal to zero. Define the node set $S = \{s\}$. Let $i := s$. Let $\pi_s := 0$.

2. (FORWARD) For each admissible forward arc originating at node $i, k(i,j)$ such that $j \notin S$, calculate the length of the path to node j through node i: $\pi_i + h_k$. If this value is less than π_j, replace π_j by $\pi_i + h_k$. Replace the back pointer for j with k.

3. (MIRROR) For each arc terminating at node $i, k(j,i)$ such that $j \notin S$, determine if the mirror arc $-k(i,j)$ is admissible. If it is, calculate the length to node j through the mirror arc: $\pi_i - h_k$. If this value is less than π_j, replace π_j by $\pi_i - h_k$. Replace the back pointer for j with $-k$.

4. (SELECT) Find $D = \min_{j \in N - S}[\pi_j]$. Let i_E be the node for which the minimum is obtained. IFIN is set to zero if any node is not yet a member of S.

5. (ADD) If $D < DM$, include i_E in S. If $i_E = t$, stop with the shortest path from s to t. If $i_E \neq t$, let $i := i_E$ and go to step 2. If $D = DM$ and all the nodes are in the set S, stop with the shortest path tree. If all the nodes are not in S, set $NP = 1$ to indicate that there is no path from s to t.

PSHRTM ALGORITHM

Purpose: Starting with a feasible basis tree and node potentials that violate the dual feasibility conditions for some nonbasic arcs, this algorithm iteratively modifies the basic tree and node potentials to obtain an optimum solution.

1. (SELECT) Find a nonbasic arc $k_E(i,j)$ for which $\pi_i + h_{k_E} < \pi_j$. If there is no such arc, stop with the optimum solution. Otherwise, let $d_E = \pi_i + h_{k_E} - \pi_j$.

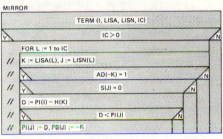

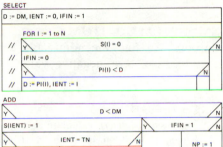

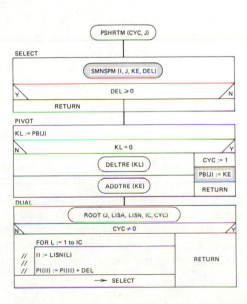

2. (PIVOT)Let $k_L(i',j)$ be the basic arc that terminates at node j. Delete this arc from the basis tree. Add arc k_E to the basis tree.

3. (DUAL) Find the nodes in the tree rooted at node j. If a negative cycle is found in this step, stop. There is no solution to the problem. Otherwise add d_E to the potentials for all nodes in the set.

SMNSPM ALGORITHM

Purpose: To find the admissible arc that has the most negative value of $\pi_i + h_k - \pi_j$.

1. Let $D:=9999$. $k_E:=0$.
2. For each arc $k(i,j)$, calculate $d_k = \pi_i + h_k - \pi_j$. If $d_k = 0$, go on to the next arc. If $d_k < 0$, arc k is admissible, and $d_k < D$, let $k_E := k$ and $D := d_k$. If $d_k > 0$, arc $-k$ is admissible, and $-d_k < D$, let $k_E := k$ and $D := -d_k$. Go on to the next arc.
3. If $k_E > 0$, set $i := o(k_E)$ and $j := t(k_E)$. If $k_E < 0$, set $i := t(-k_E)$ and $j := o(-k_E)$.

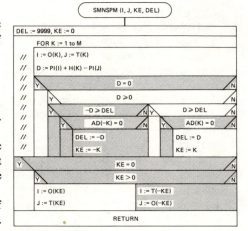

Application of DSHRTM to a marginal network will not necessarily yield a spanning tree. This could cause problems in the use of PSHRTM. Consider the simple transportation network of Figure 7.5a. The feasible solution given there yields the associated marginal network of Figure 7.5b. Clearly, nodes 1, 2, 5, and 6 cannot become members of a directed tree rooted at slack node 9.

Fortunately, a spanning forest of admissible arcs serves as well as a spanning tree in the implementation of PSHRTM. Suppose a feasible solution to a minimum cost flow problem is found and an attempt to form a spanning tree rooted at the slack node is made. If the attempt fails only the nodes in a set N_1 are included in the tree rooted at the slack node. We call this tree $D_{T1} = [N_1, M_1]$. The nodes not contained in N_1 are members of $\overline{N}_1$. Since we were unable to connect sets N_1 and $\overline{N}_1$, all arcs passing from N_1 to $\overline{N}_1$ must be inadmissible. It can be shown that they will be inadmissible for every feasible flow solution. For the example network of Figure 7.5 the tree, D_{T1}, found with the node 9 as the root consists of nodes $N_1 = \{9,4,8,3,7\}$ and arcs $M_1 = \{11,10,-8,7\}$.

To complete the spanning forest, we must choose a new root node that is a member of $\overline{N}_1$ and attempt to find a tree of admissible arcs that spans the remainder of the nodes. For the example, we arbitrarily choose node 6 and find

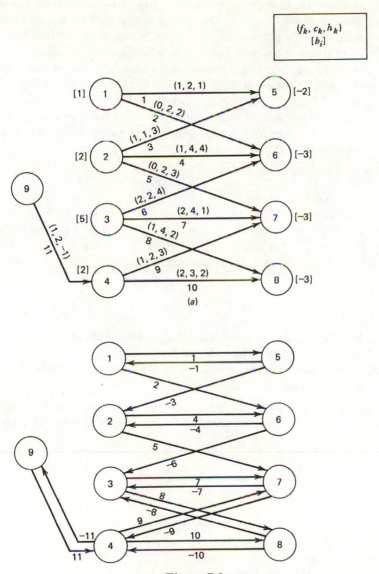

Figure 7.5
Example Network with Marginal Network

181

$D_{T2} = [N_2, M_2], N_2 = \{6, 2\}, M_2 = \{-4\}$. If the new tree completes a spanning forest, the process stops. Otherwise we choose a new root node and attempt to complete the forest. The process continues until a spanning forest is found. The result for the example appears in Figure 7.6, which consists of three trees.

Given the spanning forest thus found, we can apply the primal algorithm independently to each subnetwork of the original network defined by the node sets $N_1, N_2, N_3 \ldots$. Thus the problem is decomposed by the trees formed in this process. Rather than implement this decomposition explicitly, we assign π values so that the algorithm never finds cycles with arcs in more than one subnetwork. Thus, if r_j is the root of D_{Tj} we assign

$$\pi(r_j) = (j - 1)R$$

where R is a large number. In the example $\pi_9 = 0$, $\pi_6 = R$, $\pi_5 = 2R$ where $R = 9999$. The other π values are assigned using the trees and the usual dual condition $\pi_i + h_k = \pi_j$, where arc $k(i,j)$ is a tree arc. This artifice isolates the subnetworks. Admissible arcs $k(i,j)$ passing from N_k to $N_\ell(\ell > k)$ will always have

$$d_k = \pi_i + h_k - \pi_j > 0$$

because $\pi_i \gg \pi_j$. When d_k is positive, arc k will not be included in a cycle. There are no admissible arcs passing from N_k to $N_\ell(\ell > k)$.

This process is implemented by step 2 of algorithm SHORTM. Figure 7.6 shows the π values assigned for the example.

Let us use the network of Figure 7.7 to assist in explaining algorithm PRIMAL1. Figure 7.7 presents the flow configuration that is obtained when PHASE1 is applied to

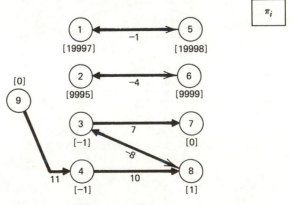

Figure 7.6
Spanning Forest for Figure 7.5

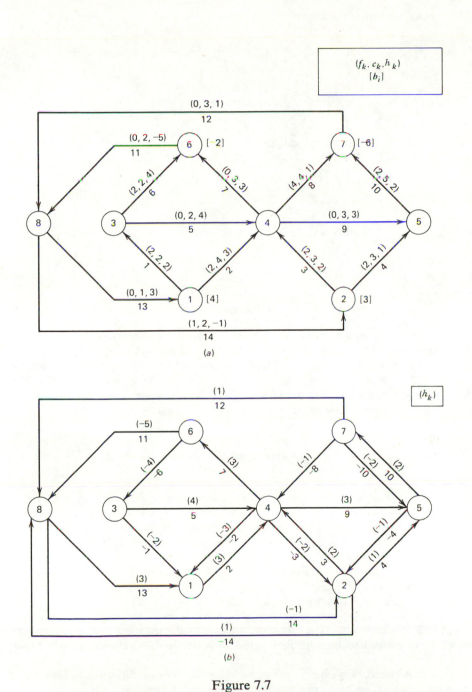

Figure 7.7
Example Application of Algorithm PRIMAL1 (*a*) Original Network
(*b*) Marginal Network (*c*) Tree from DSHRTM (*d*) Original Network
(*e*) Marginal Network (*f*) Tree from DSHRTM (*g*) Tree from TRECHG

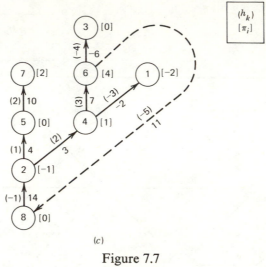

(c)

Figure 7.7
(continued)

the network of Figure 7.4. (Note that this configuration is markedly different from that given in Figure 7.4a.) Figure 7.7b pictures the marginal network defined by the flows of Figure 7.7a. Applying algorithm DSHRTM to the network of Figure 7.7b yields the spanning tree given in Figure 7.7c. The dashed arc 11(6, 8) is selected by algorithm SMNSPM as the admissible arc from Figure 7.7b with the most negative value of d_k.

It is easy to see that the entry of arc 11(6, 8) forms a negative cycle with value

$$d_{11} = \pi_6 + h_{11} - \pi_8 = 4 + (-5) - 0 = -1$$

This can be verified by summing the costs of the arcs on the cycle; that is, $d_{11} = h_{14} + h_3 + h_7 + h_{11} = -1$. To achieve a new feasible flow pattern with less cost than that given in Figure 7.7a, we need only augment the flow around the cycle. The amount of flow increase around the cycle is limited to 1 additional unit of flow, which saturates both arcs 3 and 14 at their capacities.

The resulting new flow configuration and marginal network are given in Figures 7.7d and 7.7e, respectively. Figure 7.7f presents the spanning tree DSHRTM obtains from Figure 7.7e. Arc $-1(3, 1)$ is then selected by SMNSPM.

Entry of arc $-1(3, 1)$ does not cause a negative cycle to be formed. Performing a TRECHG operation with $k_E = -1$ and $k_L = 13$ yields the spanning tree of Figure 7.7g, which is an optimal solution to the shortest path tree problem associated with Figure 7.7e.

Since no additional negative cycles can be found, the flow configuration of Figure 7.7d is an optimal solution to this example minimum cost flow problem.

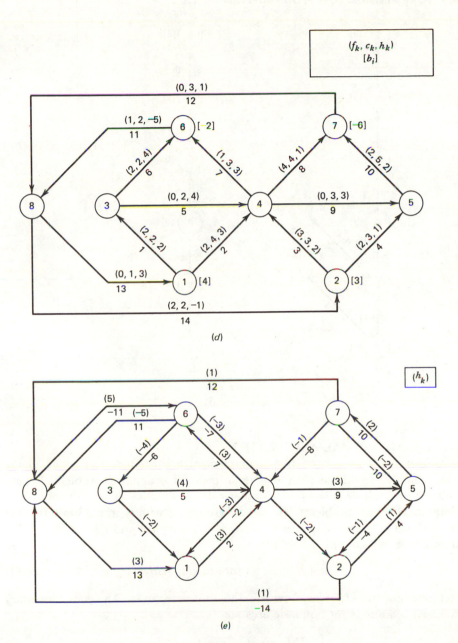

Figure 7.7
(continued)

185

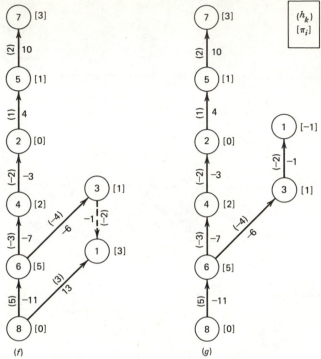

Figure 7.7
(continued)

7.4   A BASIC PRIMAL ALGORITHM

This algorithm uses a basis tree to implement a procedure analogous to the primal simplex algorithm for the general linear programming problem. Recent studies have indicated that this is perhaps the most efficient way to solve pure minimum cost flow problems. The procedure begins with a primal basic feasible flow defined on a spanning tree $D_T = [N, M_T]$ of forward and mirror arcs. Node potentials are set so that

$$\pi_i + h_k = \pi_j \qquad \text{for arc } k(i,j)\epsilon M_T \tag{1}$$

Each nonbasic arc $k(i,j)$ has flow equal to either zero or c_k. The complementary slackness conditions for nonbasic arcs are

$$\text{If } f_k = 0, \text{ then } \pi_i + h_k \geqslant \pi_j \tag{2}$$

and

$$\text{If } f_k = c_k, \text{ then } \pi_i + h_k \leqslant \pi_j \tag{3}$$

These conditions are not satisfied for all arcs until the termination of the algorithm.

The basic primal algorithm has four principal parts.

1. Select an arc that violates one of the conditions 2 or 3 and call this the *entering arc* $k_E(i_E, j_E)$.
2. Determine the cycle composed of basic arcs and the entering arc on which flow must be changed in order to bring k_E into the basis. Find the maximum flow change possible. Either the entering arc or some basic arc will be determined to leave the basis. We call this the *leaving arc* $k_L(i_L, j_L)$.
3. Modify the basis tree so that k_E enters and k_L leaves. If k_E and k_L are the same arc no change is necessary.
4. Modify the node potentials so that condition 1 is satisfied for the new tree, and go to step 1.

This process repeats until the optimality conditions are satisfied for all arcs. The details of this procedure are described in the PRIMAL algorithm.

PRIMAL ALGORITHM

Purpose: To perform the primal simplex technique for the pure network minimum cost flow problem. The algorithm begins with an initial feasible basic solution.

1. (SELECT) Select an arc $k_E(i,j)$ that violates the optimality conditions to enter the basis. If there are none, stop with the optimum solution. Let $\Delta = \pi_i + h_{k_E} - \pi_j$ if $k_E > 0$ or $-\pi_i - h_{-k_E} + \pi_j$ if $k_E < 0$.
2. (FLOW) Find the cycle, consisting of the entering arc and basic arcs, on which flow must be increased to let k_E enter the basis. Determine the maximum amount that flow can be increased in these arcs while maintaining feasibility. Change the arc flow by that amount. Select an arc which becomes inadmissible with this flow change to leave the basis, $k_L(u,v)$. If the entering arc is the same as the leaving arc, go to step 1.

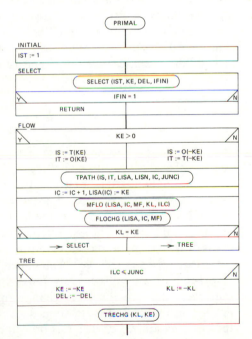

3. (TREE) If the k_L is in the forward part of the path from i to j, let $k_E := -k_E$; let $\Delta := -\Delta$. If k_L is in the reverse part of the path from i to j, let $k_L := -k_L$.

4. (POTENTIAL) Find the terminal node of k_E, r'. Determine the set of nodes in the tree rooted at r'. Add Δ to all nodes in this set. Go to step 1.

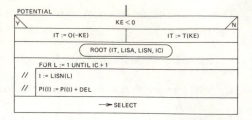

The first step of the algorithm requires the selection of an admissible arc to enter the basis. For an arc not in the basis, flow is either zero or at capacity. When $f_k(i,j) = 0$ the forward arc $k(i,j)$ is admissible and we define:

$$d_k = \pi_i + h_k - \pi_j$$

When $f_k(i,j) = c_k$ the mirror arc $-k$ is admissible and we define:

$$d_{-k} = \pi_j - h_k - \pi_i$$

The node potentials are defined so that $d_k = 0$ for basic arcs. For a nonbasic arc $d_k > 0$ if the arc satisfies the optimality condition, and $d_k < 0$ if it does not. There are a number of computational options for choosing the arc to enter the basis. The one used for the general simplex technique in most introductory texts is called "the most negative rule." This rule chooses the arc with the most negative value of d_k to enter the basis. Thus

$$d_{k_E} = \frac{\text{Min } d_k}{k\varepsilon(M^* - M_T)} \tag{4}$$

If $d_{k_E} > 0$ the solution is optimum. The time cost for each iteration is $O(m)$ since each nonbasic arc must be checked.

An alternative is the "first negative rule" in which the arc first encountered with a negative d_k is selected to enter the basis. After the basis change, the search for the new arc to enter the basis is resumed at the arc with an index one greater than the arc that just entered the basis. The advantage of this procedure is that for most iterations an entering arc is found after only a small portion of the arcs have been reviewed. The upper bound on the time cost is still $O(m)$ but in practice there will be many fewer than m arcs reviewed. This reduction is important because the basis change operations are $O(n)$ in time cost. Since $m > n$, the time cost per iteration depends strongly on the rule for determining the entering arc. Of course, this procedure does not guarantee that $d_k > 0$ for all arcs with indices less than the entering arc, and the procedure must continue to pass through the arc set until $d_k > 0$ for all nonbasic arcs.

There are a number of variants on this procedure. We describe in algorithm SELECT the "first node most negative" rule. This rule arises from transportation problem computations. There each node, or row of the transportation tableau, is reviewed in turn. When a row is found with a negative d_k, the other entries in the row are reviewed to find the most negative d_k. The cell with the most negative d_k enters the basis. After the basis change, the search for an entering cell begins in the next row. This procedure has been found to be very effective in computational studies. It is modified here for the minimum cost flow network problem by considering each node in turn and reviewing the forward and mirror arcs *originating* at the node. When a node is first encountered with an arc with $d_k < 0$, the arc with the most negative d_k that originates at that node is chosen to enter the basis. This arc is called $k_E(i_E, j_E)$. The second step in the procedure is involved with changing arc flows to bring arc k_E into the basis. In that process the arc to leave the basis, arc $k_L(i_L, j_L)$, is determined.

SELECT ALGORITHM

Purpose: To find an arc not satisfying optimality conditions using "first node most negative" rule.

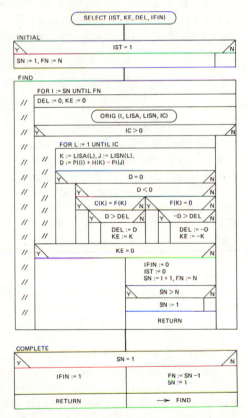

1. (INITIAL) If this is the first pass through this subroutine set search range for nodes from 1 to n.
2. (FIND) Go through the list of nodes in the search range. For each node i in the range find the set of arcs originating at the node. Check the optimality condition for each arc. If an arc is found that violates the conditions, continue to check the arcs for that node. Choose the arc for which the conditions are violated the most. Set the node search range for the next pass through this algorithm from $i+1$ to n and return. If no nonoptimum arcs are found, go to step 3.
3. (COMPLETE) If the search in step 2 ranged from node 1, there are no nonoptimum arcs and the solution is optimum. If not, set the search range from 1 to one less than the range used for step 2 and return to step 2.

Arc k_E may be either a forward or mirror arc. If $k_E > 0$, the arc is brought into the basis by increasing the flow in arc k_E of the original network. If $k_E < 0$, the arc is brought into the basis by decreasing the flow in arc $-k_E$ of the original network. Flows will be changed only in basis arcs and in arc k_E. The basis tree defines a unique path from the terminal node of k_E to its origin node. This path together with k_E defines a cycle $D_C = [N_C, M_C]$ on which flow is to be changed. Using the notation of Section 5 in Chapter 4, we may write the arcs of the cycle

$$M_C = M_F \cup M_R^- \cup k_E$$

where M_F and M_R are the forward and reverse parts of the path in the tree which connects the origin and terminal nodes of arc k_E.

The maximum flow change on an arc in this cycle is

$$\mu_k = \begin{cases} c_k - f_k & \text{if } k \in M_C \quad \text{and} \quad k > 0 \\ f_{-k} & \text{if } k \in M_C \quad \text{and} \quad k < 0 \end{cases}$$

Therefore, the maximum flow change for the cycle is μ_{k_L} the minimum of the μ_k, for the arcs in the cycle. The arc to leave the basis, $k_L(u,v)$, is the arc that enforces the minimum cycle change. The leaving arc may be k_E or any other arc in the cycle. The flows are changed in the original network by

$$f_k' = f_k + \mu_{k_L} \qquad \text{for } k \in M_C \quad \text{and} \quad k > 0$$
$$f_{-k}' = f_{-k} - \mu_{k_L} \qquad \text{for } k \in M_C \quad \text{and} \quad k < 0$$
$$f_k' = f_k \qquad \text{for } k \notin M_C$$

where f_k' is the modified flow in arc k of the original network.

With the identification of the entering and leaving arc the basis may change. If $k_L = k_E$ no change is necessary. If $k_L \neq k_E$, we must consider the two possibilities pictured in Figure 7.8. We recall that, any time an arc $k_L(i_L, j_L)$ is deleted from a basis tree, two directed subtrees are formed. One subtree, $D_1 = [N_1, M_1]$, is rooted at the slack node while the other subtree, $D_2 = [N_2, M_2]$, is rooted at j_L. When an arc $k_E(i_E, j_E)$ is added to the two subtrees, i_E must be an element of N_1 and j_E must be an element of N_2, if the resulting network is to be a directed spanning tree, that is, a basis tree. This characteristic is, of course, necessary to any basic primal algorithm.

When k_L is on the reverse path, as in Figure 7.8a, the new basis is formed by deleting k_L, reversing the arcs from j to v, and inserting k_E. When k_L is in the forward path, as in Figure 7.8b, the new basis is formed by deleting k_L, reversing the arcs from i to v, and inserting $-k_E$. Note that, since k_E may be negative, $-k_E$ may represent a forward arc.

The final step in the procedure is to determine the node potentials for the new tree. Only the potentials in the subtree rooted at the terminal node of the entering arc need be changed. Again we must consider the two cases of Figure 7.8. For the case of Figure 7.8a, let N_j be the set of nodes in the new basis tree rooted at node j_E. The new node potentials are

$$\pi_i' = \pi_i + d_{k_E} \qquad \text{for } i \in N_j$$
$$\pi_i' = \pi_i \qquad \text{for } i \in N - N_j$$

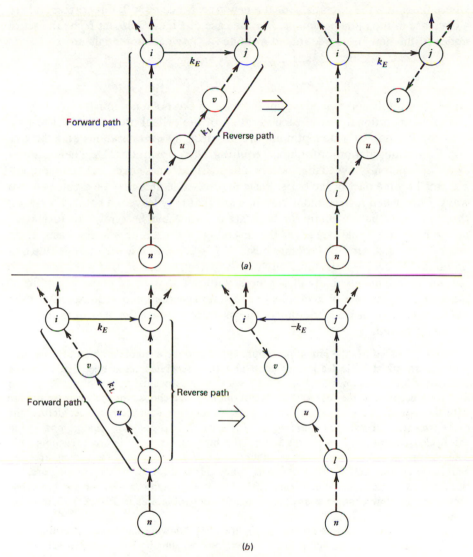

Figure 7.8
Changing the Basis Tree

where the π_i' are the potentials for the new tree. Since $d_{k_E} < 0$, this corresponds to a decrease in the node potentials. For the case of Figure 7.8b, let N_i be the set of nodes in the new basis tree rooted at node i_E. The new potentials are

$$\pi_i' = \pi_i - d_{k_E} \qquad \text{for } i \in N_i$$
$$\pi_i' = \pi_i \qquad \text{for } i \in N - N_i$$

Since $d_{k_E} < 0$ this corresponds to an increase in the node potentials.

The computational cost of each iteration of the primal procedure is $O(m+n)$, with the location of a nonoptimum arc requiring $O(m)$ operations and the tree and flow manipulation procedures requiring $O(n)$ operations. Degenerate iterations, in which no flow change occurs, are possible. These occur when inadmissible arcs become part of the basis. Basic inadmissible arcs may be created in two ways. First, when there is more than one arc that may leave the basis (a tie in μ_k) the arc or arcs that remain in the basis are inadmissible. Secondly, the formation of the basis tree usually requires the reversing of certain arcs in the basis. If an arc k is reversed, arc $-k$ becomes basic. If $f_k = 0$, arc $-k$ is an inadmissible arc. Similarly, if arc $-k$ is reversed, arc k becomes basic. If $f_k = c_k$, arc k is inadmissible. The possibility of degeneracy allows no realistic upper bound to be placed on the number of iterations the primal procedure may require (other than the total number of basic solutions). However, in practice the algorithm seems to be very efficient.

Figure 7.9 illustrates the primal procedure applied to the example problem. Figure 7.9a presents an initial basic feasible solution and the associated basis tree. The nonbasic admissible arcs are shown as dashed lines with their associated values of d_k. Since two arcs have negative d_k the first solution is not optimal. We choose for this example the arc with the most negative value, arc 5(2, 5). This arc together with the basis tree defines the cycle consisting of the arc set $M_C = \{5, -2, 3, -6\}$. The largest flow change possible in this cycle is 2, with the flow in arcs 5 and 3 reaching capacity and in arc 2 reaching zero. The changed arc flows are shown in Figure 7.9b. Of the three possible leaving arcs, we arbitrarily choose arc 5 to leave the basis. Since this is the same as the entering arc, the basic tree does not change in Figure 7.9b. The node potentials also remain the same. However, the flow changes cause the nonbasic admissible arcs in Figure 7.9b to differ from those in Figure 7.9a.

Only arc $-9(6, 3)$ has a negative d_k in Figure 7.9b. Allowing this arc to enter the basis, we find that the corresponding cycle determined by the basis is the arc set, $M_C = \{-9, 8, -2, 3\}$. This is a degenerate iteration because no flow change is possible on the cycle; that is, the flow in arc 2 is zero and the flow in arc 3 is at capacity. We arbitrarily choose arc -2 to leave the basis. Since this is in the forward part of the cycle we must reverse arc -9 as it enters the basis. Thus arc 9(3, 6) enters the basis and arc $-2(5, 1)$ leaves to form the new tree shown in Figure 7.9c. The node potentials are then changed for the nodes in the tree rooted at node 6, that is, nodes 6, 1, and 2. Since d_k is now positive for all of the admissible nonbasic arcs, an optimal solution has been obtained.

The basic algorithm has the advantage over the nonbasic algorithm because negative cycles are easy to discover and identify using the basis. Every nonbasic

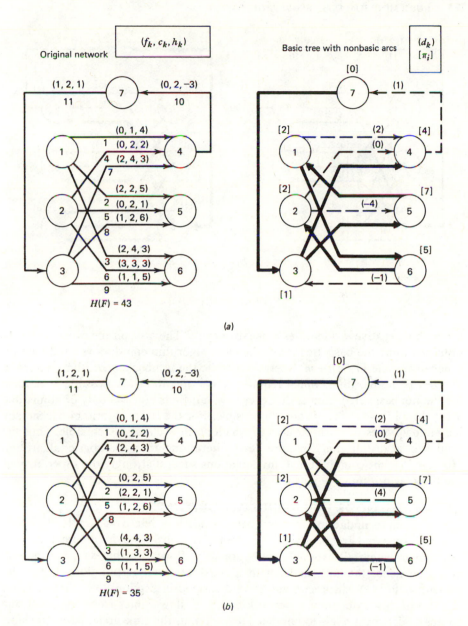

Figure 7.9
Application of Algorithm PRIMAL

193

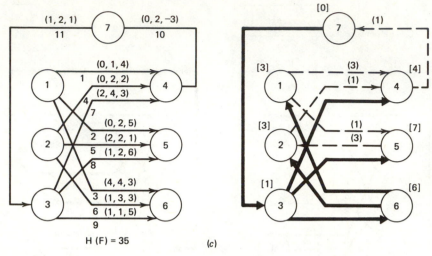

Figure 7.9
(continued)

arc with a negative d_k identifies a negative cycle. The arcs on the cycle are easily obtained using the basis tree. Thus the basic algorithm can discover and identify a negative cycle in $O(n+m)$ operations while the nonbasic algorithm requires a shortest path algorithm, which requires $O(n^3)$ operations. A possible advantage for the nonbasic algorithm is that every cycle it finds consists only of admissible arcs; thus a flow change is always possible. For the basic algorithm degeneracy may allow cycles to be discovered in which no flow change is possible. Thus the nonbasic algorithm may require fewer iterations than the basic algorithm. However, recent computational investigations suggest strongly the superiority of the basic algorithm.

7.5 A DUAL NODE INFEASIBLE ALGORITHM

The problem formulation described below leads to a relatively simple algorithm that has appeared frequently in the literature. Consider a minimum cost flow problem in which all external inputs are slack flows. Further suppose there is only one node, the sink node, with a fixed external output flow. Using the transformations of slack external flows described in Section 2 of Chapter 3, we can model this problem as a network where all flow comes from one point, the source node, and travels to another unique point, the sink node. After transformation, the problem becomes one of finding the network flow that satisfies the flow requirement into the sink node at minimum cost.

Figure 7.10a illustrates the example problem of Figure 7.1 specified in this format. The fixed external inputs for nodes 1, 2, and 3 are changed to slack inputs by using a large negative slack cost (-10). If a feasible solution exists these external flows will be

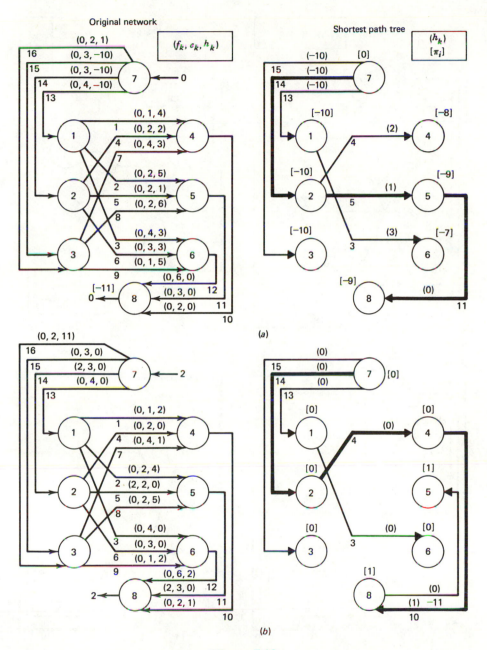

Figure 7.10
Example of the Nonbasic Incremental Algorithm

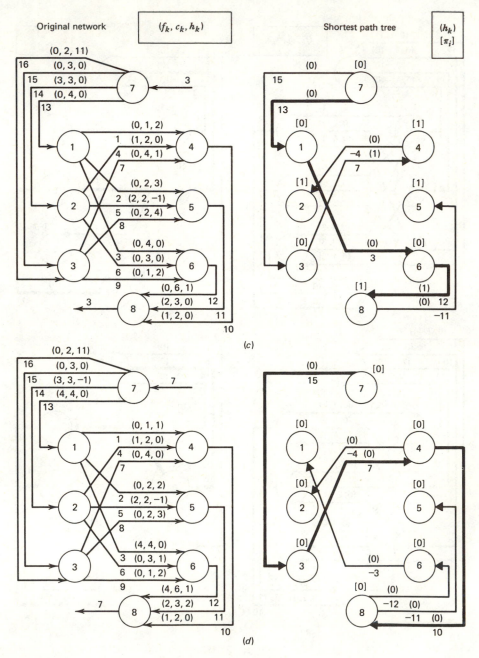

Figure 7.10
(continued)

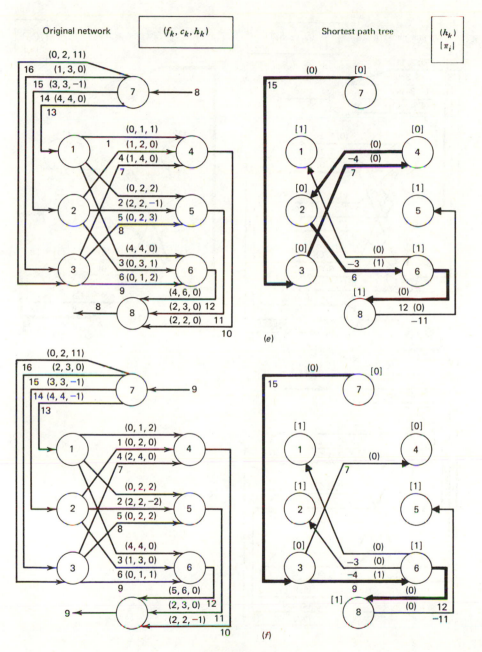

Figure 7.10
(continued)

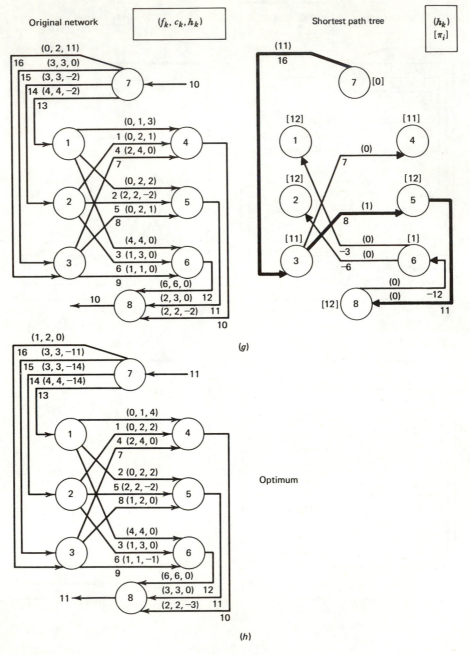

Figure 7.10
(continued)

satisfied. For simplicity, the slack external output from node 4 has been dropped. Next we create node 8, the sink node, and arcs 10, 11, and 12 with capacities equal to the external output flows on nodes 6, 5, and 4, respectively. The fixed external flow on the sink is set equal to -11, the sum of the former external flows on nodes 4, 5, and 6. The optimal solution to the new problem will be the same as the optimal solution to the problem of Figure 7.1 (less the slack output from node 4).

This algorithm consists of two iterative steps. First the least cost path from the source node to the sink node over admissible arcs is found using a shortest path algorithm. Second, the flow is augmented in this path by the maximum possible amount. The changed flows define a new marginal network and we return to the first step. The two steps are repeated until one of two stopping conditions occur: if the sink node's requirements are satisfied, the arc flows are optimum for the minimum cost flow problem. The nonexistence of a path prior to satisfying the sink node's requirements is sufficient to conclude that no feasible solutions exist.

This technique is called an *incremental approach* because the sink flow is incremented at each iteration. The resulting algorithm is a dual node infeasible algorithm because the complementary slackness conditions are satisfied throughout, but the conservation of flow at the sink node is satisfied only at termination. The general procedure is shown as the INCREME algorithm.

INCREME ALGORITHM

Purpose: To solve the minimum cost flow problem for pure networks with a flow incremental approach. Starting with zero flow, the flow is iteratively increased through the minimum cost path from source (SC) to sink (SK) until a specified sink flow (VR) is obtained.

1. (INITIAL) Set all arc flows to zero.
2. (PATH) Find the least cost path from source to sink in the marginal network.
3. (ARCLST) Obtain arcs on the least cost path using back pointers. If no path exists, set INF:= 1 and return.
4. (FLOW) Find the maximum amount the flow can be increased in the path, FM. If this flow is greater than that necessary to bring the sink flow to the required value, set the flow increase to that amount and stop after this iteration. If not, increase the flow in the arcs defined by the path and return to step 2.

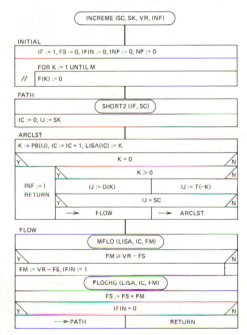

Figure 7.10 illustrates the incremental algorithm applied to the example problem described above. Each part of the figure shows the network flow existing at an iteration and also the shortest path tree for the marginal network. The flow augmenting path at each iteration is the shortest path from the source node to the sink node in the marginal network.

Algorithm SHORT2, which finds the flow augmenting path, may take several forms depending on the cost characteristics of the original network. If the original network includes both positive and negative arc costs but no negative cycles, the initial shortest path tree is found with a primal algorithm. It can be shown that if the arc costs are replaced by the following calculation

$$h'_k = \pi_i + h_k - \pi_j \qquad \text{for arc } k(i,j) \tag{5}$$

the relative costs of the paths from source to sink remain unchanged. Using the values of the node potentials obtained from the initial shortest path calculation, it may easily be shown that $h'_k > 0$ for all admissible arcs. We call h'_k the reduced cost of arc k. Since reduced arc costs are positive, every iteration after the first can use Dijkstra's algorithm to find the flow augmenting path.

SHORT2 ALGORITHM

Purpose: To find the least cost path in the marginal network for algorithm INCREME.

1. (PATH) Find the shortest path from the slack node to the sink node. If this is the first pass through, this algorithm uses sequentially Dijkstra's algorithm and a primal shortest path algorithm. If this is not the first pass, use Dijkstra's algorithm.
2. (REDUCE) Replace the cost of each arc by its reduced cost.

$$h_k := \pi_i + h_k - \pi_j \qquad \text{for arc } k(i,j)$$

Return.

```
                              SHORT2 (IF, SC)

PATH
Y ┌──────────────── IF = 1 ────────────────┐ N
  │  DSHRTM (SC, 0, NP)  │                  │
  │                      │  DSHRTM (SC, 0, NP)
  │  TREINT(N)           │                  │
  │                      │                  │
  │  PSHRTM (CYC, JJ)    │                  │
REDUCE
  ┌──────────────────────────────────────┐
  │     FOR K := 1 UNTIL M                │
// │   I := O(K), J := T(K)               │
// │   H(K) := PI(I) + H(K) − PI(J)       │
  └──────────────────────────────────────┘
                  IF := 0

                  RETURN
```

Notice that in every iteration after the first the new reduced costs may be computed directly using the reduced costs from the previous step. Therefore, the original arc costs need not be kept in memory during the solution process. The technique described above is coded in algorithm SHORT2.

When the original network consists only of arcs with nonnegative length, Dijkstra's algorithm can be used for the first iteration as well. Thus the PSHRTM algorithm could be deleted from SHORT2 for this case. It can be shown that Dijkstra's algorithm will work with some negative arcs if these arcs either originate at the source or terminate at the sink. Thus, for the example problem and also other problems where fixed external flows are replaced with arcs having large negative cost, Dijkstra's algorithm can be used at every iteration. The time cost for SHORT2 is $O(n^2 + m)$ after the first iteration— ($O(n^2)$ for DSHRTM and $O(m)$ for calculating the reduced costs).

The incremental algorithm does not guarantee that the solution remains basic. One way to make the solutions basic is to replace DSHRTM after the first iteration with a version of DUALSP that is appropriate for the marginal network. With the flow change on the augmenting path, one or more arcs on the tree become inadmissible. The algorithm would delete these arcs and replace them with admissible arcs in order that the flow solution would remain basic. Also, it is not necessary to compute the reduced costs. Indeed, if reduced costs are computed it would also be required that the node potentials be adjusted to reflect the effect of the reduced costs. Unless this were done before selection of an entering arc, an incorrect entering arc could very well be chosen. If there were no degeneracy, the time cost for each iteration would be $O(n + m)$, smaller than for DSHORT. As discussed in Section 7 of Chapter 5, degeneracy is possible, thus making the upper bound on time cost $O(n^2 + nm)$, which is greater than for DSHORT. To obtain this bound requires $n - 1$ degenerate solutions for each nondegenerate solution. A much smaller proportion of degenerate solutions seems to be the case in practical problems.

The presence of cycles with negative cost does not imply an unbounded solution of the minimal cost flow problem. Rather if a negative cycle exists the flow in the arcs of the cycle should be increased until some arc is saturated. Thus the marginal network will not contain these cycles. All negative cycles must be saturated before the incremental algorithm may proceed. This can be accomplished by preceding the first application of SHORT2 with a primal minimum cost flow algorithm to find the optimum flows with all external flows zero.

7.6   HISTORICAL PERSPECTIVE

Since the initial description of the transportation problem by Hitchcock (1941), Kantorovich (1942), and Koopmans (1947), numbers of papers have been written concerning the computational aspects of pure minimum cost network flow problems. Many of these deal with the special case of the transportation problem. A smaller number deal with the special case of the assignment problem. Rather than attempt to survey this extensive literature, we list below the principal references related to the approaches described in this chapter. Surveys of earlier work appear in Charnes and Cooper (1961), Ford and Fulkerson (1962), and Dantzig (1963). Busacker and Saaty (1965), Fulkerson (1966), Hu (1969), Elmaghraby (1970), Frank and Frisch (1971), and Golden and Magnanti (1977).

The nonbasic primal procedure discussed in Section 7.3 stems principally from Klein (1967). Subsequent papers on methods to find negative cycles of admissible arcs were by Florian and Robert (1971) and Bennington (1973). Bennington (1973) suggests using a primal simplex shortest path algorithm to find negative cycles.

The most active area of interest today is the basic simplex solution technique as discussed in Section 7.4. The early precursors to this work were Dantzig (1951) and Charnes and Cooper (1954), who applied the simplex technique to the transportation problem. Orden (1956) extended these results to the transshipment problem. More recently, several authors have utilized the special representation of basis trees to increase markedly the computational efficiency of primal simplex techniques. Srinivasan and Thompson (1972, 1973) and Glover, Karney, and Klingman (1972) utilize the triple label representation for transportation problems. Langley et al. (1974) uses the triple label method in a primal simplex algorithm for the capacitated transportation problem. Glover, Karney, and Klingman (1974a) apply the triple label technique to pure network problems. Glover, Klingman, and Stutz (1974) introduced the augmented threaded index method for the pure network problem. Bradley, Brown, and Graves (1977) provide the details for a primal algorithm using this basis representation. Several computational studies including Srinivasan and Thompson (1973), Glover, Karney, and Klingman (1974), and Glover, Karney, Klingman, and Napier (1974) attest to the superiority of the primal simplex procedure for network flow programming problems.

The incremental flow procedure discussed in Section 7.5 has as its principal precursor the report by Busacker and Gowen (1961). An incremental approach is also discussed in Ford and Fulkerson (1962). Later references appear in Edmonds and Karp (1972) and Tomizawa (1972). A related algorithm applied to the generalized minimum cost flow problem is discussed in Jensen and Bhaumik (1977).

EXERCISES

1. Outline a method, in detail, to reduce any feasible *nonbasic* solution to a minimum cost flow problem to a *basic* feasible solution. Use that method to construct a basic feasible solution from the feasible solution given in Figure. 7.2*d*.
2. Why can't FPATH2 be used instead of FPATH1 in algorithm PHASE1?
3. Starting with the initial feasible solution given in Figure 7.2*d*, use the method of PRIMAL1 to obtain the optimal minimum cost flow solution to the network given there.

4. Starting with the feasible solution given in Figure 7.4c, use the method of PRIMAL1 to obtain the optimal minimum cost flow solution to the network given there.
5. Using the method of ARTIFIC construct the initial all-artificial basic feasible solution to the network of Figure 7.4.

6. (a) Beginning with the basic solution found in exercise 5, use the method of algorithm PRIMAL to obtain the optimal minimum cost flow solution.
   (b) Starting with the initial basic feasible solution given in Figure 7.3, use the method of algorithm PRIMAL to obtain the optimal minimum cost flow solution.

7. (a) Use the method of algorithm INCREME to obtain the optimal minimum cost flow solution for the network of Figure 7.4.
   (b) Beginning with the spanning forest in Figure 7.6, use PRIMAL1 to find the optimal flows.

8. Modify algorithm DUALSP so that any admissible mirror arc K(J,I) that has $S(I) = 1$ and $S(J) = 0$ will be considered as a possible entering arc. Call this algorithm DULSPM.
9. Replacing algorithm DSHRTM after the first iteration with algorithm DULSPM, use the method of algorithm INCREME to obtain the optimal minimum cost flow solution for the network of figure 7.10.
10. For the application problems of Section 3, Chapter 2:

    (a) Use PRIMAL1 to solve the Tanglewood Chair and Hickory Dickory Loading problems. (Let $P_1 = 50$, $P_2 = 47$, and $P_3 = 60$ for the latter problem.)
    (b) Use PRIMAL to solve the Hirem-Firem and Bromwich Restaurant problems.
    (c) Use INCREME to solve the Amalgamated Grain and Joe's Tavern problems.

11. Solve the applications problems of Section 4, Chapter 2.
12. Solve the applications problems of Section 5, Chapter 2.
13. Solve the following problems as minimum cost flow problems and draw the networks giving the optimal flow configuration.

    (a) The Wimperly Fire Department (of Chapter 2).
    (b) Krumley Painting and Wall Papering (of Chapter 2).
    (c) Figure 5.2, Chapter 5.
    (d) Figure 6.1, Chapter 6.

14. One way to achieve a solution that satisfies both complementary slackness and conservation of flow for any minimum cost flow network is: (1) use DSHRTM to find the shortest path tree in the marginal network; (2) place flows on the tree arcs so that all fixed external flows are satisfied (ignore arc capacities and lower bounds). Show why the above statement is true.
15. If the technique described in exercise 14 is used, one or more arcs may have flows assigned to them that violate the arc's lower or upper bound. Outline, in detail, a *dual arc infeasible* algorithm that could then be used to achieve the optimal solution.

16. Apply the techniques described in exercises 14 and 15 to obtain the optimal solution for the network given below.

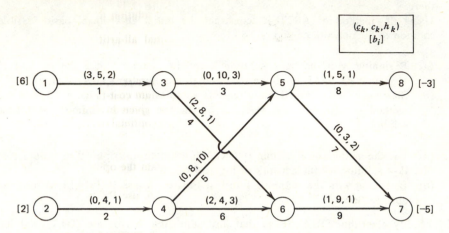

CHAPTER 8
THE OUT-OF-KILTER ALGORITHM

One of the first specialized algorithms to solve the pure minimum cost network flow problem was the Out-of-Kilter Algorithm, abbreviated OKA. It experienced wide use during the 1960s and early 1970s. In addition to its historical interest, the OKA has characteristics that make it useful today. Among its advantages are:

- It is simple to understand and thus useful for the classroom.
- It requires no special memory structures for network representation. Arc parameters are entered and stored in arbitrary order in lists.
- It uses no external node flow parameters. All such information is described by arc parameters.
- The algorithm may initiate with any set of flows that satisfies conservation of flow. Thus the algorithm is particularly useful for sensitivity analysis when arc parameters are changed.

Among the disadvantages of the OKA are:

- The use of arcs to specify node external flows often results in an unexpectedly high memory requirement.
- Solutions are not necessarily basic. This may result in slow convergence to the optimum.
- The solution procedures may require numerous passes through arc length lists, thus the OKA is not likely to be computationally efficient for large problems.

All of the algorithms presented in the previous chapters have been either primal or dual algorithms that have required either primal or dual initial feasible solutions. Conversely, the OKA is a primal–dual algorithm that may begin with

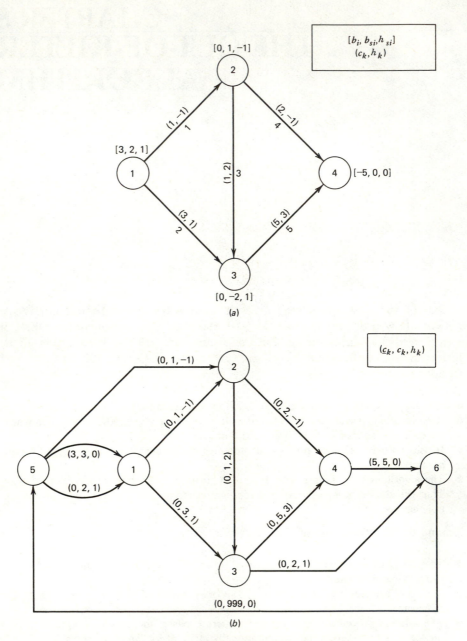

Figure 8.1
Example Network Model for the OKA

a solution that is neither primal nor dual feasible. This provides a measure of flexibility not available with other algorithms.

It is possible to use basis trees to make the OKA more efficient than the classical implementation. We discuss this idea later in the chapter.

8.1   NETWORK MODEL

The OKA network model allows three arc parameters: lower bound on flow $\underline{c}_k$; upper bound on flow, c_k; and cost per unit flow, h_k. All node external flows must be represented by arcs. Figure 8.1a illustrates a network with external flows. Its equivalent OKA representation is shown in Figure 8.1b.

The OKA model is obtained by defining a source node, node 5, and a sink node, node 6. All positive external flows for node i generate an arc from the source to node i. For a fixed external flow, the arc lower and upper bounds are both set equal to the amount of the fixed flow, and the arc cost is set to zero. For a slack external flow, the lower bound is set to zero and the upper bound to the maximum slack flow. The arc cost equals the cost of the slack flow. In a similar manner, negative external flows are modeled with arcs terminating at the sink node.

Finally, we create a "return arc" from the sink to the source with a zero lower bound, a very large upper bound, and a zero cost. The purpose of this arc is to provide conservation of flow at both the source and the sink. Since there are no external flows in the network, the amount of flow leaving the source must equal the amount entering the sink, and this same amount must pass through the return arc. The source and sink may be combined into a single node, eliminating the return arc. For many applications, it is pictorially and conceptually simpler to use both the source and the sink. The source can be thought of as the provider of all flow and the sink as the absorber of all flow.

Subroutine READOK reads the network data in the format used in the previous chapters and converts it into the OKA format. Since the classical implementation of the OKA uses an arbitrary ordering of the arcs, neither ORIGS nor TERMS is used.

READOK ALGORITHM

Purpose: To read node fixed and slack external flows and arc parameters and create a representation that can be used by the Out-of-Kilter Algorithm.

1. (INITIAL) Initialize the number of arcs to zero. Read the number of nodes and create the source and sink nodes.
2. (NODE) Read a node data item. If the item is blank, go to step 3. If the fixed flow is zero, go to step 2a. Otherwise, find the sign of the fixed flow. If it is positive, create an arc from source to the node with both bounds set to the fixed flow and cost set to zero. If it is negative, create an arc from the node to the sink with parameters as above.
2a. If the slack flow is zero, repeat step 2. If it is not, find the sign of the slack flow. If it is positive, create an arc from the source to the node with zero lower bound, upper bound equal to the slack flow, and cost equal to the slack cost. If it is negative, create an arc from the node to the sink with parameters as above. Repeat step 2.
3. (ARC) Read the arc data item. If the item is blank, the complete data set is read; create the return arc and go to step 4. Otherwise increase the arc number by one and place the parameter information in memory. Repeat step 3.
4. (BZERO) Set all B(I):=0, PI(I):=0, and F(K):=0. Return.

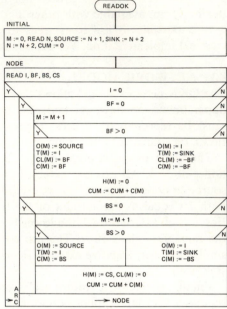

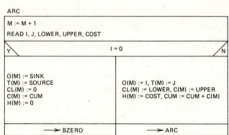

8.2 LINEAR PROGRAMMING MODEL

With all node external flows set to zero and conservation of flow required at each node, the linear programming formulation becomes

$$\text{Min.} \sum_{k=1}^{m} h_k f_k$$

Conservation of flow

$$\sum_{k\in M_{T_i}} f_k - \sum_{k\in M_{O_i}} f_k = 0 \qquad i=1,\ldots,n$$

Lower bounds

$$f_k \geqslant \underline{c}_k \qquad k=1,\ldots,m$$

Upper bounds

$$f_k \leqslant c_k \qquad k=1,\ldots,m$$

As previously discussed (Chapter 3), one of the conservation-of-flow constraints is redundant. Note that this formulation does not restrict the signs of $\underline{c}_k$ or c_k. However, for the purposes of this chapter, we will assume that $0 \leqslant \underline{c}_k \leqslant c_k$.

To form the dual of this linear programming problem, we assign π_i to the conservation of flow constraint for node i, δ_k to the upper bound constraint for arc k, and ω_k to the lower bound constraint for arc k. The dual problem then becomes

$$\text{Min.} \quad \sum_{k=1}^{m} \delta_k c_k + \sum_{k=1}^{m} \omega_k \underline{c}_k$$

$$\text{st.} \quad \pi_i - \pi_j + \delta_k - \omega_k \geqslant -h_k \qquad \text{for arc } k(i,j)\in M$$

$$\pi_i \text{ unrestricted} \qquad i\in N$$
$$\delta_k \geqslant 0 \qquad k\in M$$
$$\omega_k \geqslant 0 \qquad k\in M$$

Since one of the primal constraints is redundant, one of the π dual variables is arbitrary.

From these primal and dual statements, we can derive conditions for optimality in a manner similar to that of Chapter 3. For convenience we define

$$d_k = \pi_i - \pi_j + h_k \qquad \text{for arc } k(i,j)$$

Given a primal solution F and a partial dual solution π, we find that the two solutions are optimal for their respective problems if the following conditions exist.

1. Primal feasibility.

   (a) Conservation of flow at each node.
   (b) $f_k \geqslant \underline{c}_k$ for all arcs k
   (c) $f_k \leqslant c_k$ for all arcs k

2. Restricted dual feasibility.

  (a) $\delta_k = \text{Max} \, [0, -d_k]$
  (b) $\omega_k = \text{Max} \, [0, d_k]$

3. Complementary slackness.

  (a) $d_k = 0$ for $0 < f_k < c_k$
  (b) $f_k = \underline{c}_k$ for $d_k > 0$
  (c) $f_k = c_k$ for $d_k < 0$

The OKA algorithm requires an initial flow solution that satisfies condition 1(a) but not necessarily condition 1(b) or 1(c). The initial node potentials are arbitrary so condition 3 is usually not satisfied. Conditions 2(a) and 2(b) are not used in the algorithm but are useful for postoptimality sensitivity analysis.

The algorithm varies the flows and potentials in a manner that moves all arcs toward satisfying conditions 1(b), 1(c), and 3. The ultimate attainment of an optimal solution in a finite number of iterations is guaranteed if a feasible solution exists. If no feasible solution exists, the algorithm will discover and indicate that condition.

8.3 KILTER STATES

If the π_i's are defined for each node and flows are defined for each arc, d_k can be computed for each arc. Every arc in the network must fall in one and only one of the nine states identified in Table 8.1.

For each state, we can determine whether the conditions for optimality are satisfied. Those that do satisfy the conditions are called *in-kilter states* (α, β, ρ). Those that do not are called the *out-of-kilter states* $(\alpha_1, \beta_1, \rho_1, \alpha_2, \beta_2, \rho_2)$. An arc is identified as in-kilter or out-of-kilter on the basis of its kilter state. These kilter states are summarized in a pictorial fashion in Figure 8.2.

Table 8.1   THE KILTER STATES

State	d	f	In Kilter?	Why?
α	$d>0$	$f = \underline{c}$	Yes	Satisfies 3(b)
β	$d=0$	$\underline{c} \leqslant f \leqslant c$	Yes	Satisfies 3(a)
ρ	$d<0$	$f = c$	Yes	Satisfies 3(c)
α_1	$d>0$	$f < \underline{c}$	No	Violates 1(b), 3(b)
β_1	$d=0$	$f < \underline{c}$	No	Violates 1(b)
ρ_1	$d<0$	$f < c$	No	Violates 3(c)
α_2	$d>0$	$f > \underline{c}$	No	Violates 3(b)
β_2	$d=0$	$f > c$	No	Violates 1(c)
ρ_2	$d<0$	$f > c$	No	Violates 1(c), 3(c)

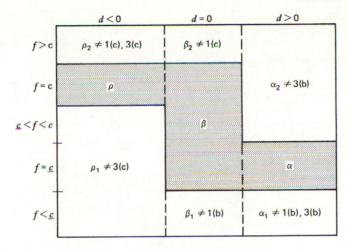

Figure 8.2
Pictorial Representation of the Kilter States

EXAMPLE 1

Consider the problem and solutions shown in Figures 8.3a and 8.3b. Determine if either solution is optimal.

Arc	Figure 8.3a			Figure 8.3b		
k	d_k	f_k	Kilter State	d_k	f_k	Kilter State
1	0	3	β	0	3	β
2	9	3	α	7	3	α
3	-4	0	ρ_1	-2	2	ρ
4	0	3	β	0	1	β
5	2	0	α	0	0	β
6	0	1	β	2	0	α
7	0	1	β	0	2	β
8	2	2	α_2	4	1	α
9	0	3	β	0	3	β

The solution shown in Figure 8.3a is not optimal because arcs 3 and 8 are out of kilter. The solution of Figure 8.3b is optimal because all arcs are in-kilter.

If, for a given solution (**f** and **π**), all the arcs are in-kilter, we are assured that the solution is optimal. If one or more arcs are out-of-kilter, the algorithm must change the flows or the node potentials in such a way that the arcs are driven toward an in-kilter solution.

We now describe a set of procedures that will yield the optimum (or an indication of infeasibility) in a finite number of steps. The overall algorithm is implemented by KILTER, which has three main parts.

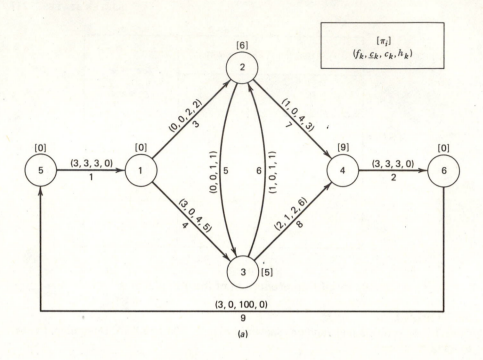

(a)

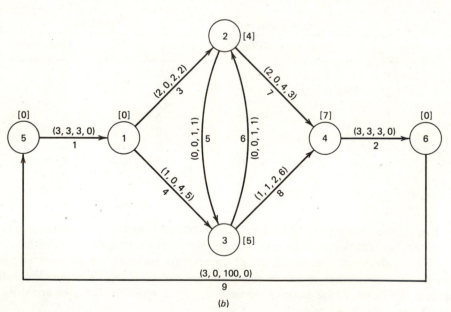

(b)

Figure 8.3
Two Solutions to a Network Example

1. Find an out-of-kilter arc.
2. Holding node potentials constant, change the flow in the network in order to drive the arc in-kilter.
3. If the flow change is unsuccessful, hold the flows constant and change node potentials in order to bring the arc in-kilter or allow additional flow changes.

KILTER ALGORITHM

Purpose: Given an initial flow that satisfies conservation of flow and arbitrary node potentials, this procedure modifies flows and node potentials to obtain an optimal solution. If there is no feasible solution, the subroutine returns INFEAS = 1.

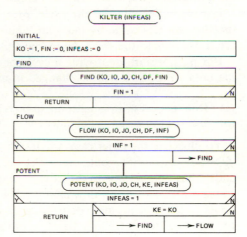

1. (INITIAL) Set k_o to 1. Set the optimum termination indicator (FIN) and the infeasibility indicator (INFEAS) to 0.
2. (FIND) Determine if arc k_o is in-kilter. If it is not, return the terminals of k_o, i_o and j_o, the direction of flow change (CH = 1 for increase, CH = −1 for decrease), and the amount of flow change (DF) required to bring arc k_o into kilter. If arc k_o is in-kilter, find an arc which is out-of-kilter. Let this be arc k_o and return the information noted above. If there are no out-of-kilter arcs (FIN = 1), return with the optimum solution.
3. (FLOW) Try to change the flow in the network to bring arc k_o into kilter. If this is possible (INF = 0), return to step 2. If this is impossible (INF = 1), go to step 4.
4. (POTENT) Try to change the node potentials to bring arc k_o into kilter or allow more flow changes. If the potentials are changed and arc k_o is brought into kilter, return to step 2. If it has not, return to step 3. If the potentials cannot be changed, there is no feasible solution to the original flow problem (INFEAS = 1). Return.

8.4 FLOW CHANGE

We first investigate a procedure which attempts to drive an arc to an in-kilter state by changing the arc flows in the network while holding all node potentials constant. The OKA assumes that flow is conserved at each node. When the initial solution satisfies this condition, all subsequent solutions will also satisfy the condition since flow is always changed on a cycle.

For example, consider the cycle, $M_c = \{3, -6, -4\}$, of Figure 8.3a. Changing the flow on this cycle by one unit yields $f_3 = 1$, $f_4 = 2$, and $f_6 = 0$ while all other arc flows remain the same. Note that this solution satisfies conservation of flow.

For purposes of identifying a cycle on which flow may be changed, we adopt the policy that flow will be changed in an arc only if the flow change:

(a) Does not cause an in-kilter arc to become out-of-kilter.
(b) Does not cause an out-of-kilter arc to become "more out-of-kilter."

To illustrate the latter case, we do not want to increase the flow in an arc with $f_k > c_k$ because that would cause the arc to move even further from an in-kilter state. This policy leads to the flow change rules specified in Table 8.2. Here only states α and ρ allow no flow change. The flow change allowed in state β may be either positive or negative depending on the arc flow. Note that an arc in any of the out-of-kilter states may be driven to an in-kilter state by an appropriate flow change. These statements are easily verified by reference to Figure 8.2.

Table 8.2 leads to a new admissibility function that specifies when an arc can be included in a flow change cycle.

Table 8.2 ALLOWABLE FLOW CHANGES IN ARCS

Kilter State	d	f	Allowable Flow Change	Maximum Amount of Change	New State
α	$d > 0$	$f = \underline{c}$	None	0	α
β	$d = 0$	$f < c$	Increase	$c - f$	β
		$f > \underline{c}$	Decrease	$f - \underline{c}$	β
ρ	$d < 0$	$f = c$	None	0	ρ
α_1	$d > 0$	$f < \underline{c}$	Increase	$\underline{c} - f$	α
β_1	$d = 0$	$f < \underline{c}$	Increase	$c - f$	β
ρ_1	$d < 0$	$f < c$	Increase	$c - f$	ρ
α_2	$d > 0$	$f > \underline{c}$	Decrease	$f - \underline{c}$	α
β_2	$d = 0$	$f > c$	Decrease	$f - \underline{c}$	β
ρ_2	$d < 0$	$f > c$	Decrease	$f - c$	ρ

A forward arc k is admissible if flow may be increased in the arc.

$$A_d(k) = 1 \qquad \text{if } \begin{cases} f_k < \underline{c}_k \\ \text{or} \\ d_k \leqslant 0 \quad \text{and} \quad f_k < c_k \end{cases}$$

$$A_d(k) = 0 \qquad \text{otherwise}$$

A mirror arc $-k$ is admissible if flow may be decreased in arc k.

$$A_d(-k) = 1 \qquad \text{if } \begin{cases} f_k > c_k \\ \text{or} \\ d_k \geqslant 0 \quad \text{and} \quad f_k > \underline{c}_k \end{cases}$$

$$A_d(-k) = 0 \qquad \text{otherwise}$$

The new admissibility definition is implemented in the ADOKA function.

ADOKA FUNCTION

Purpose: To compute the admissibility function for the Out-of-Kilter Algorithm.

If flow is less than the lower bound, the forward arc k is admissible. If flow is greater than the upper bound, the mirror arc $-k$ is admissible. If neither condition is true, compute the value of d_k for the arc. For a forward arc k, if $d_k \leqslant 0$ and $f_k < c_k$ the arc is admissible. Otherwise, it is not. For a mirror arc $-k$, if $d_k \geqslant 0$ and $f_k > \underline{c}_k$, the arc is admissible. Otherwise, it is not.

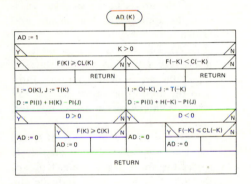

In order to find a flow change cycle, we must find an out-of-kilter arc. The procedure implemented by algorithm FIND starts at arc 1 and goes through the arcs in their indexed order until either an out-of-kilter arc is discovered or all arcs have been checked. If no out-of-kilter arc is found, the current flows are optimal.

FIND ALGORITHM

Purpose: To find an out-of-kilter arc k_o, its origin node i_o, and terminal node j_o. The direction of flow change is indicated by CH($+1$ for increase, -1 for decrease). The amount of flow change to bring arc k_o in-kilter is DF.

1. (TESTFIN) Determine if all the arcs have been checked. If so, return with FIN $= 1$. If not, find the terminal nodes of the next arc and compute d_k. If $d_k > 0$, go to step 2. If $d_k = 0$, go to step 3. If $d_k < 0$, go to step 4.
2. (ALPHA) The arc is in an α state. If $f_k = c_k$, the arc is in-kilter; go to step 1. If $f_k < c_k$, the arc is in the α_1 state. If $f_k > c_k$ the arc is in the α_2 state. Indicate the appropriate flow change and return.
3. (BETA) The arc is in a β state. If $f_k < c_k$ the arc is in the β_1 state. If $f_k > c_k$ the arc is in the β_2 state. Indicate the appropriate flow change and return. Otherwise, the arc is in-kilter. Go to step 1.
4. (RHO) The arc is in a ρ state. If $f_k = c_k$ the arc is in-kilter. Go to step 1. If $f_k < c_k$ the arc is in a ρ_1 state. If $f_k > c_k$ the arc is in a ρ_2 state. Indicate the appropriate flow change and return.

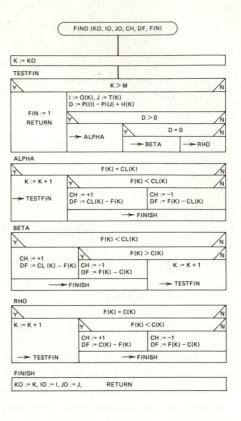

Once an out-of-kilter arc $k_o(i_o, j_o)$ is discovered, we first try to change the flows in the network in order to bring the arc into kilter. This is done in algorithm FLOW. Algorithm FIND has already determined the direction and amount of flow change required in the arc. If the flow is to be increased by an amount Δ_f, we must try to find a way to pass the flow Δ_f from node j_o to i_o through the remainder of the network using only admissible arcs. If this can be done, arc k_o can be brought to an in-kilter state without causing any other arcs to become out-of-kilter (or more out-of-kilter). The principal tool of FLOW is MAXFLO, from Chapter 6. MAXFLO performs all the manipulations required to find paths of admissible arcs and change flows. (The reader should review the MAXFLO algorithm of Chapter 6.)

FLOW ALGORITHM

Purpose: To change the flows in the network so that the flow on arc $k_o(i_o, j_o)$ can be increased or decreased while maintaining conservation of flow at each node.

If flow on arc k_o is to be increased by an amount DF, the subroutine calls MAXFLO to attempt to pass a flow DF through the network from j_o to i_o using only admissible arcs. If the maximum flow, V is less than DF, this is signaled by setting INF $= 1$ in MAXFLO. The flow on arc k_o is increased by V. If the flow on arc k_o is to be decreased, MAXFLO is used to attempt to pass a flow of DF from i_o to j_o using only admissible arcs. If the maximum flow V is less than DF this is signaled with INF $= 1$. The flow on arc k_o is decreased by V.

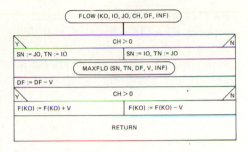

The principal subroutine used in MAXFLO is FPATH, which is used to find a path from a node s to a node t that consists only of admissible arcs. We provide a new version of this algorithm, FPATH3, which uses a labeling technique to find the desired path. FPATH3 is similar to the techniques classically employed in the OKA. Unlike the versions described in Chapter 6, FPATH3 does not depend on a specific ordering of the arcs. Note that if the arc flow is to be increased

$$s = j_o \quad \text{and} \quad t = i_o$$

If the arc flow is to be decreased

$$s = i_o \quad \text{and} \quad t = j_o$$

The use of the FPATH1 of Chapter 6 would probably make the OKA more efficient. However, it would be necessary to order the arcs as accomplished by ORIGS and to use the terminal node pointer list as determined by TERMS.

FPATH3 ALGORITHM

Purpose: FPATH3 finds a path by using a labeling procedure. The labels are indicated by the node length list S. The labeling algorithm is reported in FPATH3.

1. (INITIAL) Set all the labels to 0 except node s.
2. (LABEL) Go through the list of arcs. If an admissible arc $k(i,j)$ is found such that node i is labeled but node j is not, label node j. Let the back pointer for node j be arc k. If mirror arc $-k(j,i)$ is admissible and node j is labeled but node i is not, label node i. Let the back pointer to node i be $-k$.
3. (TEST) If node t has been labeled, a path has been found. It can be traced with the back pointers. If node t has not been labeled but some node was labeled in step 2, repeat step 2. If node t has not been labeled and no node was labeled in step 2, there is no path of admissible arcs from s to t. The labeled nodes are all the nodes that can be reached from node s through admissible arcs.

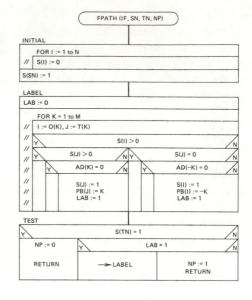

It is also necessary to provide algorithm OKAMFLO, which determines the maximum flow change in the path determined by FPATH3. This version of MFLO uses the results of Table 8.2 to obtain the maximum flow change on each arc of the path.

When no path is found in FPATH3, the event is called a *nonbreakthrough*. It signals that the maximum flow from node s to t has been found and that maximum flow is not sufficient to drive arc k_o to an in-kilter state. The set of labeled nodes $[S(I)=1]$ plays an important part in the next part of the algorithm.

OKAMFLO ALGORITHM

Purpose: OKAMFLO determines the maximum flow change allowable in the path determined by FPATH3.

1. (INITIAL) Set the initial maximum flow to a large number; set the arc indices to zero. Note that KL and ILC have no purpose here.

2. (TESTFIN) Go through arc list LISA. For each arc determine its kilter state and go to the associated part of the algorithm.

3. (ALPHA) $d_k > 0$. If $f_k \leqslant c_k$, determine if $MF > c_k - f_k$. If so, set $MF = c_k - f_k$. If $f_k > c_k$, determine if $MF > f_k - c_k$. If so, set $MF = f_k - c_k$. Go to step 2.

4. (BETA) $d_k = 0$. If $f_k > c_k$, determine if $MF > f_k - c_k$. If so, set $MF = f_k - c_k$ and go to step 2. If $f_k \leqslant c_k$, determine if $f_k < c_k$. If so, determine if $MF > c_k - f_k$. If so, set $MF = c_k - f_k$ and go to step 2. If $c_k \geqslant f_k \geqslant c_k$, go to step 5.

5. (DIRECT) If the current back pointer is a forward arc, determine if $MF > c_k - f_k$. If so, set $MF = c_k - f_k$ and go to step 2. If the current back pointer is a mirror arc, determine if $MF > f_k - c_k$. If so, set $MF = f_k - c_k$ and go to step 2.

6. (RHO) $d_k < 0$. If $f_k \leqslant c_k$, determine if $MF > c_k - f_k$. If so, set $MF = c_k - f_k$ and go to step 2. If $f_k > c_k$, determine if $MF > f_k - c_k$. If so, set $MF = f_k - c_k$ and go to step 2.

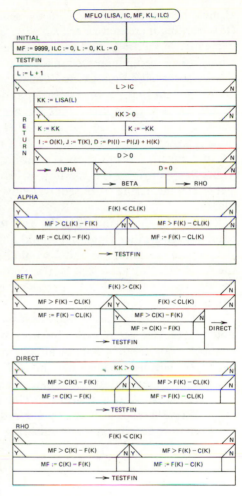

8.5 POTENTIAL CHANGE

When the flow change portion of the algorithm fails to bring arc $k_0(i_o, j_o)$ into kilter, we proceed to change the node potentials in such a way that either arc k_o is made in-kilter or more arcs are made admissible. The failure of the flow change algorithm is accompanied by the definition of two sets of nodes: N_1 is the set of nodes labeled in the search for a flow change cycle, and N_2 is the set of nodes unlabeled. The MAXFLO algorithm provides this information using list S.

$$S(I) = 1 \qquad \text{if } i \in N_1$$
$$S(I) = 0 \qquad \text{if } i \in N_2$$

It must be true that node s is in the set N_1 and t is in the set N_2. Recall that if flow is being increased in arc k_o, $s=j_o$ and $t=i_o$. Alternatively, if flow is being decreased in k_o, $s=i_o$, and $t=j_o$. Figure 8.4 schematically illustrates the case in which flow is being increased in arc k_o.

Given the situation at nonbreakthrough, we know certain characteristics about the arcs that pass between N_1 and N_2. If arc $k_1(i_1,j_1)$ originates in N_1 and terminates in N_2, we know that the arc cannot be in a kilter state that would allow a flow increase in the arc. Otherwise, the arc would be admissible and node j_1 would both be labeled and be a member of N_1. Thus, reference to Table 8.2 and Figure 8.2 will show that arc k_1 must be in one of the states listed in Table 8.3. Notice that, if the flow is less than the arc capacity for the states of Table 8.3, increasing π_j or, equivalently, *decreasing d* will make arc k_1 admissible.

If arc $k_2(i_2,j_2)$ originates in N_2 and terminates in N_1, we know that the arc cannot be in a kilter state that would allow a flow decrease in arc k. Otherwise, $-k_2$ would be admissible and i_2 would be labeled. Thus arc k_2 must be in one of the states listed in Table 8.4. Notice that, if the flow exceeds the arc lower bound for the states of Table 8.4, increasing π_i, which *increases d*, will make arc k_2 admissible.

We will change the node potentials by adding a quantity Δ_p to all the node potentials in the set N_2. The quantity Δ_p is to be chosen so that either arc k_o changes to an in-kilter state or at least one of the arcs passing from N_1 to N_2 or N_2 to N_1 becomes admissible. In addition, no change that would drive any in-kilter arc to an out-of-kilter state is allowed.

In certain cases, it is possible to change node potentials to bring arc k_o into kilter. These cases are shown in Table 8.5. If the flow in k_o is not in the range $[\underline{c},c]$, there is no change that will bring the arc to an in-kilter state. When the

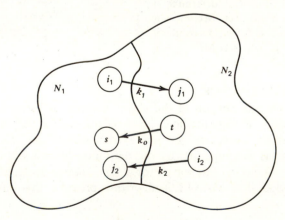

Figure 8.4
The Situation at Nonbreakthrough. Flow Is Being Increased
in the Arc k_o

Table 8.3   POSSIBLE STATES OF ARC $k(i, j)$ WHEN $i \in N_1$ AND $j \in N_2$

State	d^*	f	Increase in π_j to Make k Admissible	New State
α	$d > 0$	$f = \underline{c}$	d	β
β	$d = 0$	$f = c$	$-$	ρ
ρ	$d < 0$	$f = c$	$-$	ρ
α_2	$d > 0$	$\underline{c} < f < c$	d	β
		$f \geqslant c$	$-$	α_2, β_2, or ρ_2
β_2	$d = 0$	$f > c$	$-$	ρ_2
ρ_2	$d < 0$	$f > c$	$-$	ρ_2

$^* d_k = \pi_i - \pi_j + h_k.$

Table 8.4   POSSIBLE STATES OF ARC $k(i, j)$ WHEN $i \in N_2$ AND $j \in N_1$

State	d^*	f	Increase in π_i to Make $-k$ Admissible	New State
α	$d > 0$	$f = \underline{c}$	$-$	α
β	$d = 0$	$f = \underline{c}$	$-$	α
ρ	$d < 0$	$f = c$	$-d$	β
α_1	$d > 0$	$f < \underline{c}$	$-$	α_1
β_1	$d = 0$	$f < \underline{c}$	$-$	α_1
ρ_1	$d < 0$	$f \leqslant \underline{c}$	$-$	ρ_1, β_1, or α_1
		$\underline{c} < f < c$	$-d$	β

$^* d_k = \pi_i - \pi_j + h_k.$

Table 8.5   POTENTIAL CHANGES TO BRING ARC k_o INTO KILTER

State	d_o	f_o	Change Required
α_1	$d > 0$	$f < \underline{c}$	$-$
β_1	$d = 0$	$f < \underline{c}$	$-$
ρ_1	$d < 0$	$f < \underline{c}$	$-$
		$\underline{c} \leqslant f < c$	Increase π_i by $-d$
α_2	$d > 0$	$\underline{c} < f \leqslant c$	Increase π_j by d
		$f > c$	$-$
β_2	$d = 0$	$f > c$	$-$
ρ_2	$d < 0$	$f > c$	$-$

flow is within this range, the arc will be in either state ρ_1 or state α_2 and the changes indicated in the table will bring the arc into kilter.

Tables 8.3 and 8.4 show the values of Δ_p required to make arcs in the various states become admissible. A dash indicates that the arc cannot be made admissible with only a node potential change. For example, an arc in Table 8.3 in the β state with $f = c$ cannot become admissible because flow is already at its upper bound. Increasing π_j for this arc will cause the arc to move from the β to the ρ state. Note that for arcs passing from N_1 to N_2 only those in states α or α_2 may become admissible through simple node potential changes. For arcs passing from N_2 to N_1 only those in states ρ or ρ_1 may become admissible through node potential changes. An arc that becomes admissible always moves to the β state.

We will always choose Δ_p as the minimum value that will bring arc k_o into kilter or make an arc become admissible. Violating this rule could cause an arc to move from an in-kilter state to an out-of-kilter state. Algorithm POTENT finds the value of Δ_p and adds the value to all the node potentials in the set N_2. If there is no way to bring arc k_o into kilter or make an additional arc admissible, there is no feasible solution to the minimum cost flow problem.

On return from POTENT, the KILTER algorithm checks arc k_o to see if it has been brought into kilter. If not, the search begins again for a flow change to bring the arc into kilter. Since the potential change has introduced at least one new admissible arc, a flow change cycle may be found or at least additional nodes will be labeled. The process of flow and potential changes continues until the arc is brought into kilter or it is ascertained that the problem has no feasible solution.

POTENT ALGORITHM

Purpose: To compute the minimum value of increase to the node potentials of N_2 that will either cause arc k_o to become in-kilter or cause an arc whose terminals are in N_1 and N_2 to become admissible.

1. (INITIAL) Determine if a potential change can bring arc k_o into kilter. If so, set Δ_p to this value and set k_E to k_o. If not, set Δ_p to a large value and set k_E to 0.

2. (DUAL) Go through the list of arcs. For each arc determine if the arc has one terminal in N_1 and one terminal in N_2. If so, determine the potential change that would make the arc admissible. If this change is less than

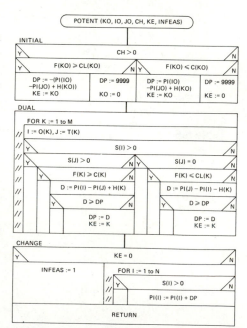

Δ_p, replace Δ_p and k_E. When all the arcs have been considered, go to step 3.

3. (CHANGE) If k_o cannot be driven into kilter and no arcs can be made admissible by a potential change ($KE = 0$), indicate infeasibility and return. If a potential change has been found, increase the node potentials on all nodes in N_2 by the amount Δ_p and return.

8.6   AN EXAMPLE APPLICATION OF THE OKA

Let us use the network of Figure 7.1, to illustrate the application of the OKA. The OKA form of the network is presented in Figure 8.5a. The flows and node potentials present in Figure 8.5a result from five flow changes after the OKA was started with all flows and node potentials set to zero.

At this point arcs 1 through 6 have been driven to an in-kilter state. Arc 7(5,8) is out-of-kilter and a positive flow change of one unit from node 5 to node 8 is required if arc 7 is to be brought into kilter by a flow change alone. Unfortunately, of those arcs touching node 8, only arc 18(8,7) is admissible and none of arcs 1, 2, 3, or 4 is admissible. Clearly, there is no path from node 8 to node 5 through admissible arcs and a change of the node potentials is called for.

At this point, $N_1 = [7, 8]$ and a search through the arcs yields $\Delta_p = 3$ from arc 6. Adding Δ_p to the node potentials of all nodes in N_2 yields the values given in Figure 8.5b. With the new values of the π_i's, arc -6 becomes admissible. Arcs -12 and -15, previously admissible, becomes accessible. There is still no path from node 8 to node 5; hence, we again have a nonbreakthrough condition.

Calling POTENT once more yields $\Delta_p = 1$ from arc 14. Adding Δ_p to the node potentials of $N_2 = [1, 5, 6]$ causes arcs -11 and 14 to become accessible. As may be observed in Figure 8.5c, there is still no path from node 8 to node 5.

An additional call to POTENT finally yields a path as seen in Figure 8.5d. This path is achieved by adding $\Delta_p = 2$ to the node potential of node 5, the single remaining member of N_2. This makes arc 16 admissible and allows the path of arcs -6, -15, and 16 to be formed. Passing the required 1 unit of flow arc in the cycle of arcs $[-6, -15, 16, 7]$ drives arc 7 to an in-kilter state and raises the total flow cost to 25. The new flows are shown in Figure 8.5e.

By observing Figure 8.5e, we note that arc 8(6,8) is in the out-of-kilter state α_1. Thus, we desire to increase its flow from the current value of 4 to its lower bound of 6. An initial attempt finds the cycle of admissible arcs $[-6, -12, 14, 8]$ shown in Figure 8.5f. Unfortunately, the flow in arcs 6 and 12 can only be reduced by 1 unit and still remain in-kilter. Passing, 1 unit of flow around the cycle causes arcs 6 and 12 to become inadmissible and leaves arc 8 still out-of-kilter. The new flows appear in Figure 8.5g and result in a cost of 29.

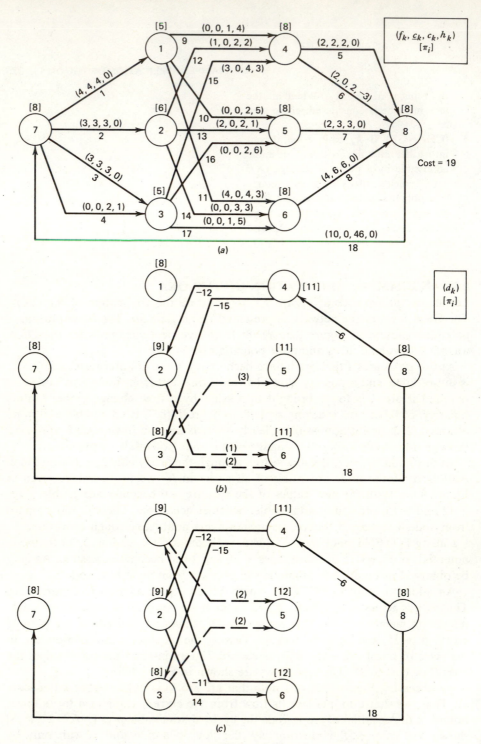

Figure 8.5
Example Application of the Out-of-Kilter Algorithm

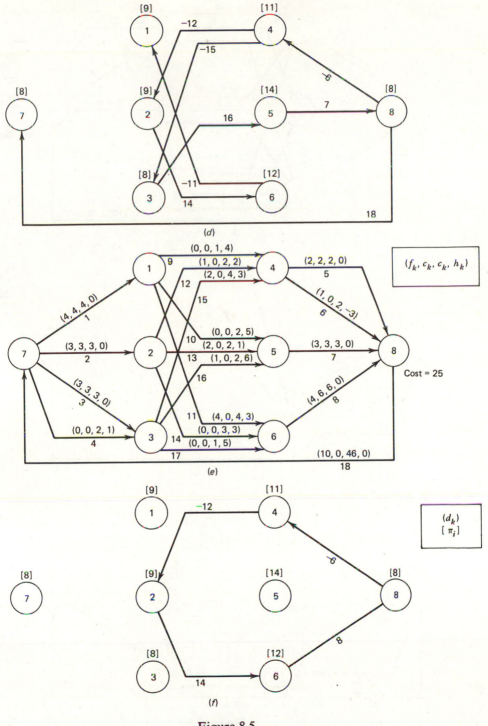

Figure 8.5
(continued)

225

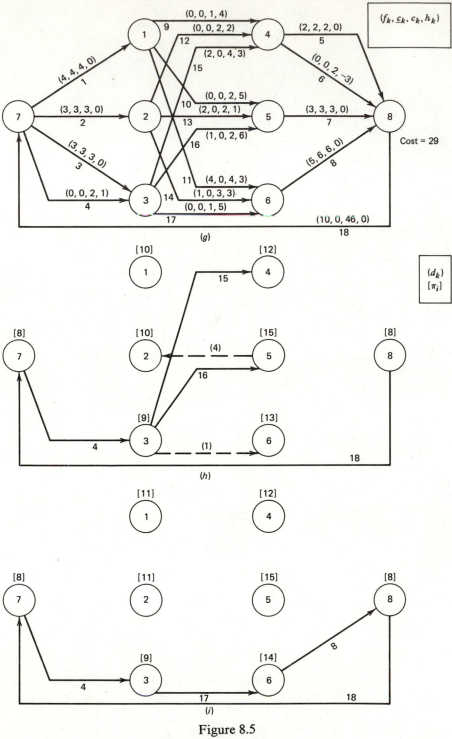

Figure 8.5
(continued)

226

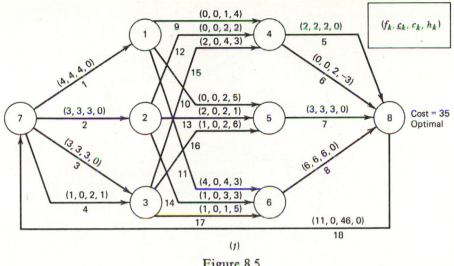

Figure 8.5
(continued)

Since no additional paths of admissible arcs exist from node 8 to node 6, we must call POTENT once again. The resultant node potentials are presented in Figure 8.5h. One more call to POTENT is required before a path from node 8 to node 6 is obtained in Figure 8.5i. Passing 1 unit of flow around the cycle of arcs [18,4,17,8] drives arc 8 to a kilter state. All the remaining arcs are also in-kilter. Therefore, Figure 8.5j describes an optimal set of flows with a total cost of 35.

8.7 HISTORICAL PERSPECTIVE

The Out-of-Kilter Algorithm was first presented in the literature by Fulkerson (1961). Ford and Fulkerson (1962) describe the algorithm in their classic book, *Flows in Networks*. Clasen (1968) gives a computer code implementing the OKA. Durbin and Kroehke (1967) and Phillips and Jensen (1974) provide tutorials.

Barr et al. (1974) developed a new version of the OKA considerably faster than the codes based on the Clasen approach. Barr et al. (1974) and Glover, Karney and Klingman (1974) show with computational tests that basic primal approaches are considerably more efficient than the OKA.

EXERCISES

1. Modify the networks given below to be consistent with the OKA model.

    (a) Maximize the flow entering at nodes 1, 2, and 3 and leaving the network at nodes 8, 9, and 10.

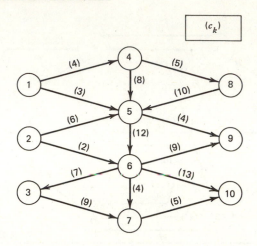

(c_k)

(b) Find the shortest path from node 1 to node 10.

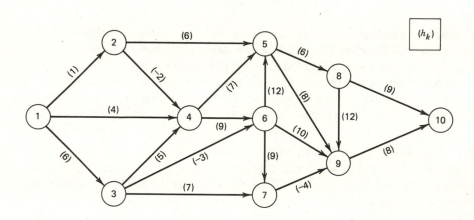

(h_k)

(c) Minimize the total flow cost of the following network while satisfying all fixed external flow requirements.

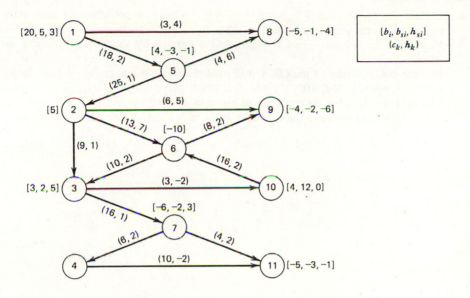

2. Give the kilter states for each arc in the network given below.

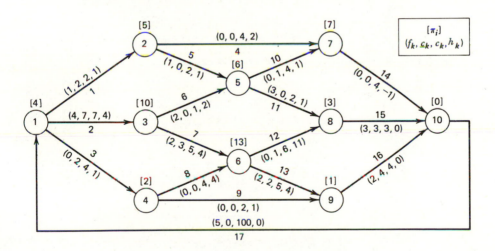

3. (a) Draw the marginal network for the network of exercise 2.
   (b) Give the minimum flow change required to bring each out-of-kilter arc into kilter.
   (c) Give the minimum change in d_k to bring each arc into kilter. If the arc cannot be brought into kilter by a change in d_k alone, indicate this fact.

4. Explain why an addition of Δ_p units to every π_i, $i \in N_2$, does not affect the kilter state of any arc that is wholly contained in either N_1 or N_2.

5. Verify that the solution given in Figure 8.5j is optimal by showing all arcs are in-kilter.

6. Perform the iterations of the OKA that yield the status given in Figure 8.5a. Begin with all $\pi_i = 0$ and all $f_k = 0$.

7. Use the OKA to solve the problems of exercises 1 and 2.

8. Use the OKA to optimize the networks given in Sections 3 through 7 of Chapter 2.

CHAPTER 9
NETWORK MANIPULATION ALGORITHMS FOR THE GENERALIZED NETWORK

9.1 INTRODUCTION

A *gain* is an additional arc parameter a_k that models a linear increase or decrease in flow as it passes through an arc. Figure 9.1 represents a typical arc $k(i,j)$ with the familiar cost parameter h_k, capacity parameter c_k, and the new gain parameter a_k. The quantity f_k is the flow in the arc as it leaves node i. The cost and capacity are as previously defined. The gain parameter represents a change in the flow value as it passes through the arc. Since f_k is the quantity of flow leaving node i, let f'_k be the quantity of flow entering node j through arc k. This quantity is

$$f'_k = a_k f_k$$

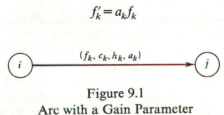

Figure 9.1
Arc with a Gain Parameter

Note if $a_k = 1$, we have no change in the flow as it passes through the arc. Therefore, if the gain is equal to 1 for all arcs, we have the pure minimum cost flow problem of Chapter 7. When $0 < a_k < 1$, flow is decreased as it passes through the arc. When $a_k > 1$, flow is increased. The results of this chapter are also valid when arcs have negative gains, although the physical interpretation is not obvious. Zero gain factors are not allowed. The terminology "generalized network" is used to indicate that some arc gain parameters will differ from a value of 1 in the network.

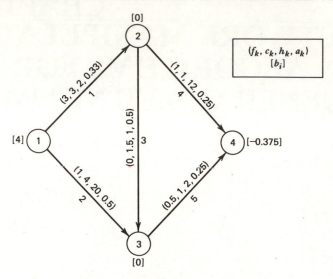

Figure 9.2
Generalized Minimum Cost Flow Problem

Figure 9.2 illustrates a minimum cost flow problem for a generalized network. The optimal solution is also shown in the figure.

Several characteristics can be observed from this example. First, note that arc flows are *no longer necessarily integer*. Also the flow into the network is no longer necessarily equal to the flow out of the network. However, conservation of flow is still required at each node.

Addition of the gain parameter expands greatly the class of models that can be handled by network programming techniques. With the gain parameter, a model can represent such things as the increase in the value of money invested in a project, the evaporation of water stored in a reservoir, and translation of one flow quantity to another. Chapter 2 illustrates several examples where the gain parameter is an important aspect of the network model.

Efficient techniques for solution of generalized network problems are available. They are not as efficient as those applicable to the pure minimum cost flow problem, but they are still considerably more efficient than solution by a general linear programming algorithm. We describe in this chapter the network manipulation algorithms that will be used in the solution algorithms of Chapter 10 to solve the generalized minimum cost flow problem. A good understanding of these algorithms will greatly facilitate the understanding and use of the materials presented in Chapter 10. In Chapter 10, two major approaches to the generalized minimum cost flow problem, a dual feasible incremental approach and a primal feasible simplex approach, will be presented. As we shall see, the former is the easiest to understand while the latter is an extension of the primal simplex approach for the pure minimum cost flow problem.

9.2   THE NETWORK WITH GAINS MODEL

In Chapters 9 and 10, we will restrict our attention to a somewhat simpler model form than was allowed for the pure network flow problem. This restriction permits simpler algorithms and does not exclude any practical class of problems. The form does, however, require more initial effort in data preparation.

We consider a network that has a single source, designated s, which can supply an unlimited quantity of flow, and a single sink, designated t, which will demand some specified quantity of flow b_t. All other nodes will have zero external flows. Each arc has three parameters: capacity, c_k; cost, h_k; and gain, a_k. The goal of a minimum cost flow algorithm is to find a set of arc flows that satisfies conservation of flow at each node except the source, has a net flow of b_t into the sink, and has minimum total cost.

Although this representation would at first seem somewhat restrictive for the modeler, it is possible to express most problems in this form. For example, consider a transportation problem with demand nodes and supply nodes as illustrated in Figure 9.3. The demand and supply nodes are connected by a network in which arcs have positive gains. The single-source–single-sink representation of this problem appears in Figure 9.4.

In Figure 9.4, arcs have been placed from the source to each supply node. The capacities on these arcs represent the amounts available at the supply nodes. Arcs from the demand nodes to the sink have capacities equal to the amounts demanded. If a minimum cost flow equal to the sum of the demands is obtained at the sink, then the original problem is solved.

For some problems it may be necessary to impose a lower bound, $\underline{c}_k$, on the flow in one or more arcs $k(i,j)$. Such a situation would be represented for this

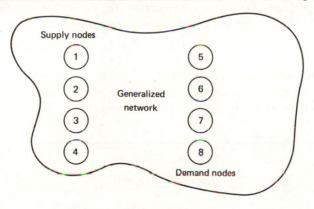

Figure 9.3
Transportation Problem

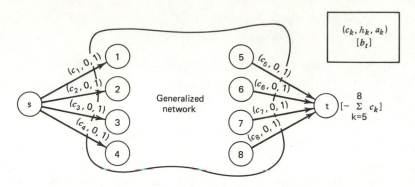

Figure 9.4
Generalized Network Model for the Transportation Problem

algorithm by two parallel arcs for each such arc. The parameters on one of the parallel arcs will be $(\underline{c}_k, -R, a_k)$ where $\underline{c}_k$ is lower bound on flow for this arc, $-R$ is a large negative cost, and a_k is the original arc gain. The second parallel arc has the parameters $(c_k - \underline{c}_k, h_k, a_k)$. Because of the large negative cost, the algorithm will saturate the first parallel arc before sending any flow on the second parallel arc.

Some problems include both positive and *negative* arc costs (negative costs model revenues). Often, no specific output flow is specified and the goal of the problem is to find the output flow that minimizes total cost, or alternatively, maximizes profit. For this situation, an additional arc is constructed from the source to the sink with parameters $(R, 0, 1)$ where R is a large number. Thus this arc transmits any amount of flow at zero cost with a gain of one. The required output flow is set to some large number that is greater than the optimum flow for the original problem. With this construction, the algorithm will deliver flow to the sink through the network only if the marginal cost is negative (or profitable).

9.3   LINEAR PROGRAMMING MODEL

The general linear programming model for the network with gains may be expressed in the following way.

$$\text{Min.} \sum_{k=1}^{m} h_k f_k$$

$$\text{st.} - \sum_{k \in M_{Ti}} a_k f_k + \sum_{k \in M_{Oi}} f_k = 0 \qquad \begin{array}{l} i \in N \\ i \neq s, t \end{array}$$

$$- \sum_{k \in M_{Tt}} a_k f_k + \sum_{k \in M_{Ot}} f_k = -b_t$$

$$0 \leq f_k \leq c_k \qquad k \in M$$

Note that no conservation-of-flow constraint is written for the source node, since this node can supply as much flow as required. For the example of Figure 9.2, this model becomes

$$
\begin{aligned}
\text{Min.} \quad & 2f_1 + 20f_2 + 1f_3 + 12f_4 + 2f_5 \\
\text{st.} \quad & -0.33f_1 + f_3 + f_4 = 0 \\
& -0.5f_2 - 0.5f_3 + f_5 = 0 \\
& -0.25f_4 - 0.25f_5 = -0.375
\end{aligned}
$$

$$
\begin{aligned}
0 \leqslant f_1 \leqslant 3 \\
0 \leqslant f_2 \leqslant 4 \\
0 \leqslant f_3 \leqslant 1.5 \\
0 \leqslant f_4 \leqslant 1 \\
0 \leqslant f_5 \leqslant 1
\end{aligned}
$$

The example illustrates that each arc, except those originating at the source, is represented twice in the constraint matrix. The flow for arc $k(i,j)$ appears in the constraint for node i with a coefficient of 1 and in the constraint for node j with the coefficient $-a_k$. Arcs originating at the source appear only once.

The basis for this linear program will consist of $n-1$ arcs chosen so that the associated $n-1$ columns are linearly independent. Recall that, for the pure network problem, such a selection formed a directed spanning tree. For the gains problem, a selection of $n-1$ linearly independent columns may result in a more general form. In particular, a basis for the gains problem may contain one or more cycles.

To illustrate, consider the example network of Figure 9.2 with the gain parameters represented by the general designation a_k. The constraint matrix formed by the conservation-of-flow equations appears as

$$
\begin{array}{c}
\text{Arcs} \\
\mathbf{A} =
\begin{array}{ccccc}
1 & 2 & 3 & 4 & 5 \\
\end{array} \\
\begin{bmatrix}
-a_1 & 0 & 1 & 1 & 0 \\
0 & -a_2 & -a_3 & 0 & 1 \\
0 & 0 & 0 & -a_4 & -a_5
\end{bmatrix}
\begin{array}{c}
2 \\
3 \\
4
\end{array}
\quad \text{Nodes}
\end{array}
$$

Our problem is to choose a set of three independent columns from this matrix. The set will be independent if the square matrix formed by the columns has a nonzero determinant. First we will choose columns whose associated arcs form a tree. For example, suppose columns 1, 3, and 4 are chosen. The basis matrix becomes

$$
\mathbf{B} =
\begin{bmatrix}
-a_1 & +1 & +1 \\
0 & -a_3 & 0 \\
0 & 0 & -a_4
\end{bmatrix}
$$

The determinant of this basis matrix is

$$|\mathbf{B}| = - a_1 a_3 a_4$$

Since the gain factors are nonzero, this determinant can never equal zero. It can be shown for the general network that any selection of columns that describes a tree can be arranged to form a matrix with diagonal elements equal to the gain factors and lower off-diagonal elements equal to zero. Thus, **B** can be arranged to appear as

$$\mathbf{B} = \begin{bmatrix} -a_{k(1)} & & \\ 0 & -a_{k(2)} & \\ 0 & 0 & -a_{k(3)} \end{bmatrix}$$

where $k(1)$ is the arc entering the node with the lowest index, $k(2)$ is the arc entering the node with the next lowest, and so on. Thus $|B|$ can always be written as

$$|\mathbf{B}| = \pm \left(\prod_{k \in M_B} a_k \right)$$

where M_B is the set of arcs in the basis tree. Since $|\mathbf{B}|$ can never equal 0, the choice of a tree as a basis is acceptable.

In order to form a tree in the manner just shown, one of the arcs chosen must originate at the source node. Otherwise it is impossible to arrange the arcs so that the first column has only one nonzero element. Pictorially, we will represent such a basis as a directed tree rooted at the source. Where it is necessary, we will use mirror arcs to obtain the directed tree. Figure 9.5 shows several basis trees for the example problem.

Next, consider the case in which the arcs chosen form a cycle. For example, select arcs 3, 4, and 5 so that

$$\mathbf{B} = \begin{bmatrix} 1 & 1 & 0 \\ -a_3 & 0 & 1 \\ 0 & -a_4 & -a_5 \end{bmatrix}$$

and

$$|\mathbf{B}| = a_4 - (a_3 \cdot a_5)$$

Note that if all the gains are unity, as for the pure network, the determinant becomes zero. Thus, a cycle is not an acceptable component of a basis for a pure

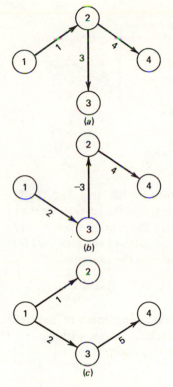

Figure 9.5
Example Spanning Trees

network. If, however,

$$|\mathbf{B}| = a_4 - (a_3 \cdot a_5) \neq 0$$

then the cycle is an acceptable basis. Rewriting this expression in an alternative form, we note that the cycle forms a basis if and only if

$$a_3 \cdot a_5 \neq a_4$$

or, equivalently

$$\frac{a_3 \cdot a_5}{a_4} \neq 1$$

The latter expression has a graphical interpretation that we will find useful. Note that the cycle appears as in Figure 9.6. Starting at node 2 and passing

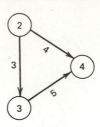

Figure 9.6
Basic Cycle

around the cycle, we encounter the arcs 3, 5, and -4. The -4 in the cycle indicates that arc 4 is traversed in the reverse, or mirror, direction. The gain of a mirror arc $-k$ as $1/a_k$. Thus the gain of arc -4 is $1/a_4$. The gain of the cycle is defined as the product of the gains of the arcs that form the cycle. We assign the symbol β to the cycle gain. For the example

$$\beta = \frac{a_3 a_5}{a_4}$$

A cycle can be an acceptable component of the basis if β does not equal one.

In general, a basis may include a cycle with various trees originating from it as in Figure 9.7. The basis for this example is

Arcs

	1	2	3	4	5	6	7	
	1	0	0	$-a_4$	0	0	0	2
	$-a_1$	1	1	0	0	0	0	3
	0	$-a_2$	0	1	1	0	0	4
$\mathbf{B} =$	0	0	$-a_3$	0	0	0	0	5 Nodes
	0	0	0	0	$-a_5$	1	1	6
	0	0	0	0	0	$-a_6$	0	7
	0	0	0	0	0	0	$-a_7$	8

Rearranging the rows and columns of this basis slightly we obtain

Arcs

	1	2	4	3	5	6	7	
	$-a_1$	1	0	1	0	0	0	3
	0	$-a_2$	1	0	1	0	0	4
	1	0	$-a_4$	0	0	0	0	2
$\mathbf{B} =$	0	0	0	$-a_3$	0	0	0	5 Nodes
	0	0	0	0	$-a_5$	1	1	6
	0	0	0	0	0	$-a_6$	0	7
	0	0	0	0	0	0	$-a_7$	8

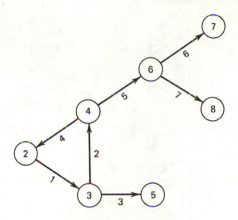

Figure 9.7
A Cycle Originating Trees

In this form, we note that the basis has the same form as that described by a tree except for the existence of a single nonzero element in the lower subdiagonal portion of the matrix. It is easy to show that

$$|\mathbf{B}| = a_3 a_5 a_6 a_7 - (a_1 a_2 a_3 a_4 a_5 a_6 a_7)$$

For this basis, the condition

$$|\mathbf{B}| = a_3 a_5 a_6 a_7 - a_1 a_2 a_3 a_4 a_5 a_6 a_7 \neq 0$$

is equivalent to the condition

$$a_1 a_2 a_4 \neq 1$$

Again we observe that the cycle gain may not equal unity if the arcs are to be linearly independent. In general, it can be shown that for a basis component with arcs M_B that includes cycle arcs, M_C, $(M_C \subset M_B)$ the determinant of the associated matrix will be proportional to

$$\prod_{k \in M_B} a_k - \prod_{k \in M_B - M_C} a_k$$

(See exercise 3 of this chapter for an example with a mirror arc in a cycle.)

We represent a basis component that includes a cycle as a directed graph with all noncycle arcs directed as if the cycle were the root (as in Figure 9.7). It is convenient to think of these arcs as a tree rooted at a cycle. The arcs in the cycle are oriented to form a directed cycle. There are two ways that the cycle can be directed, as shown in Figure 9.8. Either orientation is acceptable.

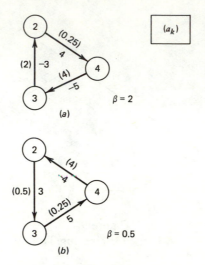

(a_k)

$\beta = 2$

(a)

$\beta = 0.5$

(b)

Figure 9.8
Representations of a Cycle

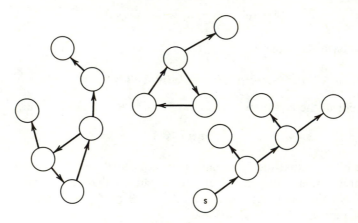

Figure 9.9
Basis with Several Components

240

A particular basis may be a combination of components as illustrated in Figure 9.9. Since there is only one source node there can be only one tree component. However, there may be several components rooted at cycles.

9.4   THE DUAL OF THE LINEAR PROGRAM

The dual for the generalized minimum cost flow problem can be written as follows.

$$\text{Min.} \quad -\pi_t b_t + \sum_{k=1}^{m} \delta_k c_k$$

$$\text{st.} \quad +\pi_i - a_k \pi_j + \delta_k \geq -h_k \quad \text{for each arc } k(i,j) \in M \quad i \neq s$$

$$-a_k \pi_j + \delta_k \geq -h_k \quad \text{for each arc } k(s,j) \in M$$

If we arbitrarily define $\pi_s = 0$, we can write the dual constraints in the form

$$\pi_i - a_k \pi_j + \delta_k \geq -h_k \quad \text{for each arc } k(i,j) \in M$$

It can be shown that, since $c_k > 0$, at optimality

$$\delta_k = \text{Max}(0, -h_k - \pi_i + a_k \pi_j)$$

This leads to the following complementary slackness conditions for optimality.

1. $+\pi_i - a_k \pi_j = -h_k$ for $0 < f_k < c_k$.
2. $f_k = 0$ for $\pi_i - a_k \pi_j > -h_k$.
3. $f_k = c_k$ for $\pi_i - a_k \pi_j < -h_k$.

These conditions have intuitive interpretations when we identify π_i as the cost of obtaining one additional unit of flow at node i and π_j as the cost of obtaining one additional unit of flow at node j. Consider the general arc $k(i,j)$ as in Figure 9.10.

If one additional unit of flow is to be obtained at node j through arc k, the additional flow leaving node i in arc k must be $1/a_k$. The cost of bringing this

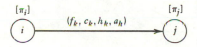

Figure 9.10
A General Arc with Dual Node Variables

flow to node i and passing it through arc k is

$$\left(\frac{1}{a_k}\right)(\pi_i + h_k)$$

If $(1/a_k)\ (\pi_i + h_k) > \pi_j$, which is equivalent to condition (2 above), then f_k should be set equal to zero; that is, any increase in f_k above a value of zero *increases* the total cost of the flows in the network.

Similarly, if $(1/a_k)\ (\pi_i + h_k) < \pi_j$, which is equivalent to condition (3), this implies that for an optimal solution $f_k = c_k$.

If $0 < f_k < c_k$, neither condition 2 nor 3 can exist for otherwise it would be beneficial to either increase or decrease flow in arc k.

These conditions are used extensively in Chapter 10 to aid the process of finding and proving optimality.

9.5   REPRESENTATION OF BASIS NETWORKS

As we did for the pure minimum cost flow problems, we use basis subnetworks extensively in the computational algorithms that we apply to the generalized minimum cost flow problems. These will be represented by the triple labeling scheme of Chapter 4. Some modification and reinterpretation of the basis manipulation algorithms will be necessary to account for the possibility of cycles in the basis. We will make extensive use of Figure 9.11 as an illustrative example network throughout the following discussion of those modifications.

Clearly, if the basis is a spanning tree rooted at the source node, the triple labeling scheme, as described in Chapter 4, is a satisfactory method of representation. Consider the subnetwork of Figure 9.12, which consists of a subtree rooted at the source and a subtree rooted at a cycle. One of the several triple

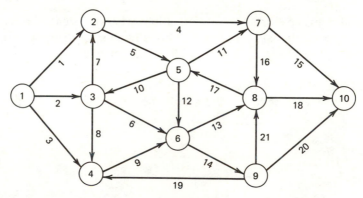

Figure 9.11
Example Network

pointer representations is also shown in Figure 9.12. Note that the presence of the cycle does not present any problems for the triple pointer representation. Each node still has a unique back pointer arc $p_B(i)$. When a node has no successors, a zero is indicated in the forward pointer $p_F(i)$ (i.e., nodes 2,5,7,10). A node with a single immediate successor has the associated node indicated in its forward pointer (i.e., nodes 1,4,6,8). A node with more than one immediate successor (i.e., nodes 3 and 9) has its additional immediate successors indicated in the right pointers of those immediate successor nodes. For example, to find the immediate successors of node 3, we find the first one (node 5) in its forward pointer and its second one (node 2) as the right pointer of node 5. The immediate successors of node 9 are node 10 $[p_F(9)=10]$, node 8 $[p_R(10)=8]$, and node 4 $[p_R(8)=4]$. Thus the cycle presents no difficulty for the triple labeling scheme.

To observe the pointer representation for a more complex subnetwork, consider the example shown in Figure 9.13, which gives only one of several alternative pointer representations of that subnetwork. Note that the graphical designations such as *left-most successor* and *next-right successor* do not have meaning when cycles are present. Also, the concept of depth in a tree is ambiguous when we allow subnetworks that include cycles. (What is the depth of the nodes on a cycle?) The economies obtained with the depth parameter for

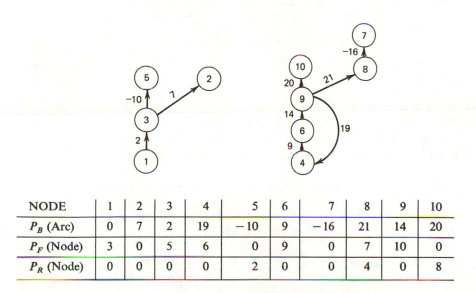

NODE	1	2	3	4	5	6	7	8	9	10
P_B (Arc)	0	7	2	19	−10	9	−16	21	14	20
P_F (Node)	3	0	5	6	0	9	0	7	10	0
P_R (Node)	0	0	0	0	2	0	0	4	0	8

Figure 9.12
Basis Network with a Cycle

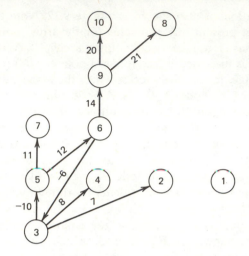

NODE	1	2	3	4	5	6	7	8	9	10
P_B (Arc)	0	7	−6	8	−10	12	11	21	14	20
P_F (Node)	0	0	5	0	7	9	0	0	10	0
P_R (Node)	0	0	0	2	4	0	6	0	3	8

Figure 9.13
Complex Subnetwork Rooted at a Cycle

the pure problem are not in fact available for the problem with gains. Therefore, the depth characteristic is not used in this chapter.

In the remainder of this section, we describe three operations that are to be performed on the basis subnetworks for generalized networks.

1. Find the unique directed trail, originating at a source node or cycle, to a specified node.
2. Find the set of nodes in that portion of the subnetwork rooted at some node.
3. Delete an arc from a subnetwork and add another.

Trail-Finding Algorithm

For the pure network problems, the path finding algorithm determined a unique tree path between a pair of nodes i and j. For the gains algorithms, this capability is not required. Rather, we address the question of finding the unique directed trail to some node i from the source node or from a cycle. For this process we simply follow the back pointers of node i and its predecessors until the source node is found or a node is encountered a second time. The latter situation indicates the presence of a cycle. A node encountered twice in the search process is the "junction" node. For example, in Figure 9.13 the trail to node 10 is

$$M_P = (20, 14, 12, -10, -6)$$
$$N_P = (10, 9, 6, 5, 3, 6)$$

Here the junction node is node 6. Note that we have included node 6 in N_P twice to indicate that it is encountered twice in the trail. Also note that the arcs and nodes are listed in the order in which they are encountered when the path is traversed backward from node 10. Two other illustrative trails are given in Table 9.1 for nodes in Figure 9.13.

The algorithm also computes a node parameter γ_j that will be used a great deal in later applications. Let j be some node on the trail that terminates at node i. Let P_j be the set of arcs on the trail from j to i.

If node j is not on a cycle

$$\gamma_j = \frac{1}{\displaystyle\prod_{k \in P_j} a_k}$$

If node j is on a cycle

$$\gamma_j = \frac{\beta/(\beta-1)}{\displaystyle\prod_{k \in P_j} a_k}$$

where β is the cycle gain. Also, $\gamma_j = 1$ if $j = i$.

Thus for the network of Figure 9.13 using $i = 10$

$$\gamma_{10} = 1$$

$$\gamma_6 = \frac{1}{a_{20}a_{14}}$$

$$\gamma_3 = \left(\frac{a_{-10}a_{12}a_{-6}}{a_{-10}a_{12}a_{-6} - 1} \right)\left(\frac{1}{a_{20}a_{14}a_{12}a_{-10}} \right)$$

Table 9.1

Node	M_P	N_P	Junction Node
7	$(11, -10, -6, 12)$	$(7, 5, 3, 6, 5)$	5
3	$(-6, 12, -10)$	$(3, 6, 5, 3)$	3

The usefulness of γ_j will become apparent in later sections. The formal algorithm that performs the above computations is called TRAIL. Its coding follows.

TRAIL ALGORITHM

Purpose: To find the trail in a basis network that originates at the source node or at a cycle and terminates at a specified node IT.

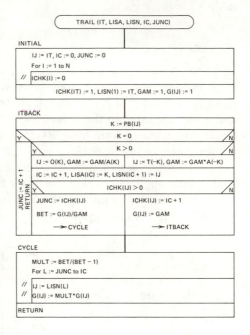

1. (INITIAL) Initialize the node pointer to IT. Set the check list to zero. Check node IT and put IT on the node list.
2. (ITBACK) Find the back pointer for the current path node. If it is zero, the source node has been reached; thus, return. If it is not zero, find the origin node of the arc and put the arc and its origin node on the arc and node lists. Test if the node has already been visited. If so, this is the junction node. Compute β and go to step 3. If not, check the node and repeat this step.
3. (CYCLE) Adjust node gains on the cycle by the factor $\beta/\beta - 1$.

Notes: The lists LISN and LISA contain the nodes and arcs on the path, respectively. They are listed in order starting from node IT and progressing backwards, through the back pointers. There will always be one more entry in LISN than in LISA. JUNC is the first index of the junction node in the node list. If it is IC + 1, the path originates at the source. ICHK is a working list used to determine if a node is encountered twice. If a node is encountered twice in the backtrack operation a cycle is indicated. IC is the number of entries in the arc list.

Finding a Subnetwork Rooted at a Particular Node

The activity described below is similar to the procedure used for pure networks and implemented by the ROOT algorithm of Chapter 4. Our goal is to identify that portion of the basis that is rooted at a specified node. We will use

Table 9.2

Root Node	M_T	N_T
1	$(2, -10, 7)$	$(1, 3, 5, 2)$
8	(-16)	$(8, 7)$
9	$(20, 21, -16, 19, 9, 14)$	$(9, 10, 8, 7, 4, 6, 9)$

the basis of Figure 9.12 as an example. Table 9.2 illustrates the subnetworks defined by several different root nodes.

These examples illustrate that the existence of a cycle adds a new aspect to the problem. As in the pure network case, when the root node is not on the cycle, the subnetwork defined is a tree. This is illustrated in Table 9.2 for the root nodes 1 and 8. When the root node is on a cycle, the subnetwork defined is the entire network component that includes the cycle. This component has the same number of nodes as arcs, while the tree has one more node than arc. Note, however, that, when the root node is on a cycle, algorithm ROOTG causes its index to appear twice on the node list N_T. Table 9.2 illustrates this case for root node 9. The ROOT algorithm of Chapter 4 was designed to terminate when a cycle was found. The ROOTG algorithm is a simple modification of the ROOT algorithm that does not terminate in the event of a cycle but rather continues the search.

ROOTG ALGORITHM

Purpose: To find the list of arcs (LISA) and the list of nodes (LISN) that are in the subnetwork rooted at the node IROOT. If the pointers indicate a cycle returning to IROOT, CYC is set to 1.

1. (INITIAL) Let II = IROOT; store IROOT in LISN.
2. (FORWARD) If node II has a forward pointer node, go to step 4. Otherwise, if node II is not equal to IROOT go to step 3. Else, the subtree consists of a single node, IROOT; RETURN.
3. (RIGHT) If node II has no right pointer, we must backtrack; go to step 5. Otherwise, go to step 4.
4. (ADDLST) Store the arc and node encountered on the latest search. If node II is equal to IROOT a cycle returning to IROOT exists; go to step 3.

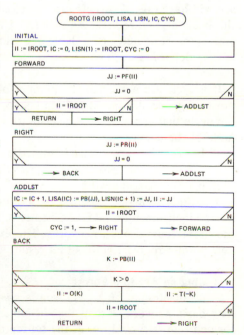

5. (BACK) Backtrack from node II using
    its back pointer arc. If the node II is
    not equal to IROOT, go to step 3.
    Otherwise RETURN.

Note that in ROOTG a cycle is indicated in the ADDLST operation when the search process encounters a forward or right pointer (JJ) that is equal to the root node (IROOT). Consider Figure 9.13 with IROOT=3. The values of M_T and N_T shown below correspond to the list LISA and LISN constructed by ROOTG.

$$IROOT = node\ 3$$
$$M_T = (-10, 11, 12, 14, 20, 21, -6, 8, 7)$$
$$N_T = (3, 5, 7, 6, 9, 10, 8, 3, 4, 2)$$

After node 8 is added to N_T, the procedure backtracks to node 9. Since $p_R(9)=3$, node 3 is added, once more, to N_T and the cycle indication is set. At this point, instead of returning as in ROOT, ROOTG continues through two backtracks to reach node 5. Since $p_R(5)=4$, node 4 is added to N_T. Since $p_R(4)=2$, node 2 is added.

Deleting an Arc from the Basis Network

Although the basis network for the gains problem may consist of several components, there will always be one tree rooted at the source. Deletion of an arc will have one of two effects. We illustrate these effects by deleting arcs from the basis of Figure 9.12. As illustrated in Figure 9.14, deleting a noncycle arc causes the basis network to have one additional tree rooted at the terminal node of the deleted arc. (The changed elements of the pointer representations are shaded.)

If a cycle arc is deleted the number of components of the basis network will not change. As illustrated in Figure 9.15, the component that had been rooted at the cycle that contained the deleted arc will now be a tree rooted at the terminal node of the deleted arc.

Either of these cases is an intermediate stage in the process of creating a new basis network. The discussion below will detail how an arc is to be added to obtain a new basis network.

The algorithm that changes the pointer representation to reflect the deleted arc is algorithm DELTRE of Chapter 4. No modification is necessary to handle the more general basis networks. The example Figures 9.14 and 9.15 show the new pointer representations that result.

Adding an Arc to a Forest

After deleting arc $k_L(i_L, j_L)$ from a basis, we must add arc $k_E(i_E, j_E)$ to form a new basis network. Recall that in either case of the deletion process a tree is formed that is rooted not at the source but at j_L. We will restrict attention here

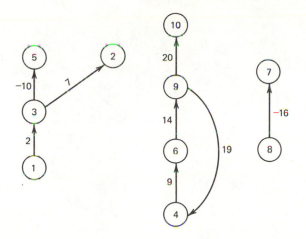

Node	1	2	3	4	5	6	7	8	9	10
P_B (Arc)	0	7	2	19	-10	9	-16	0	14	20
P_F (Node)	3	0	5	6	0	9	0	7	10	0
P_R (Node)	0	0	0	0	2	0	0	0	0	4

Figure 9.14

The Basis Network of Figure 9.12 with Arc 21 Deleted

to the case in which the arc to be added terminates at j_L; that is, $j_E = j_L$. The next section describes how a more general situation is handled.

If $j_E = j_L$, there are two possible cases for arc k_E.

1. Node i_E lies in the tree rooted at j_E, and thus k_E forms a cycle when it is added.
2. Node i_E lies in some other tree (not rooted at j_E). The addition of k_E ties the two trees together thus reducing the number of basis components by one.

Case 1 is illustrated by Figure 9.16 where arc $-13(8,6)$ is added to the tree of Figure 15. The second case is illustrated by Figure 9.17 where arc $12(5,6)$ is added to the forest of Figure 9.15. The new pointer representations are also shown in each case. The shaded cells indicate changes from Figure 9.15.

The ADDTREG algorithm accomplishes the desired computer implementation. ADDTREG is the same as the ADDTRE algorithm of Chapter 4 except that all references to depth pointers are deleted. Notice that the ADDTREG algorithm is given the same name as the ADDTRE algorithm, SUBROUTINE ADDTRE. This facilitates the computer implementation of the generalized network algorithms.

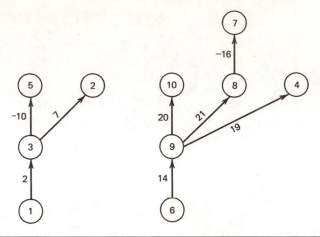

Node	1	2	3	4	5	6	7	8	9	10
P_b (Arc)	0	7	2	19	−10	0	−16	21	14	20
P_F (Node)	3	0	5	0	0	9	0	7	10	0
P_R (Node)	0	0	0	0	2	0	0	4	0	8

Figure 9.15
The Basis Network of Figure 9.12 with Arc 9 Deleted

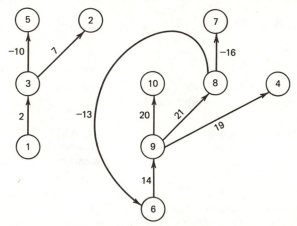

Node	1	2	3	4	5	6	7	8	9	10
P_B	0	7	2	19	−10	−13	−16	21	14	20
P_F	3	0	5	0	0	9	0	6	10	0
P_R	0	0	0	0	2	7	0	4	0	8

Figure 9.16
The Network of Figure 9.15 with Arc −13 Added

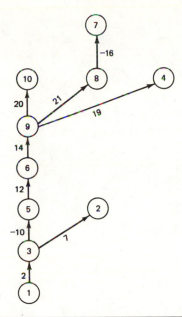

Node	1	2	3	4	5	6	7	8	9	10
P_B	0	7	2	19	-10	12	-16	21	14	20
P_F	3	0	5	0	6	9	0	7	10	0
P_R	0	0	0	0	2	0	0	4	0	8

Figure 9.17
The Network of Figure 9.15 with Arc 12 Added

ADDTREG ALGORITHM

Purpose: To add arc $k_E(i_E, j_E)$ to a disconnected directed subnetwork. Node j_E must be the root of a tree.

1. (FORWARD) If the forward pointer of node i_E is zero, go to step 2. Otherwise, let the right pointer of j_E equal the forward pointer of i_E and go to step 2.

2. (BACK) Let the back pointer of j_E equal k_E and let the forward pointer of i_E equal j_E.

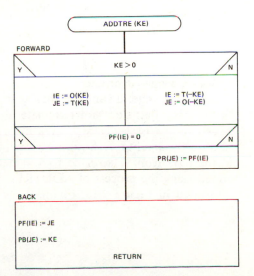

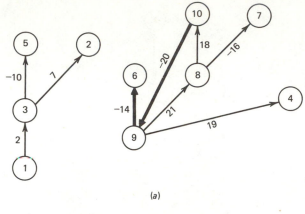

(a)

(a) $k_L = 12(5, 6)$, $k_E = 18(8, 10)$

	1	2	3	4	5	6	7	8	9	10
P_B	0	7	2	19	−10	−14	−16	21	−20	18
P_F	3	0	5	0	0	0	0	10	6	9
P_R	0	0	0	0	2	8	0	4	0	7

Figure 9.18
Examples of the Basis Change Operation

Basis Change Algorithm

In a more general situation than that considered above, an arc may be added to the basis that does not terminate at node j_L. Say the arc $k_E(i_E, j_E)$ is to be added to the basis. Here j_E is a node in the tree rooted at j_L but j_E is not necessarily equal to j_L. A valid basis network is formed by first reversing the arcs in the path from j_E to j_L to form a new tree rooted at j_E. Now the arc k_E is added using the process of ADDTREG. Several examples of this procedure are illustrated in Figure 9.18. Figure 9.18a shows the basis that results when arc

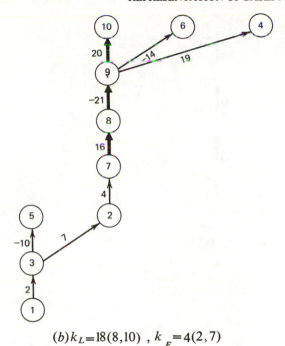

$(b) k_L = 18(8,10)$, $k_E = 4(2,7)$

	1	2	3	4	5	6	7	8	9	10
P_B	0	7	2	19	− 10	− 14	4	16	− 21	20
P_F	3	7	5	0	0	0	8	9	10	0
P_R	0	0	0	0	2	4	0	0	0	6

Figure 9.18
(continued)

12(5,6) is deleted from the basis of Figure 9.17 and arc 18(8, 10) is added. Figure 9.18b is the result of deleting arc 18(8, 10) from Figure 9.18a and adding arc 4(2, 7). Finally, Figure 9.18c results from the deletion of arc 4(2, 7) from Figure 9.18b and the addition of arc 3(1, 4). The reversed arcs, in each figure, are shown with heavy lines. The operation of changing basis networks is accomplished by the TRECHG algorithm of Chapter 4.

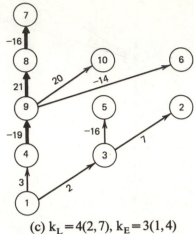

(c) $k_L = 4(2,7)$, $k_E = 3(1,4)$

	1	2	3	4	5	6	7	8	9	10
P_B	0	7	2	3	−10	−14	−16	21	−19	20
P_F	4	0	5	9	0	0	0	7	8	0
P_R	0	0	0	3	2	0	0	10	0	6

Figure 9.18
(continued)

9.6   FLOW FOR GENERALIZED NETWORKS

Flow on Paths

Consider the simple directed path of Figure 9.19 where only the gain parameters are defined and the amount of flow leaving node 6 is V. The set of equations that defines the flows for this system is

$$
\begin{aligned}
-a_1 f_1 \;+f_2 &= 0 && (1)\\
-a_2 f_2 \;+f_3 &= 0 && (2)\\
-a_3 f_3 \;+f_4 &= 0 && (3)\\
-a_4 f_4 \;+f_5 &= 0 && (4)\\
-a_5 f_5 &= -V && (5)
\end{aligned}
$$

$$\text{①} \xrightarrow[1]{(a_1)} \text{②} \xrightarrow[2]{(a_2)} \text{③} \xrightarrow[3]{(a_3)} \text{④} \xrightarrow[4]{(a_4)} \text{⑤} \xrightarrow[5]{(a_5)} \text{⑥} \rightarrow V$$

Figure 9.19
Directed Path with Gains ($s = 1, t = 6$)

This set of equations may be solved by first solving for f_5 in equation (5) and substituting its value into equation (4). This allows us to solve directly for f_4 in equation (4). Continuing this process through all five equations yields

$$f_5 = \frac{V}{a_5} \tag{5a}$$

$$f_4 = \frac{f_5}{a_4} = \frac{V}{a_4 a_5} \tag{4a}$$

$$f_3 = \frac{f_4}{a_3} = \frac{V}{a_3 a_4 a_5} \tag{3a}$$

$$f_2 = \frac{f_3}{a_2} = \frac{V}{a_2 a_3 a_4 a_5} \tag{2a}$$

$$f_1 = \frac{f_2}{a_1} = \frac{V}{a_1 a_2 a_3 a_4 a_5} \tag{1a}$$

We recall from our discussion of algorithm TRAIL that the node parameter γ_i is equal to the inverse of the product of the arc gains on the trail from node i to the last node on the trail. Thus for the example of Figure 9.19

$$\gamma_6 = 1, \qquad \gamma_5 = \frac{1}{a_5}, \qquad \gamma_4 = \frac{1}{a_4 a_5}, \qquad \gamma_3 = \frac{1}{a_3 a_4 a_5}$$

$$\gamma_2 = \frac{1}{a_2 a_3 a_4 a_5}, \qquad \gamma_1 = \frac{1}{a_1 a_2 a_3 a_4 a_5}$$

Substituting these relations into equations (1a) to (5a), we see that the flow on each arc in any directed path may be expressed in terms of the γ_j as

$$f_k = \frac{V \gamma_j}{a_k} \qquad \text{for arc } k(i,j) \in M_P \tag{6}$$

We use this equation instead of the simpler $f_k = V \gamma_i$ because a node in a basis network will terminate only one arc but may originate more than one arc.

EXAMPLE

For the network of Figure 9.19, let $a_1 = 2$, $a_2 = 0.9$, $a_3 = 1.5$, $a_4 = 0.7$, and $a_5 = 0.3$, and determine the value of the arc flows if $V = 3$.

Solution

Compute

$$\gamma_6 = 1 \qquad \gamma_5 = 3.33 \qquad \gamma_4 = 4.76$$
$$\gamma_3 = 3.17 \qquad \gamma_2 = 3.53 \qquad \gamma_1 = 1.76$$

using equation (6).

$$f_1 = 5.29 \qquad f_2 = 10.58 \qquad f_3 = 9.52$$
$$f_4 = 14.28 \qquad f_5 = 10$$

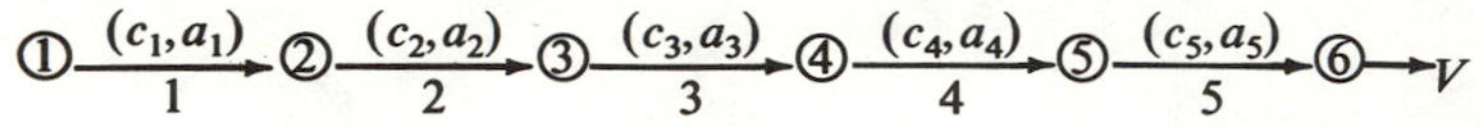

Figure 9.20
Directed Path with Arc Gains and Capacities

Figure 9.20 has a capacity defined for each arc. We now address the problem of finding the maximum flow from node 1 to node 6.

Equation (6) specifies each arc flow as a function of V. The capacity constraints require that

$$f_k \leqslant c_k \qquad k \in M_P$$

Equivalent expressions of this restriction are

$$\frac{V}{a_k} \gamma_j \leqslant c_k \qquad \text{for } k(i,j) \in M_P$$

$$V \leqslant \frac{(c_k a_k)}{\gamma_j} \qquad \text{for } k(i,j) \in M_P$$

Thus the maximum flow V on the directed path is

$$V = \operatorname*{Min}_{k \in M_P} \frac{(c_k a_k)}{\gamma_j} \tag{7}$$

EXAMPLE
If for the previous example

$$c_1 = 3, \qquad c_2 = 5, \qquad c_3 = 2, \qquad c_4 = 7, \qquad c_5 = 6$$

the maximum value of V, from equation (7), is

$$V = \operatorname{Min}[1.701, 1.4175, 0.63, 1.47, 1.8] = 0.63$$

The associated arc flows are computed from Equation (6) as

$$f_1 = 1.11, \qquad f_2 = 2.22, \qquad f_3 = 2, \qquad f_4 = 3, \qquad f_5 = 2.1$$

Flow-Generating Cycles

Consider now the simple network consisting of a directed path and a directed cycle as in Figure 9.21. Here $M_P = [5, 4, 3, 2, 1]$. We identify the path as $M_H = \{4, 5\}$ and the cycle as $M_C = \{3, 2, 1\}$.

Here node 1, the source, stands alone, and we consider only the part of the network consisting of arcs 1 through 5. The set of equations that defines the set of flows on this network is

$$
\begin{aligned}
-a_1 f_1 \quad +f_2 \quad &= 0 \\
-a_2 f_2 \quad +f_3 \quad &= 0 \\
+f_1 \quad -a_3 f_3 \quad +f_4 \quad &= 0 \\
-a_4 f_4 \quad +f_5 &= 0 \\
-a_5 f_5 &= -V
\end{aligned}
$$

Again, we can solve this set of five equations for the five arc flows.

$$
f_5 = \frac{V}{a_5}
$$

$$
f_4 = \frac{f_5}{a_4} = \frac{V}{a_4 a_5}
$$

We must revise the solution process because of the off-diagonal term in the third equation. Rewriting the first three equations we obtain

$$
\begin{aligned}
-a_1 f_1 \quad +f_2 \quad &= 0 \\
-a_2 f_2 \quad +f_3 &= 0 \\
f_1 \quad -a_3 f_3 &= -f_4
\end{aligned}
$$

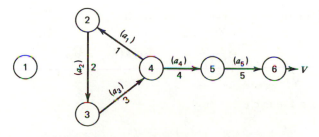

Figure 9.21
Example Network with a Flow-Generating Cycle

Solving for f_3 by Cramer's rule yields

$$f_3 = \frac{\begin{vmatrix} -a_1 & 1 & 0 \\ 0 & -a_2 & 0 \\ 1 & 0 & -f_4 \end{vmatrix}}{\begin{vmatrix} -a_1 & 1 & 0 \\ 0 & -a_2 & 1 \\ 1 & 0 & -a_3 \end{vmatrix}} = \frac{-a_1 a_2 f_4}{1 - a_1 a_2 a_3}$$

Since the gain around the cycle, β, is $a_1 a_2 a_3$,

$$f_3 = \frac{-a_1 a_2 f_4}{1 - \beta}$$

Simultaneously multiplying and dividing the right side by β and substituting for f_4 yields

$$f_3 = \left(\frac{-\beta}{1-\beta} \right) \frac{f_4}{a_3}$$

$$= \left(\frac{\beta}{\beta - 1} \right) \frac{V}{a_3 a_4 a_5}$$

When all $a_k > 0$, it is apparent that f_3 can be positive only if the cycle gain is greater than 1. Cycles with $\beta > 1$ are called *flow-generating cycles* because flow is generated without any contribution from external flows.

Completing the solution of the flow equations yields

$$f_2 = \frac{f_3}{a_2}$$

$$= \left(\frac{\beta}{\beta - 1} \right) \frac{V}{a_2 a_3 a_4 a_5}$$

Again recall the definition of γ_j for arcs on the path

$$\gamma_j = \frac{1}{\underset{k \in P_j}{\Pi} a_k} \qquad \text{for } k(i,j) \in M_H \tag{8}$$

and for arcs on the cycle

$$\gamma_j = \left(\frac{\beta}{\beta - 1} \right) \frac{1}{\underset{k \in P_j}{\Pi} a_k} \qquad \text{for } k(i,j) \in M_C \tag{9}$$

The general expression for flow on the arcs can then be written

$$f_k = \frac{V\gamma_j}{a_k} \qquad \text{for } k(i,j) \in M_H \cup M_C \qquad (10)$$

Although this is the same as the expression developed for the path case, note that γ_j is adjusted for arcs on the cycle.

EXAMPLE

Find the arc flows in Figure 9.21 if $V=3, a_1=0.5, a_2=3, a_3=1, a_4=0.3, a_5=0.7$. Clearly, $\gamma_6=1, \gamma_5=1.43, \gamma_4=14.29, \gamma_3=14.29, \gamma_2=4.76, \beta=a_1a_2a_3=1.5$.
From equation (10) above, we obtain

$$f_1=28.57, \qquad f_2=14.29, \qquad f_3=14.29, \qquad f_4=14.29, \quad \text{and} \quad f_5=4.29$$

The solution is presented in a graphical manner below.

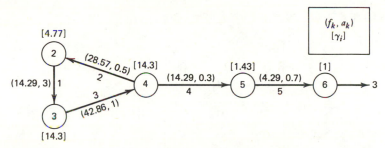

It is instructive to study this figure and note that conservation of flow is indeed satisfied at each node. In particular, note that at node 4

$$-f_1 + a_3 f_3 - f_4 = -28.57 + 42.85 - 14.29 = 0$$

When the arcs have capacities, an appropriate question is, What is the maximum flow that can be generated by a flow-augmenting cycle–path combination and what are the corresponding arc flows? With arc capacities defined, the maximum flow that can pass through each arc is

$$f_k = \frac{V\gamma_j}{a_k} \leqslant c_k \qquad \text{for } k \in M_P \qquad (11)$$

Therefore, the maximum flow change is

$$V = \operatorname*{Min}_{k \in M_P} (a_k c_k / \gamma_j) \qquad (12)$$

Once the maximum value of V is arrived at through equation (12), the corresponding arc flows are obtained by means of equation (10).

Flow-Absorbing Cycles

Just as a cycle can generate flow if its gain is greater than unity, a cycle can absorb flow if its gain is less than unity. Consider the network of Figure 9.22. The equations for conservation of flow are

$$
\begin{aligned}
f_1 && &= V \\
-a_1 f_1 &+ f_2 && = 0 \\
& -a_2 f_2 + f_3 && -a_5 f_5 = 0 \\
&& -a_3 f_3 + f_4 && = 0 \\
&& -a_4 f_4 &+ f_5 = 0
\end{aligned}
$$

Again solving for the flows in the arcs of Figure 9.22, we obtain

$$
f_1 = V
$$

$$
f_2 = a_1 f_1 = V a_1
$$

Using Cramer's rule to solve for f_3, we find

$$
f_3 = \frac{\begin{vmatrix} a_2 f_2 & 0 & -a_5 \\ 0 & 1 & 0 \\ 0 & -a_4 & 1 \end{vmatrix}}{\begin{vmatrix} 1 & 0 & -a_5 \\ -a_3 & 1 & 0 \\ 0 & -a_4 & 1 \end{vmatrix}} = \frac{a_2 f_2}{1 - a_3 a_4 a_5}
$$

Defining β as the cycle gain and substituting for f_2 gives

$$
f_3 = \frac{a_1 a_2 V}{1 - \beta}
$$

It is apparent that, for f_3 to be positive, β must be less than 1. We call such a cycle a flow-absorbing cycle.

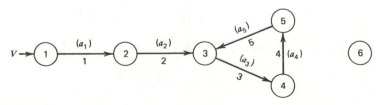

Figure 9.22
Example Network with a Flow-Absorbing Cycle

The values for f_4 and f_5 follow directly.

$$f_4 = a_3 f_3 = \frac{a_1 a_2 a_3 V}{1 - \beta}$$

$$f_5 = a_4 f_4 = \frac{a_1 a_2 a_3 a_4 V}{1 - \beta}$$

EXAMPLE

Let $a_1 = 2, a_2 = 1, a_3 = 0.5, a_4 = 1, a_5 = 0.4$. Solve for the flows when $V = 3$. Note that $\beta = (0.5)(1)(0.4) = 0.2 < 1$, so this is a flow-absorbing cycle. The computed arc flows are shown in the figure below.

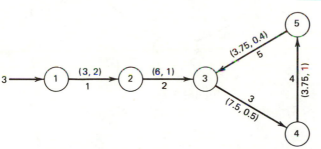

(f_k, a_k)

When some arc gains are negative, the cycle gain may also be negative (when an odd number of arcs with negative gains are in the cycle). In this case, the cycle must be a flow absorbing cycle.

The Expanded Network

Any iterative algorithm for the minimum cost flow problem will operate to modify arc flows in such a way that the optimal solution will ultimately be obtained. Just as it was in our discussion of pure networks, it is convenient to define an expanded network to assist in keeping track of the effects of such changes upon a given set of arc flows in a generalized network. As before, the expanded network replaces each arc of the original network with two arcs, the forward and mirror arcs. Forward arcs are designated with positive integers and mirror arcs with negative integers. This construct is illustrated in Figure 9.23, where the original arc is given in Figure 9.23a and the associated pair of arcs in the expanded network is given in Figure 9.23b. If the expanded network parameters are denoted with asterisks, their general definitions may be written as

Forward arcs

$$h_k^* = h_k$$
$$a_k^* = a_k$$

(13)

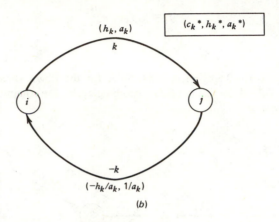

Figure 9.23
Transformation to Expanded Network

Mirror arcs

$$h^*_{-k} = \frac{-h_k}{a_k}$$

$$a^*_{-k} = \frac{1}{a_k} \tag{14}$$

Flows in the expanded network imply changes in the flows of the original network by the relation

$$\Delta_k = f^*_k - (f^*_{-k})(a^*_{-k}) = f^*_k - \frac{f^*_{-k}}{a_k} \tag{15}$$

where Δ_k is the change in flow in the original network. The procedures will assure that forward and mirror arcs will never simultaneously have nonzero flow. Thus, at most, one of the two terms on the right side of equation (15) will be nonzero while the other term will have value zero.

In this text, the expanded network is used primarily as a conceptual tool to aid in the description of the manipulations of the algorithms. The expanded network parameters and flows are never used explicitly in the algorithms and computer programs. We do frequently reference mirror arcs, however, through the use of a negative arc index.

Flow Augmentation Trails in the Expanded Network

The flow modification algorithms to be described in Chapter 10 define flow augmentation trails in terms of the expanded network. Thus the directed trail may include both forward and mirror arcs. The general case is illustrated in Figure 9.24. In the remainder of this section, we repeat the flow equations that have been developed above in their appropriate form for the expanded network.

Let M_P be the set of arcs in the flow augmenting trail: $M_P = (k_1, k_2, \ldots, k_p)$. The arcs in this set are ordered so that k_1 terminates at the sink, k_2 terminates at the origin node of k_1, and so on. Node 1 in Figure 9.24 is called the sink in this discussion. In applications, this may be any node of the network.

The value of γ_i is computed for each node on the trail, but not on a cycle as

$$\gamma_i = \frac{1}{\displaystyle\prod_{\ell=1}^{i-1} a_{k_\ell}^*} \tag{16}$$

For the sink node γ_i is defined as 1. If the trail originates at a cycle, γ_i is computed for nodes on the cycle using the cycle gain β as

$$\gamma_i = \left(\frac{\beta}{\beta-1}\right)\left[\frac{1}{\displaystyle\prod_{\ell=1}^{i-1} a_{k_\ell}^*}\right] \tag{17}$$

Note that γ_i can also be defined recursively. If arc $k_i(i+1, i)$ is an arc on the trail, then if $i+1$ is not the junction node for the cycle

$$\gamma_{i+1} = \frac{\gamma_i}{a_k^*}$$

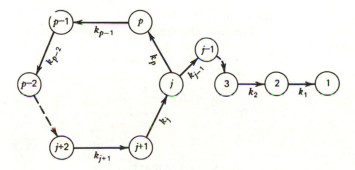

Figure 9.24
The General Case of a Flow-Augmenting Trail

If node $i+1$ is the junction node

$$\gamma_{i+1} = \left(\frac{\beta}{\beta-1}\right)\left(\frac{\gamma_i}{a_k^*}\right)$$

where β is the cycle gain.

In terms of the parameters of the original network, the γ's are computed using the following recursive formulae.

$$\gamma_1 = 1$$

When node $i+1$ is not the junction node

$$\gamma_{i+1} = \begin{cases} \dfrac{\gamma_i}{a_{k_i}} & k_i > 0 \\ \\ \gamma_i a_{-k_i} & k_i < 0 \end{cases} \tag{18}$$

When node $i+1$ is the junction node

$$\gamma_{i+1} = \begin{cases} \left(\dfrac{\beta}{\beta-1}\right)\dfrac{\gamma_i}{a_{k_i}} & k_i > 0 \\ \\ \left(\dfrac{\beta}{\beta-1}\right)\gamma_i a_{k_i} & k_i < 0 \end{cases} \tag{19}$$

The flows in the expanded network determined by a flow increment of V at the sink are

$$f_{k_i}^* = \frac{\gamma_i V}{a_{k_i}^*} \qquad \text{for } k(i,j) \in M_P$$

Note that the marginal flows may be either positive or negative, depending on the signs of γ_i, V, and a_{ki}^*. The flow change in the original network implied by a flow increment of V is, for $k_i > 0$,

$$\Delta_{k_i} = f_{k_i}^* = \frac{\gamma_i V}{a_{k_i}} \tag{20}$$

for $k_i < 0$

$$\Delta_{-k_i} = (f_{k_i}^*)(a_{k_i}^*)$$
$$= -\gamma_i V \tag{21}$$

Now let us determine the maximum positive flow increment V that can be obtained at the sink. Assume there are initial flows in the arcs of the trail identified as f_{k_i} or f_{-k_i}, depending on whether k_i is positive or negative. We define V_{k_i} as the increase in sink flow that will drive the flow in arc k to one of its bounds. Consider first the case where k_i is a forward arc ($k_i > 0$). The flow in arc k_i after the flow increment is

$$f_{k_i} + \Delta_{k_i} = f_{k_i} + \frac{\gamma_i V}{a_{k_i}}$$

Since V is positive, the direction of flow change depends on the sign of γ_i / a_{k_i}. If this factor has a positive sign, the arc flow increases and the value of V, which drives it to capacity, is determined by

$$f_{k_i} + \frac{\gamma_i V_{k_i}}{a_{k_i}} = c_{k_i}$$

or
$$V_{k_i} = \frac{(c_{k_i} - f_{k_i}) a_{k_i}}{\gamma_i} \qquad \text{for} \begin{cases} k_i > 0 \\ \gamma_{\frac{i}{ak_i}} > 0 \end{cases} \qquad (22)$$

where V_{k_i} is the value of V that drives arc k to its bound. When the factor γ_i / a_{k_i} is negative, the flow in arc k_i decreases, and the flow increase that drives it to zero is determined by

$$f_{k_i} + \frac{\gamma_i V_{k_i}}{a_{k_i}} = 0$$

or
$$V_{k_i} = \frac{-f_{k_i} a_{k_i}}{\gamma_i} \qquad \text{for} \begin{cases} k_i > 0 \\ \dfrac{\gamma_i}{a_{k_i}} < 0 \end{cases} \qquad (23)$$

Now we consider the case where arc k_i is a mirror arc ($k_i < 0$). The flow in arc k_i after the increment is

$$f_{-k_i} + \Delta_{-k_i} = f_{-k_i} - \gamma_i V$$

Since V is positive, the direction of the flow change depends only on the sign of γ_i. When γ_i is positive, flow in arc $-k_i$ decreases, and the flow increment that drives it to zero is determined by

$$f_{-k_i} - \gamma_i V_{k_i} = 0$$

or
$$V_{k_i} = \frac{f_{-k_i}}{\gamma_i} \qquad \text{for} \begin{cases} k_i < 0 \\ \gamma_i > 0 \end{cases} \qquad (24)$$

For γ_i negative, the flow increment increases the flow in the arc $-k_i$, and the increment that drives the flow to capacity is determined by

$$f_{-k_i} - \gamma_i V_{k_i} = c_{-k_i}$$

$$V_{k_i} = \frac{f_{-k_i} - c_{-k_i}}{\gamma_i} \qquad \text{for} \begin{cases} k_i < 0 \\ \gamma_i < 0 \end{cases} \tag{25}$$

For each of the four cases studied above, V_{k_i} is positive. The maximum flow increment for the trail is the flow increment that will drive the flow on one or more of the arcs to zero or its capacity without causing any other flow to be infeasible. Thus,

$$V = \underset{k_i \in M_P}{\text{Min}} \left[V_{k_i} \right] \tag{26}$$

where the V_{k_i} is determined as one of the four cases above. The MFLOG algorithm is used to find the maximum flow change in a trail, given the γ values and lists of arcs and nodes on the trail. The FLOWG algorithm modifies arc flows for a given trail and a given value of V.

MFLOG ALGORITHM

Purpose: To find the maximum flow augmentation possible in a trail originating at the source or a cycle.

Notes: The list of arcs in the path are stored in the list LISN. IC is the number of arcs. MF is the maximum flow change. KL is the arc which becomes saturated with the flow change. ILC is the index of KL.

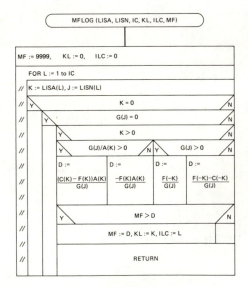

FLOWG ALGORITHM

Purpose: To change the flow in each arc an amount prescribed by MF and G(J).

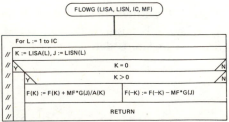

EXAMPLE

Consider the portion of a larger original network given in Figure 9.25a. The flows shown on the figure do not satisfy conservation of flow because this example has been extracted from a hypothetical network and not all arcs are shown. A flow-augmenting path selected from the expanded network for this example appears in Figure 9.25b.

For the path of Figure 9.25b, we compute the values of γ shown on the figure. Using equations (22), (24), and (26), we find that the maximum flow change in the path is

$$V = \text{Min} (2, 3, 9, 13.5, 6.75) = 2$$

Increasing the flow by 2 at node 6 yields from equation (20) for the marginal network

$$f_1^* = 0.296, \quad f_{-2}^* = 0.592, \quad f_3^* = 0.444, \quad f_{-4}^* = 0.67, \quad f_5^* = 2$$

The marginal flows modify the flows in the original network via equation (20) and (21) to obtain

$$f_1 = 2 + 0.296 = 2.296$$
$$f_2 = 3 - (0.592)(0.75) = 2.556$$
$$f_3 = 0 + 0.444 = 0.444$$
$$f_4 = 3 - (0.67)(3) = 0.99$$
$$f_5 = 4 + 2 = 6$$

When flow is to be decreased at the sink rather than increased, the foregoing discussion is correct if the signs of the γ_i are reversed. The γ_i are then defined as the increase of flow out of node i per unit of flow *decrease* at the sink. This is easily accomplished by assigning $\gamma = -1$ for the sink, and then using the recursive equations (18) and (19) to find the other values of γ for nodes on the trail.

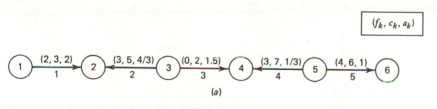

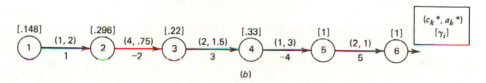

Figure 9.25
A Path in the Expanded Network

EXAMPLE

Given the flow absorbing cycle of Figure 9.22, compute the arc flows to obtain a reduction of three units of flow at node 1. Here node 1 is the sink. The trail to the sink is $M_P = \{-1, -2, 5, 4, 3\}$, and $N_P = \{1, 2, 3, 5, 4, 3\}$. The flow is to decrease at the sink by 3. Assume the arc gains:

$$a_1 = 2, \qquad a_2 = 1, \qquad a_3 = 0.5, \qquad a_4 = 1, \qquad a_5 = 0.4$$

Since flow is to be decreased, we assign the value $\gamma_1 = -1$. We then use equations (18) and (19) to compute $\gamma_2 = -1/a_{-1} = -a_1 = -2$. We compute the cycle gain as

$$\beta = (0.5)(1.)(0.4) = 0.2$$

$$\gamma_3 = \left(\frac{\beta}{\beta - 1}\right)\frac{\gamma_2}{a_{-2}} = \frac{0.2}{-0.8}(-2)(1) = 0.5$$

$$\gamma_5 = \frac{\gamma_3}{a_5} = \frac{0.5}{0.4} = 1.25$$

$$\gamma_4 = \frac{\gamma_5}{a_4} = \frac{1.25}{1} = 1.25$$

Using equation (21) with $V = 3$ and the initial flows equal to zero, we compute

$$f_1 = -V\gamma_1 = -(3)(-1) = 3 \qquad \text{for } -1 \in M_P$$
$$f_2 = -V\gamma_2 = -(3)(-2) = 6 \qquad \text{for } -2 \in M_P$$

Using equation (20), we compute

$$f_5 = \frac{V\gamma_3}{a_5} = \frac{(3)(0.5)}{0.4} = 3.75 \qquad \text{for } 5 \in M_P$$

$$f_4 = \frac{V\gamma_4}{a_4} = \frac{(3)(1.25)}{1} = 3.75 \qquad \text{for } 4 \in M_P$$

$$f_3 = \frac{V\gamma_4}{a_3} = \frac{(3)(1.25)}{0.5} = 7.5 \qquad \text{for } 3 \in M_P$$

9.7 THE NODE POTENTIALS

The primal–dual theory of linear programming provides the following conditions for optimality for the generalized minimum cost flow problem.

(a) $\pi_i - a_k\pi_j = -h_k$ for f_k basic. $\qquad\qquad\qquad\qquad$ (27)
(b) $\pi_i - a_k\pi_j \geqslant -h_k$ for $f_k = 0$ and nonbasic.
(c) $\pi_i - a_k\pi_j \leqslant -h_k$ for $f_k = c_k$ and nonbasic.
(d) f_k is feasible.

Condition (a) provides a convenient means of determining the dual variables π, given a basis of the primal problem. Recall that a basis of this problem is a selection of $n-1$ arcs of the network such that $|\mathbf{B}| \neq 0$, where $\mathbf{B}$ is the set of columns of the LP matrix associated with the basis arcs. The set of equations indicated by condition (a) for the basic arcs can be written in vector–matrix form as

$$\boldsymbol{\pi}\mathbf{B} = -\mathbf{H}_B \tag{28}$$

where $\boldsymbol{\pi}$ is the vector of node potentials and $\mathbf{H}_B$ is the vector of arc costs for the basic arcs. When the basis network consists of more than one component, one component will be a tree rooted at the source node and the other components will be trees rooted at cycles. In this case, the rows and columns of the basis can be rearranged such that the basis matrix has a block diagonal form.

$$\mathbf{B} = \begin{bmatrix} \mathbf{B}_T & 0 & 0 \\ 0 & \mathbf{B}_{C1} & 0 \\ 0 & 0 & \mathbf{B}_{C2} \end{bmatrix}$$

Here $\mathbf{B}_T$ is the basis tree associated with the tree rooted at the source and $\mathbf{B}_{C1}$ and $\mathbf{B}_{C2}$ are the basis matrices associated with the trees rooted at the cycles.

EXAMPLE

Consider the basis network defined by Figure 9.12. The basis matrix for this example with rows and columns rearranged is

$$\begin{array}{ccccccccc}
2 & 7 & -10 & 19 & 9 & 14 & 21 & -16 & 10
\end{array}$$

$$\left[\begin{array}{ccc|cccccc}
-a_2 & 1 & 1 & & & & & & \\
 & -a_7 & & & & & & & \\
 & & -a_{(-10)} & & & & & & \\ \hline
 & & & -a_{19} & 1 & & & & \\
 & & & & -a_9 & 1 & & & \\
 & & & 1 & & -a_{14} & 1 & & 1 \\
 & & & & & & -a_{21} & 1 & \\
 & & & & & & & -a_{(-16)} & 1 \\
 & & & & & & & & -a_{20}
\end{array}\right]\begin{array}{c}3\\2\\5\\4\\6\\9\\8\\7\\10\end{array}$$

With the basis matrix partitioned in this way, one can partition equation (28) above into

$$\boldsymbol{\pi}_T\mathbf{B}_T = -\mathbf{H}_T$$
$$\boldsymbol{\pi}_C\mathbf{B}_C = -\mathbf{H}_C$$

In this partition, $\boldsymbol{\pi}_T$ and $\mathbf{H}_T$ are, respectively, the node π values and basic arc costs for the tree, and $\boldsymbol{\pi}_C$ and $\mathbf{H}_C$ have similar definitions for the cycle. For the example of Figure 9.12, these equations are

$$[\pi_3, \pi_2, \pi_5] \begin{bmatrix} -a_2 & 1 & 1 \\ 0 & -a_7 & 0 \\ 0 & 0 & -a_{(-10)} \end{bmatrix} = [-h_2, -h_7, -h_{(-10)}]$$

and

$$[\pi_4, \pi_6, \pi_9, \pi_8, \pi_7, \pi_{10}] \begin{bmatrix} -a_{19} & 1 & 0 & 0 & 0 & 0 \\ 0 & -a_9 & 1 & 0 & 0 & 0 \\ 1 & 0 & -a_{14} & 1 & 0 & 1 \\ 0 & 0 & 0 & -a_{21} & 1 & 0 \\ 0 & 0 & 0 & 0 & -a_{(-16)} & 0 \\ 0 & 0 & 0 & 0 & 0 & -a_{20} \end{bmatrix}$$

$$= [-h_{19}, -h_9, -h_{14}, -h_{21}, -h_{(-16)}, -h_{10}]$$

In order to determine the π values, one must solve these sets of equations. For the tree portion of the basis, one can always rearrange the rows and columns to obtain an upper diagonal form as shown in the example. This allows the π values to be determined in an iterative manner. To illustrate, we write the equations in algebraic form.

$$-a_2\pi_3 = -h_2$$
$$\pi_3 - a_7\pi_2 = -h_7$$
$$\pi_3 - a_{(-10)}\pi_5 = -h_{(-10)}$$

or

$$\pi_3 = \frac{h_2}{a_2}$$

$$\pi_2 = \frac{(\pi_3 + h_7)}{a_2}$$

$$\pi_5 = \frac{(\pi_3 + h_{(-10)})}{a_{(-10)}}$$

First π_3 is determined and then it is used to determine π_2 and π_5. In general, we note that for an arc in the tree $k(i, j)$

$$\pi_j = \frac{(\pi_i + h_k)}{a_k} \tag{29}$$

By first assigning $\pi_s = 0$, we can determine the values of π_j for each tree node using equation (29). This process is implemented in the DUAL Algorithm.

DUAL ALGORITHM

Purpose: To compute the dual variables for the basis network rooted at a given node II. The value of PI(II) is assumed available.

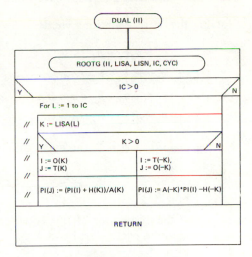

A component of the basis network that includes a cycle provides a somewhat more complex problem because of the single unity element in the lower diagonal portion of the matrix $\mathbf{B}_C$. Consider the four-arc cycle of Figure 9.26. If the four arcs are basic, they must all satisfy condition (a) of equation (25); that is

$$
\begin{aligned}
\pi_1 - a_1\pi_2 && = -h_1 \\
\pi_2 - a_2\pi_3 && = -h_2 \\
\pi_3 - a_3\pi_4 &= -h_3 \\
-a_4\pi_1 && \pi_4 = -h_4
\end{aligned}
$$

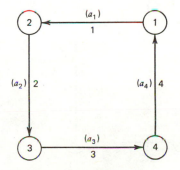

Figure 9.26
A Cycle of Four Basic Arcs

Solving the set of equations yields

$$\pi_4 = a_4\pi_1 - h_4$$
$$\pi_3 = a_3\pi_4 - h_3 = a_3 a_4\pi_1 - a_3 h_4 - h_3$$
$$\pi_2 = a_2\pi_3 - h_2 = a_2 a_3 a_4\pi_1 - a_2 a_3 h_4 - a_2 h_3 - h_2$$
$$\pi_1 = a_1\pi_2 - h_1 = a_1 a_2 a_3 a_4\pi_1 - a_1 a_2 a_3 h_4 - a_1 a_2 h_3 - a_1 h_2 - h_1$$

Noting that $a_1 a_2 a_3 a_4 = \beta$, the cycle gain, we obtain

$$\pi_1 = \frac{h_1 + a_1 h_2 + a_1 a_2 h_3 + a_1 a_2 a_3 h_4}{\beta - 1} \tag{30}$$

Note that the numerator of this equation is just the cost of passing one unit of flow around the cycle starting at arc 1.

Node i	Flow Starting at Node i	Cost of Flow Starting at Node i
1	$f_1 = 1$	$h_1 f_1 = h_1$
2	$f_2 = a_1$	$h_2 f_2 = a_1 h_2$
3	$f_3 = a_1 a_2$	$h_3 f_3 = a_1 a_2 h_3$
4	$f_4 = a_1 a_2 a_3$	$h_4 f_4 = a_1 a_2 a_3 h_4$

The flow at the terminal node of arc 4, that is, node 1, is $a_1 a_2 a_3 a_4 = \beta$. With these flows, $\beta - 1$ units of flow can be withdrawn at node 1. Thus the cost per unit of flow that can be obtained at node 1 is given by equation (30). These concepts are illustrated in Figure 9.27.

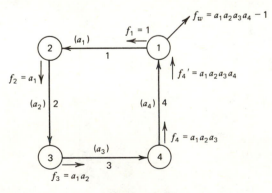

Figure 9.27
Flows in a Cycle

In general, if $M_C = (1, 2, \ldots, k_c)$ the π value at node 1 is

$$\pi_1 = \frac{h_1 + \sum_{k=2}^{k_c} h_k \prod_{j=1}^{k-1} a_j}{\beta - 1} \qquad (31)$$

We call the numerator of this expression the *unit cost* of the cycle. Although this seems a rather complex equation, it is operationally very easy to implement in the computer algorithms. Using equation (31) to evaluate the π value for any one node in the cycle, we can easily determine all others with equation (29). Subroutine CYCLE is provided to compute the unit cost and gain of a cycle.

CYCLE ALGORITHM

Purpose: To compute the gain of a cycle defined by the back pointers and the cost of circulating one unit of flow around the cycle.

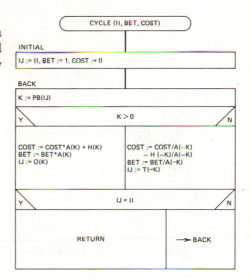

For the example of Figure 9.12, we choose arbitrarily to evaluate π_4 (any node in the cycle would do as well). The cycle gain is $\beta = a_9 a_{14} a_{19}$. The cost of passing one unit of flow around the cycle starting at node 4 is

$$h_9 + a_9 h_{14} + a_9 a_{14} h_{19}$$

Thus,

$$\pi_4 = \frac{h_9 + a_9 h_{14} + a_9 a_{14} h_{19}}{a_9 a_{14} a_{19} - 1}$$

With this value, we can determine in order

$$\pi_6 = \frac{\pi_4 + h_9}{a_9}$$

$$\pi_9 = \frac{\pi_6 + h_{14}}{a_{14}}$$

$$\pi_8 = \frac{\pi_9 + h_{21}}{a_{21}}$$

$$\pi_7 = \frac{\pi_8 + h_{(-16)}}{a_{(-16)}}$$

$$\pi_{10} = \frac{\pi_9 + h_{20}}{a_{20}}$$

EXAMPLE

For the basis of Figure 9.12, determine the π values when

Arc	2	7	10	9	14	19	21	16	20
h_k	8	4	2	1	-3	6	10	2	2
a_k	0.8	0.85	0.8	0.8	1	1.5	0.8	1	1

Since two arcs are mirror arcs, we first use equation (14) to compute

$$h_{-10} = \frac{-h_{10}}{a_{10}} = \frac{-2}{0.8} = -2.5$$

$$a_{-10} = \frac{1}{a_{10}} = 1.25$$

$$h_{-16} = \frac{-h_{16}}{a_{16}} = -2$$

$$a_{-16} = \frac{1}{a_{16}} = 1$$

Then, assuming $\pi_1 = 0$, equations (29) and (31) are used to obtain

$$\pi_3 = \frac{h_2}{a_2} = \frac{8}{0.8} = 10$$

$$\pi_2 = \frac{\pi_3 + h_7}{a_7} = \frac{10 + 4}{0.85} = 16.47$$

$$\pi_5 = \frac{\pi_3 + h_{-10}}{a_{-10}} = \frac{10 - 2.5}{1.25} = 6.0$$

$$\pi_4 = \frac{1 + (0.8)(-3) + (0.8)(1)(6)}{(0.8)(1)(1.5) - 1} = 17$$

$$\pi_6 = \frac{17 + 1}{0.8} = 22.5$$

$$\pi_9 = \frac{22.5 - 3}{1} = 19.5$$

$$\pi_8 = \frac{19.5 + 10}{0.8} = 36.875$$

$$\pi_7 = \frac{36.875 - 2}{1} = 34.875$$

$$\pi_{10} = \frac{19.5 + 2}{1} = 21.5$$

9.8 HISTORICAL PERSPECTIVE

We reserve until Chapter 10 consideration of the literature related to generalized network flow problems.

EXERCISES

1. For the network pictured below, draw the transformed network and write the linear programming model in the form described by Sections 9.2 and 9.3.

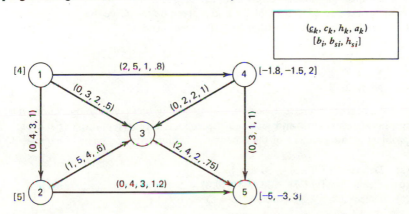

2. (a) Evaluate $|\mathbf{B}|$ for each of the spanning trees of Figure 9.5.
   (b) Verify that $|\mathbf{B}| = a_3 \cdot a_5 \cdot a_6 \cdot a_7 - a_1 \cdot a_2 \cdot a_3 \cdot a_4 \cdot a_5 \cdot a_6 \cdot a_7$ is correct for basis of Figure 9.7.

3. For the basis presented below:

   (a) Compute $|\mathbf{B}|$ for the subnetwork containing the cycle.
   (b) Construct the triple label pointer representation of the basis that would be constructed by TREINT.

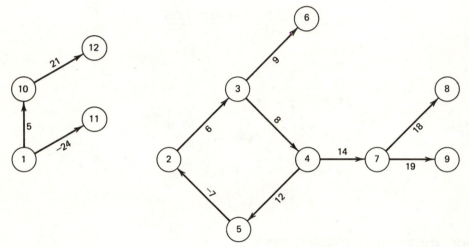

4. Give an alternative triple pointer representation for each of Figures 9.12 and 9.13.
5. For the basis of exercise 3, give the trails for the following nodes.

   (a) Node 9.
   (b) Node 3.
   (c) Node 12.

6. For the basis of exercise 3, give the list of nodes and the list of arcs contained in subnetworks rooted at each of the nodes given below. Give the lists in the order that they would appear if algorithm ROOTG were used.

   (a) Node 5.
   (b) Node 7.
   (c) Node 2.

7. For the basis of exercise 3, perform the following basis change operations. Draw the resulting subnetwork and give the triple pointer representation. (Each part of the problem is independent of the others and begins with the basis of exercise 3.)

   (a) Delete arc 12(4,5). Add arc 4(1,5).
   (b) Delete arc 14(4,7). Add arc 17(6,8).
   (c) Delete arc 6(2,3). Add arc 13(4,6).

8. In regard to basis subnetworks in generalized minimum cost flow problems of the form described in Chapter 9, why must there never be a basis component with more than 1 cycle? (i.e. why must node j_E always be an element of the tree rooted at node J_L?)

9.

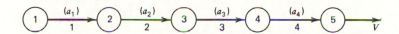

For the path presented above:

(a) If $a_1 = 2.5$, $a_2 = 0.6$, $a_3 = 0.8$, and $a_4 = 1.2$, determine the flows in each arc if $V = 5$.

(b) For the same arc gains as part (a), determine the maximum flow in the path if $c_1 = 3$, $c_2 = 9$, $c_3 = 6$, and $c_4 = 7$. Also, obtain the flows in the arcs when this maximum flow is present.

10.

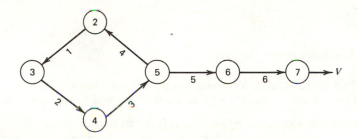

For the subnetwork above:

(a) If $a_1 = 1.5$, $a_2 = 2.5$, $a_3 = 1$, $a_4 = 0.5$, $a_5 = 0.8$, $a_6 = 1.5$, and $V = 4$, determine the flows on each arc.

(b) If the arc gains are as in part (a), determine the maximum flow possible leaving node 7 if $c_1 = 6$, $c_2 = 8$, $c_3 = 7$, $c_4 = 12$, $c_5 = 16$, and $c_6 = 18$. Also determine the flows in the arcs under the maximum flow conditions.

11. Find the flow on each arc if $V = 5$.

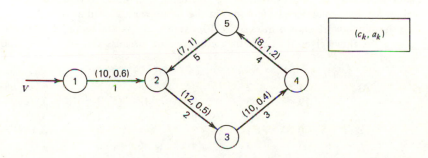

12. Draw the associated expanded network for the original network given below.

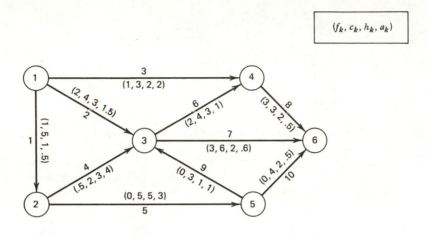

(f_k, c_k, h_k, a_k)

13. For the network of exercise 12:

   (a) Find the maximal flow increase through the path formed by arcs 1, 5, 7, and 9.
   (b) Find the maximal flow augmentation over the arcs 1, −2, 4, 5, −8, and 10.

14. Verify that the expressions in equation (27) of this chapter are correct.
15. For the data given in the table below compute the Π values for the basis of Figure 9.16.

Arc	2	7	10	13	14	16	19	20	21
h_k	8	4	2	4	−3	2	6	2	10
a_k	0.8	0.85	0.8	0.75	1	1	1.5	1	0.8

16. In the basis of Figure 9.12, verify that the choice of a node on the cycle may be made arbitrarily by first evaluating π_6 and using that value to obtain the same π_i values for the remainder of the nodes in the subnetwork rooted at the cycle.
17. For the subnetwork shown in exercise 10, assume the parameters and flows shown below. The flows given do not satisfy conservation because this is part of a larger network.

Arc	1	2	3	4	5	6
c_k	10	6	8	12	4	8
h_k	1	2	5	−2	1	−3
a_k	−0.2	1	−3	−1	2	−0.4
f_k	5	3	4	6	3	4

(a) Find the new flows if the flow out of node 7 is increased by 1.
(b) Find the maximum flow increase that can be obtained at node 7 and the corresponding arc flows.
(c) Find the maximum flow decrease that can be obtained at node 7 and the corresponding arc flows.
(d) Compute the π values for the nodes of the subnetwork, assuming it is part of a basis.

CHAPTER 10
GENERALIZED MINIMUM COST FLOW PROBLEMS

In this chapter, we present two methods for solving the generalized minimum cost flow problem, a flow augmentation approach and a primal approach. Because the first of these techniques makes extensive use of results from the generalized shortest path problem, we will discuss the generalized shortest path problem first.

10.1 THE GENERALIZED SHORTEST PATH PROBLEM

In this section, we define and solve the shortest path problem for the network with gains. The algorithm obtained will have usefulness for the general minimum cost flow problem algorithm of Section 10.6. The decision not to devote an entire chapter to the generalized shortest path problem was due primarily to the fact that it has little practical interest except as an adjunct to the generalized minimum cost flow problem. Our problem here is to find, for a selected node t, the minimum cost path through which flow can be augmented to obtain one unit of flow at the node t. This problem can be written as follows.

$$\text{Min.} \sum_{k=1}^{m} h_k f_k$$

$$\text{st.} - \sum_{k \in M_{Ti}} a_k f_k + \sum_{k \in M_{Oi}} f_k = 0 \qquad \begin{cases} i \in N \\ i \neq s \\ i \neq t \end{cases} \tag{1}$$

$$- \sum_{k \in M_{Ti}} a_k f_k + \sum_{k \in M_{Oi}} f_k = -1$$

$$f_k \geq 0 \qquad k \in M$$

The similarity of this problem statement to that of the shortest path problem of Chapter 5 earns it the name of the *generalized shortest path problem*. In this section we assume that all gain factors are positive.

Assigning the dual values π_i to each node i, we write the dual of this problem as

$$\text{Min.}\left[-\pi_t\right]$$
$$\text{st.} \quad \pi_i - a_k \pi_j \geqslant -h_k \qquad \text{for each arc } k(i,j) \in M \tag{2}$$
$$\pi_s = 0$$

From linear programming theory, we know that the solution to the primal problem will have at most $n-1$ nonzero arc flows. Complementary slackness conditions for an optimal solution of this problem are

(a) $\pi_i - a_k \pi_j = -h_k \qquad \text{for } f_k > 0.$
(b) $f_k = 0 \qquad \text{for } \pi_i - a_k \pi_j > -h_k.$

These conditions can be rewritten in a more intuitive fashion.

(a) If arc $k(i,j)$ is included in the shortest path basis, then $(\pi_i + h_k)/a_k = \pi_j$.
(b) If arc $k(i,j)$ is not included in the shortest path basis, then $(\pi_i + h_k)/a_k \geqslant \pi_j$.

For this problem, the value of π_i represents the cost of supplying one unit of flow to node i. The cost of bringing one unit of flow to node i and beginning its passage through arc $k(i,j)$ is

$$\pi_i + h_k$$

As the flow passes through the arc, the single unit of flow is multiplied by the gain factor a_k; thus the flow reaching node j is a_k. In order to obtain one unit of flow at node j, exactly $1/a_k$ units of flow must be delivered to node i and begin passage over arc k. The cost for this amount of flow is

$$\frac{(\pi_i + h_k)}{a_k}$$

Clearly this is also the cost of obtaining a single unit of flow at node j; that is

$$\pi_j = \frac{(\pi_i + h_k)}{a_k}$$

If an arc $k(i,j)$ that violates condition (b) is not in the basis, it should replace the basis arc currently terminating at node j. We recall that solution techniques for the pure shortest path problem could be partitioned according to whether all arc costs were positive. A similar though somewhat more complex division can be made for generalized shortest path problems. We will consider two cases.

1. When all arc costs are positive and all arc gains are less than or equal to one.
2. When any arc cost is negative or when any arc gain is greater than one.

In the following sections, we will present three algorithms.

1. Algorithm DSHRTG is very similar to Dijkstra's algorithm, DSHORT, and works with case 1 above.
2. Algorithm PSHRTG is a primal algorithm similar to PSHORT for the pure problem and works with case 2.
3. Algorithm DULSPG is similar to DULSPM and is used to update shortest paths for the flow augmentation approach to the generalized minimum cost flow problem.

10.2 ALL ARC COSTS POSITIVE—ALL GAINS LESS THAN OR EQUAL TO ONE

If all costs are positive and all arcs have gains less than or equal to one, Dijkstra's (DSHORT) algorithm of Chapter 5 is applicable if one slight change in the FORWARD portion of the algorithm is made. We need only change the pure network condition

$$D := PI(I) + H(K)$$

to the network with gains condition

$$D := (PI(I) + H(K))/A(K)$$

The change is shaded in the DSHRTG algorithm.

DSHRTG ALGORITHM

Purpose: To derive the shortest path from node s to node t in a generalized network when all arc costs are positive and all arc gains are $\leqslant 1$. (Only forward arcs are considered.) If $t = 0$, the algorithm derives the shortest path tree.

1. (INITIAL) Set $\pi_i = R$ (R some large number) for all nodes except s. Set all back pointers equal to zero. Define the node set $S = \{s\}$. Let $i := s$. Let $\pi_s := 0$.
2. (FORWARD) For each forward arc originating at node $i, k(i, j)$, such that $j \notin S$, calculate the length of the path

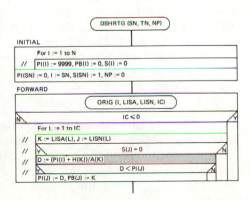

to node j through node $i, (\pi_i + h_k)/a_k$. If this value is less than π_j, replace π_j by $(\pi_i + h_k)/a_k$. Replace the back pointer for j with k.

3. (SELECT) Find $D = \mathrm{Min}_{j \in N - S}[\pi_j]$. Let i_E be the node for which the minimum is obtained. IFIN is set to zero if any node is not yet a member of S.

4. (ADD) If $D < R$, include i_E in S. If $i_E = t$, stop with the shortest path from s to t. If $i_E \neq t$, let $i := i_E$ and go to step 2. If $D = R$ and all the nodes are in the set S, stop with the shortest path tree. If all the nodes are not in S, set NP equal to 1 to indicate that there is no path from s to t.

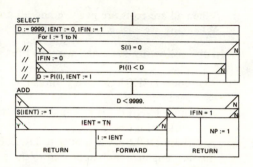

When all gains are less than or equal to 1, there can be no flow-generating cycle. This statement arises from the fact that no cycle gain can exceed a value of 1 when only forward arcs are admissible and all arc gains are less than or equal to 1. Also since all costs are positive and all gains are no more than one, it can be shown that, once a node is added to the set S by the Dijkstra procedure, there is no shorter path through nodes not yet in the set S.

Figures 10.1a and 10.1b illustrate an example of Dijkstra's algorithm applied to the generalized problem. Figure 10.1a shows the original network. In this example SN and TN have been set equal to 1 and 10, respectively. Adding node 1 to the set S, we set π_1 to zero and assign tentative node potential values to the terminal nodes of arcs 1, 2, and 3. This yields $\pi_2 = 20$, $\pi_3 = 10$, and $\pi_4 = 11.76$. Selecting the minimal π_i, for $i \notin S$, we add node 3 to S.

Moving to node 3, we note that arcs 6, 7, and 8 originate at node 3 and terminate at nodes 6, 2, and 4, respectively. Thus, π_6 is tentatively set to 13.3, π_2's old value of 20 is updated to the lesser value of 16.47, and π_4 remains at a value of 11.76, because a new value—$(\pi_3 + h_8)/a_8 = 33.85$—is larger.

Continuing in this fashion, node 10 is added to the set S with the addition of the sixth arc, arc 20, to the tree. The final optimal solution is given in Figure 10.1b where the solid arcs are numbered in the order they were added to the tree. The dotted lines are temporary back pointers found in step 2 of the algorithm.

Note that we have not manipulated flows with this algorithm. Rather we have used the primal–dual optimality conditions to directly determine the π values for each node. The path from source to sink can be found using the back pointers and starting from the sink. The flow solution to the problem could be found by successively calling subroutines MFLOG and FLOWG.

DSHRTG actually determines the shortest path tree if the algorithm is allowed to proceed until all nodes have been labeled.

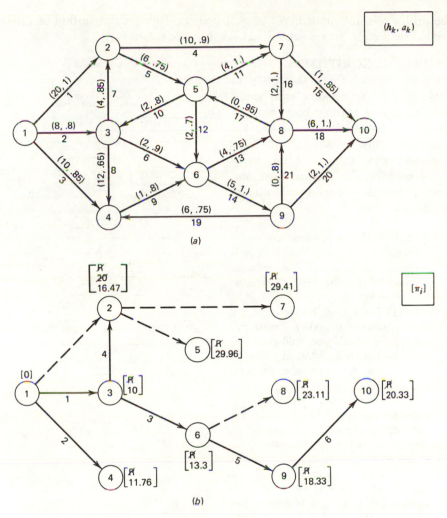

Figure 10.1
Example of Dijkstra's Algorithm Applied to the Generalized Network

10.3 NEGATIVE ARC COSTS—ARC GAINS GREATER THAN ONE

When either one or both of the following two conditions exist, Dijkstra's algorithm will not necessarily provide an optimal solution.

1.   One or more arc costs are negative.
2.   One or more arc gains exceed the value one.

Here we describe a primal algorithm, PSHRTG, which can start with any feasible basis. It will either converge to the optimum or give an indication that the problem is unbounded. We suggest that the Dijkstra's algorithm of the last section be used to provide an initial basic feasible solution.

PSHRTG ALGORITHM

Purpose: To solve the shortest path problem for the network with gains. Starting with a feasible basis network and node potentials that violate the dual feasibility condition for some nonbasic arcs, this algorithm iteratively modifies the basis network and node potentials to obtain the optimum solution.

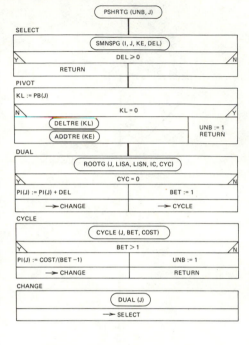

1. (SELECT) Find a nonbasic arc $k_E(i,j)$ for which $(\pi_i + h_{k_E})/a_{k_E} < \pi_j$. If there is no such arc, stop with the optimum solution. Otherwise, let $d_E = (\pi_i + h_{k_E})/a_{k_E} - \pi_j$.

2. (PIVOT) Let $k_L(i',j)$ be the basic arc that terminates at node j. Delete this arc from the basis tree. Add arc k_E to the basis tree. If $k_L = 0$, then j must be the source node. This implies the solution is unbounded.

3. (DUAL) Find the nodes in the network rooted at node j. If the network is a tree, first increment π_j, then go to step 5. If the network includes a cycle, go to step 4.

4. (CYCLE) Find the cycle gain β and new π_j. If β is <1, the solution is unbounded. If the gain is greater than one, go to step 5.

5. (CHANGE) Modify the dual variables in the tree rooted at node j.

With an initial basis tree defined, one need only satisfy the dual feasibility conditions to assure an optimum. The general procedure used by PSHRTG follows the scheme of the PSHORT algorithm of Chapter 5 for the pure problem; however, some variations are required due to the presence of the gain parameters. The algorithm follows the following rough outline.

1. Start with an initial basis network.
2. Go through the set of arcs and find an arc k_E that violates dual feasibility; that is,

$$(\pi_{i_E} + h_{k_E})/a_{k_E} < \pi_{j_E}$$

Call this arc the *entering arc*. If there are no such arcs, stop with the optimum solution.

3. Delete the basis arc that currently terminates at node j_E. Call this the *leaving arc* k_L. Add k_E to the basis.
4. Change the dual variables to account for the addition of k_E and the deletion of k_L. Return to step 2.

Step 2, above, is accomplished for the purposes of this chapter by Algorithm SMNSPG. Algorithm SMNSPG is very similar to Algorithm SMNSP of Chapter 5. The single difference between the algorithms is shaded in the coding of SMNSPG. The kinds of answers PSHRTG obtains are quite different from those of PSHORT due to the possibility of flow-generating cycles in the solution.

SMNSPG ALGORITHM

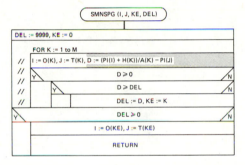

Purpose: To find the arc with the most negative value of $(\pi_i + h_k)/a_k - \pi_j$.

1. Let $D = 9999$ and $k_E = 0$.
2. For each arc $k(i,j)$, calculate $d_k = (\pi_i + h_k)/a_k - \pi_j$. If $d_k \geqslant 0$, go on to the next arc. If $d_k < 0$ and $d_k < D$, let $k_E = k$, and $D = d_k$ and go on to the next arc.
3. After completing arc scan, if search has found a $d_k < 0$, let $i = o_{k_E}$ and $j = t_{k_E}$ and return.

One of the primary differences is that the shortest path to provide one unit of flow to the sink may not originate at the source, but rather at a flow-generating cycle. Clearly, at optimality, the shortest path will be completely contained within a single component of the basis. Also the method for determining the dual variables, in step 4 above, requires some modifications due to the presence of the gain parameters.

To illustrate the selection process, consider the example of Figure 10.2*a*. Two nonbasic arcs are shown with dashed arrows. The dual feasibility condition

$$(\pi_i + h_k)/a_k \geqslant \pi_j \qquad \text{for nonbasic arcs } k(i,j)$$

can be stated in the alternative form.

$$d_k = (\pi_i + h_k)/a_k - \pi_j \geqslant 0$$

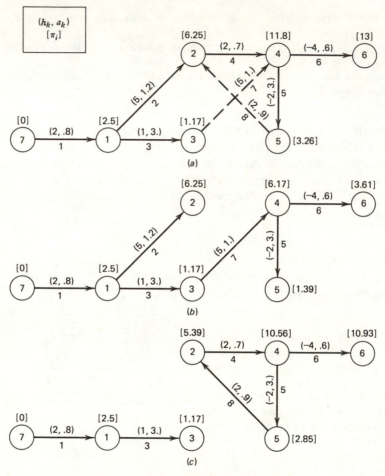

Figure 10.2
Illustrative Basis Changes for the Primal Shortest Path Algorithm

Computing this quantity for arcs 7(3,4) and 8(5,2) we obtain

$$d_7 = \frac{(\pi_3 + h_7)}{a_7} - \pi_4 = (1.17 + 5)/1 - 11.8 = -5.63$$

$$d_8 = \frac{(\pi_5 + h_8)}{a_8} - \pi_2 = (3.26 + 2)/0.9 - 6.25 = -0.406$$

Therefore both arcs do not satisfy the dual conditions and could enter the basis.

First, let arc 7 enter the basis, which forces arc 4 to be the leaving arc. Changing the basis yields Figure 10.2b. The dual variables are changed in the tree rooted at node 4, j_E.

Using primes to indicate the new dual variables, the dual variables are updated using the methods of algorithm DUAL of Chapter 9 as follows.

$$\pi_4' = \frac{(\pi_3 + h_7)}{a_7} = \pi_4 + d_7 = 11.8 - 5.63 = 6.17$$

$$\pi_5' = \frac{(\pi_4' + h_5)}{a_5} = 1.39$$

$$\pi_6' = \frac{(\pi_4' + h_6)}{a_6} = 3.61$$

Suppose arc 8 were allowed to enter the basis of Figure 10.2a instead of arc 7. Arc 2 leaves the basis network and Figure 10.2c results. The basis network now includes a flow-generating cycle. The value of π_2 shown in Figure 10.2c may be computed using the method of algorithm CYCLE in Chapter 9.
Therefore

$$\pi_2 = \frac{h_4 + a_4 h_5 + a_4 a_5 h_8}{a_4 a_5 a_8 - 1} = \frac{2 + (0.7)(-2) + (0.7)(3)(2)}{(0.7)(3)(0.9) - 1}$$

$$\pi_2 = 5.39$$

The other π values are updated using the methods of algorithm DUAL.

Three questions arise with respect to the example of Figure 10.2c.

1. Does $d_k < 0$ indicate that a reduction will be made in π_{j_E} when arc k_E enters the basis?
2. Is there a way to compute the value of π_{j_E} from d_k without the necessity of going through the complex computation above?
3. Is it possible that a flow-absorbing cycle will be created rather than a flow-generating cycle?

We will answer these questions using the more general model of Figure 10.3. The basis network of Figure 10.3 does not include a cycle; however, adding the arc 4 will form a cycle with arcs 1, 2, and 3. The criterion for entering arc 4 is

$$d_4 = \frac{(\pi_4 + h_4)}{a_4} - \pi_1 < 0 \qquad (3)$$

Since arcs 1, 2, and 3 are basic, we may state the following equations.

$$\pi_2 = \frac{(\pi_1 + h_1)}{a_1}$$

$$\pi_3 = \frac{(\pi_2 + h_2)}{a_2} = \frac{\pi_1 + h_1 + a_1 h_2}{a_1 a_2}$$

$$\pi_4 = \frac{(\pi_3 + h_3)}{a_3} = \frac{\pi_1 + h_1 + a_1 h_2 + a_1 a_2 h_3}{a_1 a_2 a_3} \qquad (4)$$

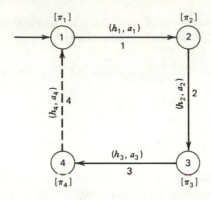

Figure 10.3
Entering Arc that Forms a Cycle

Combining equations (3) and (4) gives

$$d_4 = \frac{\pi_1 + h_1 + a_1 h_2 + a_1 a_2 h_3 + a_1 a_2 a_3 h_4}{a_1 a_2 a_3 a_4} - \pi_1$$

Recognizing that $a_1 a_2 a_3 a_4 = \beta$ and performing some algebraic manipulation yields

$$d_4 = \pi_1 \left(\frac{1}{\beta} - 1 \right) + \left[\frac{h_1 + a_1 h_2 + a_1 a_2 h_3 + a_1 a_2 a_3 h_4}{\beta} \right]$$

or

$$d_4 = \frac{(\beta - 1)}{\beta} \left[\frac{h_1 + a_1 h_2 + a_1 a_2 h_3 + a_1 a_2 a_3 h_4}{(\beta - 1)} - \pi_1 \right] \tag{5}$$

Note that the first term in the large bracket is the new value of π_1, π_1', should arc 4 be added to form the cycle. Hence

$$d_4 = \frac{(\beta - 1)}{\beta} \left[\pi_1' - \pi_1 \right] \tag{6}$$

Note that, with $\beta > 1$, $d_k < 0$ implies

$$\pi_1' - \pi_1 < 0$$

Thus a negative value of d_k does indicate a reduction in the π values of the cycle.

Equation (6) also provides the answer to the second question. With a small manipulation, we find that

$$\pi_1' = \left[\frac{\beta}{\beta-1} \right] d_4 + \pi_1$$

or in general

$$\pi_{j_E}' = \left[\frac{\beta}{\beta-1} \right] d_{k_E} + \pi_{j_E}$$

if k_E causes a cycle to be formed in the new basis.

For the example of Figure 10.2c, $\beta = 1.89$ and we may compute

$$\pi_2' = \frac{(1.89)}{0.89}(-0.406) + 6.25 = 5.39$$

The same value as determined with algorithm CYCLE.

To answer the third question, we note that for a flow-absorbing cycle $\beta < 1$. Reference to equation (5) above indicates that d_k can be negative only if

$$h_1 + a_1 h_2 + a_1 a_2 h_3 + a_1 a_2 a_3 h_4 < (\beta-1)\pi_1 \tag{7}$$

If all arc costs are positive, the left side of equation (7) and the value of π_1 both must be positive. Thus, for this case, the entering arc forming a cycle will always form a flow-generating cycle ($\beta > 1$).

In the case where one or more arc costs are negative, it is possible that the inequality will be satisfied for $\beta < 1$. This situation indicates an unbounded solution to the generalized shortest path problem. In the algorithms, we check the equivalent condition that $d_k < 0$ and $\beta < 1$. In this event, the algorithm terminates with an indication of an unbounded solution. This condition is analogous to that of the pure shortest path problem, where all $a_k \equiv 1$ and $\beta \equiv 1$. Satisfaction of equation (7) above for the pure problems would imply

$$h_1 + h_2 + h_3 + h_4 < 0$$

that is, the existence of a negative cycle.

An example of the application of PSHRTG is given in Figure 10.4. Figure 10.4a presents the tree returned by algorithm DSHRTG for the network of Figure 10.1a when the following changes are made.

$$h_{11} = -4, \qquad h_{17} = -2, \quad \text{and} \quad a_{11} = 2.5$$

Clearly, DSHRTG fails to obtain the optimal solution in the example since arc 15(7, 10) and arc 16(7, 8) each violate the dual feasibility condition with $d_{15} = -10.58$ and

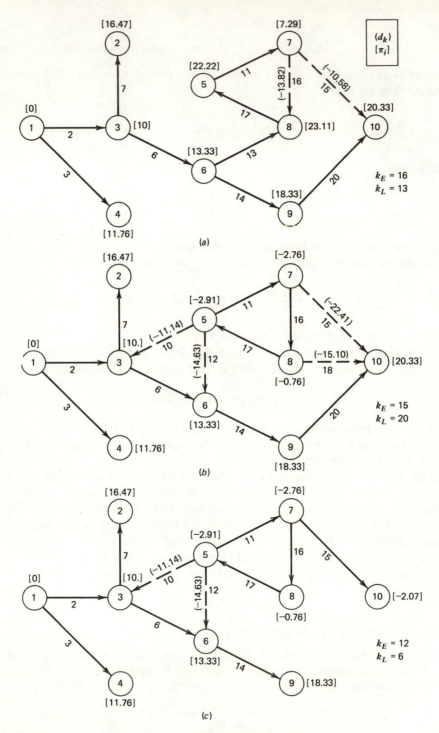

Figure 10.4
Example Application of Algorithm PSHRTG

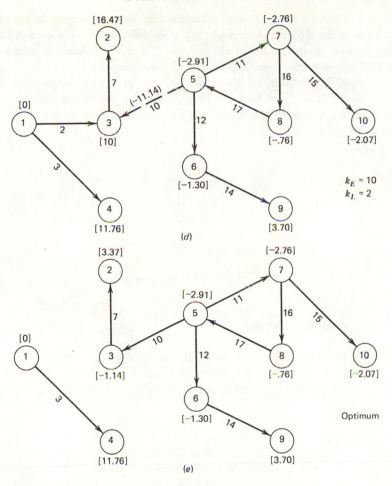

Figure 10.4
(continued)

$d_{16} = -13.82$. Since $d_{16} < d_{15}$, $k_E = 16$ and $k_L = 13$; arc 13 is the basis arc terminating at node 8.

The indicated degenerate change of basis produces the network of Figure 10.4b. Notice that the basis now consists of two components: a flow-generating cycle with $\beta = 2.375$ formed by arcs 11, 16, and 17, and a tree rooted at the source.

Continuing in this fashion, the optimal solution pictured in Figure 10.4e is obtained in three more iterations. Notice the required 1 unit of flow is optimally supplied to the sink at a total cost of -2.07 by the flow generating cycle identified in Figure 10.4b.

10.4   DUAL GENERALIZED SHORTEST PATH ALGORITHM

The previous sections were concerned with determining a basis that solved the generalized shortest path problem for a specified sink node. This section starts with a solution to that problem that was optimal before one or more arcs on the shortest path to the sink were made inadmissible. The problem is to find a new shortest path to the sink consisting of only admissible arcs. This is called a *dual algorithm* because it starts with a dual feasible but primal infeasible solution and iterates to a solution that is both primal and dual feasible.

The principal application of this algorithm is in a flow augmentation approach to the minimum cost flow problem. Thus the algorithm is to be applied to the marginal network that admits both forward and mirror arcs. The following admissibility function is used.

Forward arcs $(k > 0)$

$$AD(k) = \begin{cases} 1 & \text{if } c_k - f_k > 0 \\ 0 & \text{if } c_k - f_k = 0 \end{cases} \tag{8}$$

Mirror arcs $(k < 0)$

$$AD(k) = \begin{cases} 1 & \text{if } f_{-k} > 0 \\ 0 & \text{if } f_{-k} = 0 \end{cases} \tag{9}$$

where f_k is a flow function defined on the original network. A verbal interpretation of the admissibility function is:

> a forward arc is admissible if the flow in the original arc is less than its capacity and a mirror arc is admissible if the flow in the original arc is greater than zero.

The Leaving Arc

Recall that the shortest path to the sink lies entirely in only one component of the basis network. For example, in Figure 9.16 of Chapter 9, the sink, node 10, lies in the component defined by the node set $N = \{4, 6, 7, 8, 9, 10\}$. In Figure 9.17 of Chapter 9, the component consists of all nodes of the network. Deletion of an arc from the shortest path divides the nodes of the network into two sets of N_1 and N_2. Then N_2 is the set of nodes that still remains connected to node t after the deletion of arc k_L. The set N_1 is the complementary set of nodes; that is, $N_1 = N - N_2$. Let D_2 be the basis subnetwork containing N_2 and let D_1 be the basis subnetwork containing N_1. Then D_2 always forms a tree rooted at the terminal node of k_L. Thus D_1 consists of one or more components that always include the source node and any flow-generating cycles that remain after the deletion of arc k_L. Two example basis partitions are illustrated in Figure 10.5. Figure 10.5*a* shows the original basis network and Figures 10.5*b* and 10.5*c* give two examples of results when an arc is deleted from a basis.

With arc k_L deleted, there no longer exists a flow augmentation path to node 10. Thus the current problem is to find an admissible arc k_E to add to the basis

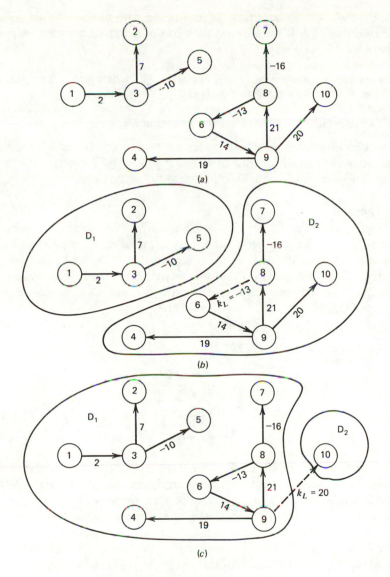

Figure 10.5
Example Basis Partitions

network that will reestablish a flow augmentation path to node 10 while assuring that the resulting path is the shortest path in the marginal network. A path can be reestablished in two ways.

1. If the arc k_E originates in N_1 and terminates in N_2, the new path will originate at the source or at some flow-generating cycle in D_1.
2. If the arc k_E both originates and terminates in N_2, the new path will originate at a new flow-generating cycle that has been formed in D_2.

Figure 10.6 illustrates three new bases that might be formed from the two basis subnetworks pictured in Figure 10.5b. Figure 10.6a illustrates the first case above while Figures 10.6b and 10.6c illustrate the second case.

Node Gains

It is helpful, at this point, to define node again γ_i, which was first encountered in Chapter 9, specifically in terms of the generalized shortest path problem. In this context, γ_i is only defined for the nodes in the set N_2 and is the inverse of the product of the arc gains on the unique path from node i to node t in the network D_2. Note that there is a unique path because D_2 is always a tree. As illustrations from Figure 10.5b, we observe that

$$\gamma_6 = \frac{1}{a_{14}a_{20}}$$

$$\gamma_7 = \frac{1}{a_{16}a_{-21}a_{20}}$$

and

$$\gamma_4 = \frac{1}{a_{-19}a_{20}}$$

(Note the path from 7 to 10 is $M_p = \{-(-16), -21, 20\}$.) As we have seen, the node gain γ_i is a useful quantity. Recall that, if one unit of flow starts at node i and travels through D_2 to t, the quantity of flow arriving at t is $1/\gamma_i$. Alternatively, if one unit of flow is to arrive at t, the quantity γ_i must begin at node i. The node gains also allow the computation of the cycle gain β for a cycle that would be formed by the addition of an arc k to the tree D_2. For example, adding arc $9(4,6)$ to Figure 10.5b forms the cycle $M_c = \{9, 14, 19\}$ with cycle gain $\beta = a_9 a_{14} a_{19}$. Alternatively, note that the value of β could also be determined by

$$\beta = a_9(\gamma_4/\gamma_6) = \frac{a_9 a_{14} a_{20}}{a_{-19}a_{20}}$$

$$= \frac{a_9 a_{14}}{a_{-19}}$$

$$= a_9 a_{14} a_{19}$$

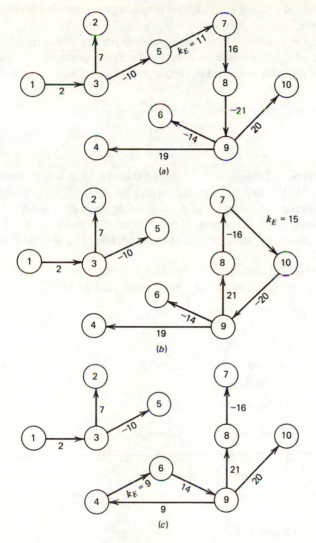

Figure 10.6
Examples of New Bases

In general, adding an arc $k(i,j)$ such that $i \in N_2$ and $j \in N_2$ will always form a cycle with a gain of

$$\beta = a_k \frac{\gamma_i}{\gamma_j} \tag{10}$$

This equation is very useful in the algorithms to be discussed because it allows the computation of cycle gains without requiring an explicit identification of the arcs in the cycle.

The node gains γ_i can be easily computed from the structure of the basis tree and the value of γ_{j_L}, the node gain of the terminal node of the leaving arc k_L. For any arc $k(i,j)$ that is included in the tree D_2, we have

$$\gamma_j = \gamma_i a_k \qquad \text{for } k \in M_2$$
$$i = o(k)$$
$$j = t(k) \tag{11}$$

Node j_L is the root of the tree D_2. Assume that we have listed the arcs of M_2 in predecessor order (i.e., an arc appears in the list after all its predecessors). Now starting with a given value of γ_{j_L} and progressing through the ordered list M_2 we can use equation (11) to compute node gains for each member of N_2. Our algorithm will use algorithm TRAIL to compute γ_{j_L} and ROOTG to obtain the ordered list M_2.

As an example, consider Figure 10.5b with $k_L = -13(8,6)$. Here $j_L = 6$. Assume the value of γ_6 has already been computed as

$$\gamma_6 = \frac{1}{a_{14}a_{20}}$$

The ordered arc set is $M_2 = \{14, 20, 21, -16, 19\}$. From this set we compute

From arc 14: $\gamma_9 = \gamma_6 a_{14} = \dfrac{1}{a_{20}}$

From arc 20: $\gamma_{10} = \gamma_9 a_{20} = 1$

From arc 21: $\gamma_8 = \gamma_9 a_{21} = \dfrac{a_{21}}{a_{20}}$

From arc -16: $\gamma_7 = \gamma_8 a_{-16} = \dfrac{\gamma_8}{a_{16}} = \dfrac{a_{21}}{a_{16}a_{20}}$

From arc 19: $\gamma_4 = \gamma_9 a_{19} = \dfrac{a_{19}}{a_{20}}$

The Entering Arc

Next let us investigate the problem of choosing an admissible arc k_E to enter the basis in order to make π'_t, the new length of the path to node t, as small as possible. The previous basis defines some value of π_t. With the deletion of an arc k_L and the addition of arc k_E, we obtain some new value π'_t. It must be true that $\pi'_t \geqslant \pi_t$ because the previous basis was optimal prior to the time that one or more of its member arcs became inadmissible. When some admissible arc $k(i,j)$ is inserted to form a new basis, its effect on π_t depends on whether k connects D_1 and D_2, or whether both ends of k lie in N_2.

We consider first the case in which $i \in N_1$ and $j \in N_2$ as illustrated in Figure 10.7a. Note that the path formed has three parts: P_1 from the source node or from a flow-generating cycle to node i; P_2 which is the arc $k(i,j)$; and P_3 from node j to node t. The values of π from the previous basic solution may be used in computing the costs on these paths. Then π'_t is the cost to obtain one unit of flow at node t through the path defined by P_1, P_2, and P_3. For one unit of flow at t, the flow at nodes j and i must be respectively γ_j and γ_j/a_k. The cost contributions are then

$$P_3 : \pi_t - \pi_j \gamma_j$$

$$P_2 : \frac{h_k \gamma_j}{a_k}$$

$$P_1 : \frac{\pi_i \gamma_j}{a_k}$$

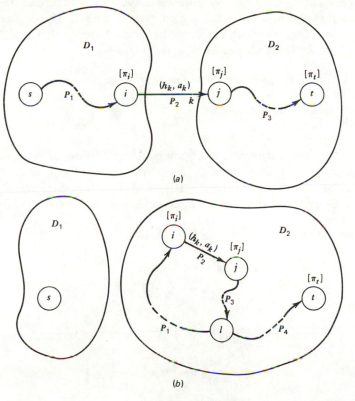

(a)

(b)

Figure 10.7

Paths Formed When an Arc Is Added to a Basis

Adding these three contributions obtains

$$d_k = \pi'_t - \pi_t = \left[\frac{\pi_i + h_k}{a_k} - \pi_j \right] \gamma_j \qquad \text{for } k(i,j) \text{ admissible}$$

$$i \in N_1 \quad \text{and} \quad j \in N_2 \qquad (12)$$

For an admissible mirror arc $-k(j,i)$ such that $j \in N_1$ and $i \in N_2$, we obtain

$$d_{-k} = \left[\frac{(\pi_j + h_{(-k)})}{a_{(-k)}} - \pi_i \right] \gamma_i$$

Substituting

$$h_{(-k)} = \frac{-h_k}{a_k} \qquad a_{(-k)} = \frac{1}{a_k}$$

$$d_{-k} = (\pi_j a_k - h_k - \pi_i)\gamma_i \qquad (13)$$

When $i \in N_2$ and $j \in N_2$, arc $k(i,j)$ forms a cycle as in Figure 10.7b. We must first insure that the new cycle has a gain greater than 1. This is checked with equation (10). Next we must evaluate the cost of the path formed with the addition of arc k. The path has four parts.

P_1: the path from some junction node ℓ to node i.
P_2: the added arc k.
P_3: the path from j to the junction node ℓ.
P_4: the path from ℓ to t.

Again using flow arguments, to obtain one unit of flow at node t requires a flow of γ_ℓ entering P_4 from node ℓ. Therefore, the flow into node ℓ from the circuit leaving P_3 must be $[\beta/(\beta-1)]\gamma_\ell$ and the flow entering P_1 must be $\gamma_\ell/(\beta-1)$. Clearly, the flow out of node j must then be $[\beta/(\beta-1)]\gamma_j$ and the flow out of node i is $[\beta/(\beta-1)](\gamma_j/a_k)$.

With these flows, the cost on the four parts is

$$P_4 = \pi_t - \pi_\ell \gamma_\ell$$

$$P_3 = \frac{\beta}{(\beta-1)} \left[\pi_\ell \gamma_\ell - \pi_j \gamma_j \right]$$

$$P_2 = \frac{\beta}{(\beta-1)} \left(\frac{h_k \gamma_j}{a_k} \right)$$

$$P_1 = \pi_i \frac{\beta}{(\beta-1)} \left[\frac{\gamma_j}{a_k} \right] - \pi_\ell \left[\frac{\gamma_\ell}{(\beta-1)} \right]$$

$$= \frac{\beta}{(\beta-1)} \left[\frac{\pi_i - \pi_\ell \gamma_\ell}{\beta \gamma_i} \right] \frac{\gamma_j}{a_k}$$

Adding these four terms, we obtain

$$d_k = \frac{\beta}{(\beta-1)} \left[\frac{(\pi_i + h_k)}{a_k} - \pi_j \right] \gamma_j \tag{14}$$

where

$$\beta = \frac{a_k \gamma_i}{\gamma_j} \qquad \text{for } k(i,j), i \in N_2, j \in N_2$$

Notice that the value of d_k obtained for arcs that form cycles differs from the value obtained for noncycle arcs only by the factor $(\beta/\beta-1)$.

For admissible mirror arcs $-k(j,i)$ the value becomes

$$d_{-k} = \frac{\beta}{(\beta-1)} (\pi_j a_k - h_k - \pi_i) \gamma_i \tag{15}$$

where

$$\beta = \frac{\gamma_j}{a_k \gamma_i} \qquad \text{for } -k(j,i) \qquad j \in N_2, i \in N_2$$

Since d_k represents the increase in the value of π_t caused by the addition of arc k to the basis, we require a dual iteration to find the arc with the smallest value of d_k. This will be the entering arc and receive the designation k_E. Let M' be the subset of admissible arcs from the expanded network such that

> if $k(i, j) \in M'$ either:
> $i \in N_1$ and $j \in N_2$
> or $i \in N_2$ and $j \in N_2$ and $a_k \gamma_i / \gamma_j > 1$

Then the entering arc is determined by

$$d_{k_E} = \frac{\text{Min}(d_k)}{k \in M'}$$

It is clear that d_{k_E} is nonnegative because the process begins with a dual feasible solution.

Statement of the Algorithm

The algorithm implementing these operations is DULSPG. An overview of the algorithm is as follows.

1. Start with an initial dual feasible solution with a basis tree defined.
2. Find the inadmissible arc on the shortest path to the sink that is furthest removed from the sink node. Let this arc be k_L. If there are none, stop with the optimal solution.
3. Identify the set N_2 and compute γ_i for each member of this set.
4. Go through the set of all admissible arcs that terminate in N_2 and choose the one with the smallest value of d_k. Let this arc be k_E. If there are no such arcs, terminate; there is no feasible solution.
5. Change the basis by deleting arc k_L and adding arc k_E. Recompute the π values for all nodes in the set N_2. Return to step 2.

DULSPG ALGORITHM

Purpose: To obtain a new optimal shortest path network when one or more arcs on the trail to the sink node, TN, are made inadmissible. The algorithm begins with a basis network with node potentials satisfying dual feasibility and complementary slackness conditions.

1. (INITIAL) List the arcs in the trail to the sink node. The arcs are listed starting from the sink and progressing backward through the path. Algorithm TRAIL computes the values of γ_i for arcs on the trail.
2. (LEAVE) From the inadmissible arcs on the path to the sink, select the one furthest from the sink (in terms of number of arcs) as the arc that is to leave the basis, k_L. Node j is the terminal node of arc k_L. If there is no inadmissible arc in the trail, the shortest path has been found. Return.
3. (TREE) Find the tree (D_2) rooted at the terminal node of k_L, (i.e., node j). Test if k_L is on a cycle. If not, go to step 4. If k_L is on the cycle, delete the second reference to node j from LISN, and delete k_L from LISA. Adjust the value of $\gamma(j)$ to account for the effects of the cycle. Go to step 4.
4. (NEWGAM) Compute and assign the values of γ_i to the nodes in the set N_2. Nodes in the set are indicated by setting $S(i)=1$ if $i \in N_2$. Initialize DEL and k_E for the search process.

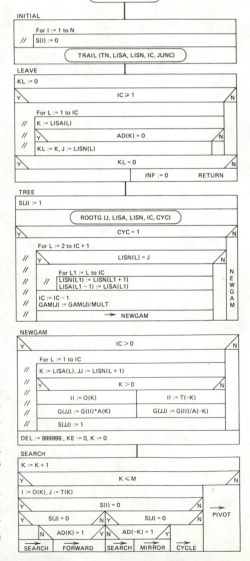

5. (SEARCH) Go through the complete set of arcs. When all the arcs have been considered, go to step 12. If arc k originates in $N_1[S(I) = 0]$ and terminates in $N_2[S(J) = 1]$ test the forward arc for admissibility. If it is admissible, go to step 6; otherwise try the next arc. If arc k originates in N_2 and terminates in N_1, test the mirror arc for admissibility. If it is admissible, go to step 7; otherwise try the next arc. If arc k both originates and terminates in N_2, go to step 8.

6. (FORWARD) Evaluate d_k for the forward arc and go to step 11.

7. (MIRROR) Evaluate d_k for the mirror arc and go to step 11.

8. (CYCLE) Compute the gain of the cycle caused by inserting the forward arc. If it is 1, try the next arc. If the cycle gain is > 1, test the forward arc for admissibility. If it is admissible, go to step 9. Otherwise try the next arc. If the gain is less than 1, test the mirror arc for admissibility. If it is admissible, invert the cycle gain and go to step 10. Otherwise try the next arc.

9. (FORCYC) Compute d_k for the forward arc that forms the flow-generating cycle and go to step 11.

10. (MIRCYC) Compute d_k for the mirror arc that forms the flow-generating cycle and go to step 11.

11. (COMPARE) Test d_k against the *smallest value* found to this point. If d_k is smaller, call this arc the *entering arc* k_E. Try the next arc.

12. (PIVOT) Test to see if an entering arc has been found. If not, there is no path of admissible arcs to the sink. Otherwise change the basis network by deleting k_L and adding k_E. Compute the π values for the new basis network. Return to step 1 to see if the new path has all admissible arcs.

FORWARD

$D := (((PI(I) + H(K))/A(K)) - PI(J))*G(J)$
$KK := K$

→ COMPARE

MIRROR

$D := (PI(J)*A(K) - H(K) - PI(I))*G(I)$
$KK := -K$

→ COMPARE

CYCLE

$BET := A(K)*G(I)/G(J)$

	BET = 1		N
Y	BET > 1		N
Y			
N	AD(K) = 1	Y N	AD(−K) = 1	Y
			BET := 1/BET
SEARCH	FORCYC	SEARCH	→ MIRCYC

FORCYC

$D := (BET/(BET − 1))(((PI(I) − H(K))/A(K)) − PI(J))*G(J)$
$KK := K$

→ COMPARE

MIRCYC

$D := (BET/(BET − 1))(PI(J)*A(K) − H(K) − PI(I))*G(I)$
$KK := -K$

→ COMPARE

COMPARE

| Y | D < DEL | N |
| $DEL := D, KE := KK$ | | |

→ SEARCH

PIVOT

N	KE = 0		Y
	TRECHG (KL, KE)		INF := 1
	For L := 1 to IC + 1		RETURN
//	I := LISN (L)		
//	PI(I) := PI(I) + DEL/G(I)		

→ INITIAL

It is possible that the algorithm may have to pass through step 2 more than once to find an optimum solution. The basis change operation of step 5 may cause the reversal of some arcs, causing inadmissible arcs to enter the solution. The method used for the selection of k_L (furthest removed from the sink) guarantees that the set N_2 will be reduced by at least one node at each iteration. Thus, DULSPG will terminate in at most n iterations.

Figure 10.8 gives an illustration of the use of algorithm DULSPG. Figure 10.8a gives the optimal basis spanning tree that results from applying DSHRTG to the network of

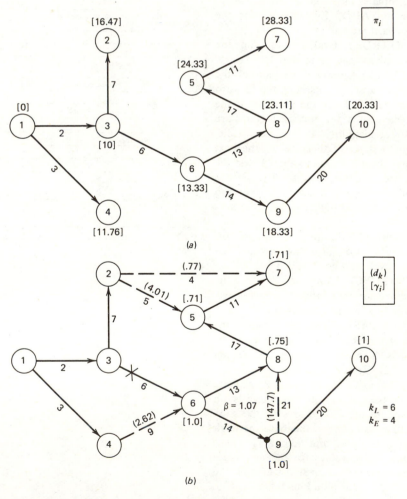

(a)

(b)

Figure 10.8
Example Application of DULSPG

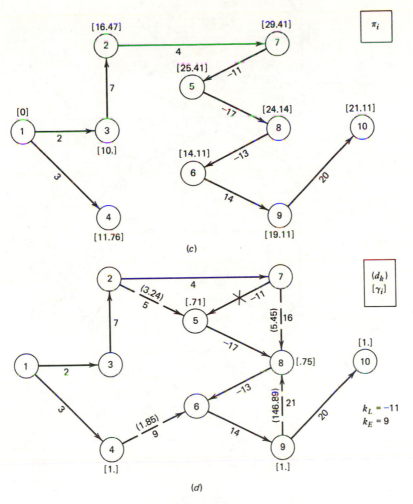

Figure 10.8
(continued)

Figure 10.1 with TN set to a value of zero. At this point, let us suppose that arc $6(3,6)$ and arc $20(9,10)$ are declared inadmissible. In the context of the generalized shortest path problem, all other forward arcs remain admissible, while all mirror arcs are inadmissible.

Applying DULSPG to Figure 10.8*a* yields Figure 10.8*b*. TRAIL has identified arcs 20, 14, 6, and 2 as the set of arcs through which flow is provided to node 10 at the least cost. Since arc 6 is the furthest inadmissible arc from node 10, it is designated as the *leaving arc*; that is, $k_L = 6$. This designation implies that $N_2 = [6, 9, 10, 8, 5, 7]$ and a new γ_i is computed for each member of N_2.

Next, the arcs are searched to determine the entering arc k_E that will add the least to the cost of providing one unit of flow to node 10. The candidate arcs are indicated by dotted lines in Figure 10.8*b*. Since arc $4(2,7)$ has $d_4 = 0.77$, which is the smallest d_k of those arcs, $k_E = 4$. Notice that candidate arc $21(9,8)$, joined with arcs 13 and 14, forms a

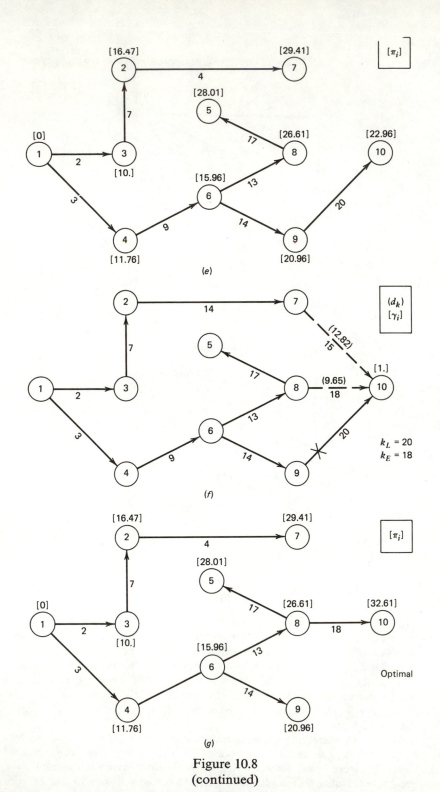

Figure 10.8
(continued)

flow-generating cycle with $\beta = 1.07$. However, arc 21 is not selected because $d_{21} = 147.7$.

Upon performing the change of basis, we obtain Figure 10.8c. Notice that the introduction of arc 4 has caused three mirror arcs to be present in the basis. Since each of these are inadmissible, they must be removed. Since arc -11 is furthest removed from node 10, $k_L = -11$, and $N_2 = [5, 8, 6, 9, 10]$.

Examination of candidate arcs yields $k_E = 9$ with $d_9 = 1.85$. The change of basis yields Figure 10.8e and fortuitously all mirror arcs in the basis during the previous step have been reversed. Thus, it remains only to replace inadmissible arc 20 with arc 18 and the optimal solution presented in Figure 10.8g is obtained.

One portion of the algorithm requires additional explanation. When k_L lies on a cycle, as does arc $-13(8, 6)$ in Figure 10.5b, subroutine TRAIL does not compute the proper value of γ_{j_L}. Because the arc is on the cycle, the quantity computed by TRAIL is

$$\left(\frac{\beta}{\beta - 1} \right) \gamma_{j_L}$$

The proper value is obtained by dividing this quantity by $(\beta / \beta - 1)$. This is accomplished in the TREE portion of DULSPG. The position of k_L on a cycle also affects the lists LISA and LISN obtained by subroutine ROOTG. For the example of Figure 10.56 (see Figure 4.16, Chapter 4 for the three-pointer representation), we obtain

$$\text{LISN} = [6, 9, 10, 8, 6, 7, 4]$$

$$\text{LISA} = [14, 20, 21, -13, -16, 19]$$

$$\text{IC} = 6 \quad \text{and} \quad \text{CYC} = 1$$

In order to proceed with the computation of the node gains, we must delete the second reference to j_L in LISN and the reference to k_L in LISA. For the example, we then obtain

$$\text{LISN} = [6, 9, 10, 8, 7, 4]$$

$$\text{LISA} = [14, 20, 21, -16, 19]$$

$$\text{IC} = 5$$

Again these modifications are accomplished in the TREE portion of the algorithm.

10.5 GENERALIZED MINIMUM COST FLOW ALGORITHMS

In the following sections, we use extensively the algorithms developed in the earlier part of this chapter and in Chapter 9 to solve the minimum cost flow problem for the network with gains. Conceptually, the network with gains problem is, in many ways, far removed from the pure network problem. The algorithms we describe below, however, are quite similar to those developed for the pure problem.

Two approaches to the minimum cost problem will be presented. First, we will consider a dual flow augmentation approach, which is an application of the generalized shortest path algorithms of the last section. The main values of this approach are its conceptual simplicity and the fact that it does not require an initial primal feasible solution.

Finally, we will describe a primal approach. This technique requires an initial basic feasible solution and uses an iterative primal procedure to achieve an optimal solution.

10.6   THE FLOW AUGMENTATION APPROACH

Given a network with arcs having capacity, cost, and gain parameters with source and sink nodes defined, and with a required quantity of flow specified for the sink, our problem is to determine the arc flows with the minimum total cost. The flows must satisfy conservation of flow except at the source and the sink. While net flow into the sink must equal the required value, the source may supply any amount of flow. In our usual format, the problem is

$$\text{Min.} \sum_{k=1}^{m} h_k f_k \tag{16}$$

$$\text{st.} - \sum_{k \in M_{Ti}} a_k f_k + \sum_{k \in M_{Oi}} f_k = 0 \qquad i = 1, \dots, n; \; i \neq s, t \tag{17}$$

$$- \sum_{k \in M_{Tt}} a_k f_k + \sum_{k \in M_{Ot}} f_k = - F_t \tag{18}$$

$$0 \leqslant f_k \leqslant c_k \qquad k \in M \tag{19}$$

Here F_t is a given amount of output flow that is to be obtained. We require in this section that the gain factors be positive.

An assignment of flows to the arcs that satisfies equations (17) and (19) will be called F. An F that satisfies equation (16) for some value of flow at the sink less than F_t will be called an *intermediate optimum*. The flow F that satisfies equation (16) to (19) is called the *optimum*.

The general approach taken to obtain a solution will be similar to that of algorithm INCREME for the pure network problem. The procedure begins with an intermediate optimum F_0 for some value of output flow at node t. Next, a shortest path network is constructed, D_s^0, that determines for each node the minimum cost per unit of additional flow to the node and the path over which that flow may be obtained. The output flow is increased in D_s^0 along the minimum cost trail defined for the sink until one or more arcs in the path become saturated. This flow augments F_0 to obtain F_1. Then F_1 is also an intermediate optimum. A new augmenting network is constructed, D_s^1, and the output flow is again augmented to obtain F_2. This process continues iteratively until the desired output flow or the maximum flow is obtained. At every step F_k is an intermediate optimum; hence, at termination the flow pattern must be the optimum or the maximum flow through the network if F_t is greater than the maximum flow.

The shortest path algorithms of the last section are used to find the shortest path networks at each step. For the first iteration, DSHRTG followed by PSHRTG obtains the shortest path. Augmenting the flow through this path causes one or more of the arcs in the path to become inadmissible. Then DULSPG is used to establish a new shortest path in the marginal network and the process continues.

Throughout the algorithm, because the values of the π_i are determined by a shortest path algorithm, the solution will be dual feasible. The solution is always basic in that at most $n-1$ arcs will have flows strictly between their bounds, and these arcs will form a basis network as previously described. The solution will satisfy conservation of flow and complementary slackness. The one requirement that is violated until the final iteration is the requirement for flow at the sink. The sink flow starts at zero and iteratively increases until it finally reaches its required value. At that point, the algorithm stops with the optimum solution.

An important benefit of this technique is that in most cases it is very easy to obtain an initial flow solution. When the network satisfies the conditions for boundedness of the shortest path problem, an initial flow of zero is acceptable. This is similar to the no negative cycle requirement for the pure network problem. We assume in the remainder of this section that a zero flow is an intermediate optimum.

For an example of the algorithm, consider the network of Figure 10.9a. The required flow at the sink for this example is $F_t = 10$. Initially, the flows are taken to be zero. Figure 10.9b shows D_s^0 determined for the initial flows using DSHRTG and PSHRTG. The augmenting path to the sink is used to increment the flows as shown in Figure 10.9c. These flows cause arc 6 to become saturated. Figure 10.9d shows an intermediate stage in the DSHRTG algorithm. Note the entering arc is 4. This network yields a degenerate iteration because the path to node 10 includes inadmissible arcs; specifically, arcs -11, -17, and -13 are inadmissible. They are the result of the basis change operation when arc 6 leaves the basis and arc 4 enters. The next iteration of DULSPG causes arc -11 to leave and the resultant path includes all admissible arcs. Subsequent shortest path networks are shown in Figures 10.9e, 10.9g, 10.9i, 10.9k, and 10.9m and the corresponding flows in the original network are shown in Figures 10.9f, 10.9h, 10.9j, and 10.9l. Figure 10.9n shows the optimum flows. Note in these figures the potential of the sink steadily increases as the algorithm progresses. An interesting augmenting network appears in Figure 10.9k. This was formed from Figure 10.9i by the removal of arc 3 and the addition of arc -6. The resultant network includes a flow-generating cycle.

Algorithm INCREMG implements the flow augmentation approach to the generalized minimum cost problem.

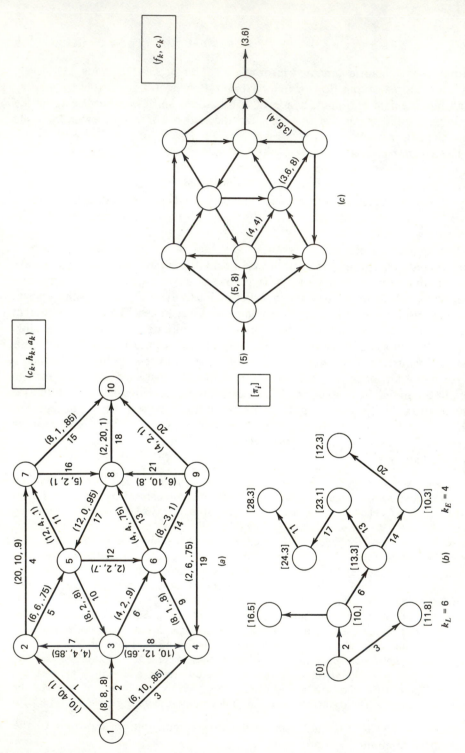

(f_k, c_k)

(c_k, h_k, a_k)

$[\pi_i]$

$k_L = 6$ $k_E = 4$

(a)

(b)

(c)

Figure 10.9
Example Application of INCREMG

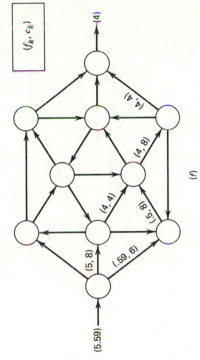

$[\pi_i]$

(d)

(e)

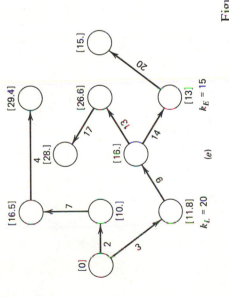

(f_k, c_k)

(f)

Figure 10.9
(continued)

311

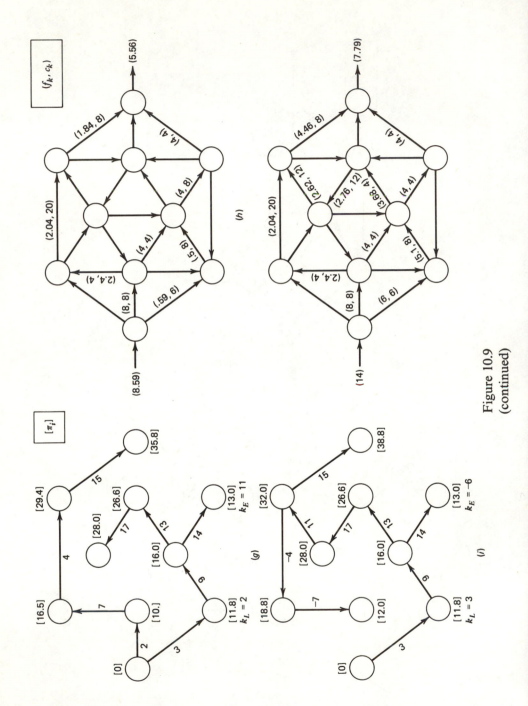

(f_k, c_k)

$[\pi_i]$

Figure 10.9
(continued)

312

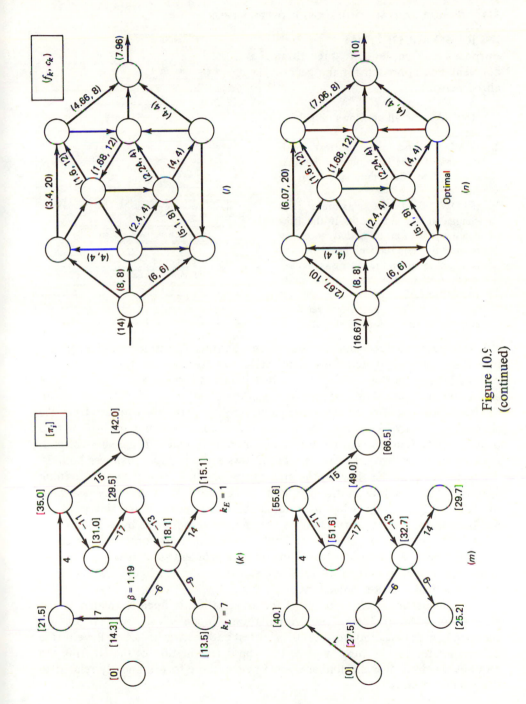

Figure 10.9
(continued)

313

INCREMG ALGORITHM

Purpose: To implement the flow augmentation approach for the generalized network.

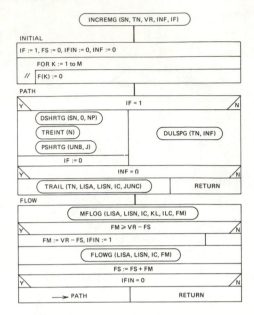

1. (INITIAL) Set all arc flows initially to zero.
2. (PATH) Find the shortest path network. If there is no path to the sink, return. Else, find the path into the sink.
3. (FLOW) Find the maximum flow change in the sink path. If the flow change is greater than that necessary to meet the required sink flow, the flow changes to meet the necessary increment. Change the flow in the path. Increment the sink flow. If the required sink flow has been reached, stop. Otherwise, return to step 2.

If there were no degeneracy, it is apparent that this algorithm would terminate in a finite number of iterations. Since arc flows are all initially at their lower bounds (zero), the flows of D_s^0 at the first iteration form a basic solution. The arcs in the basis are those that are in the shortest path network D_s^0. Arcs not in the basis have zero flow. Each subsequent iteration may delete an arc from the shortest path network only when the flow in the corresponding arc in D is zero or at capacity. Thus for all iterations, arcs not represented in the shortest path network by a forward or mirror arc have flows at their upper or lower bounds. At each iteration, the flow to the sink increases; thus, since there is a finite number of basic solutions, the algorithm must be finite. Degeneracy occurs in the DULSPG algorithm when an inadmissible arc appears in the sink path. As has been shown, the number of times this can occur is limited by n.

At each iteration, the simplex optimality condition and constraints of equations (17) and (19) of this chapter of the original problem statement are satisfied. Constraint equation (18), however, is not satisfied until the final iteration. Thus we have a dual solution procedure.

The computational effort required for an iteration is linearly related to the numbers of nodes and arcs. The determination of node gains, flow changes, and the leaving arc is accomplished on the shortest path network; hence, the effort is linearly related to the number of nodes. The determination of the entering arc requires a search over all nonbasic arcs; hence, the effort is linearly related to the number of arcs.

10.7 THE PRIMAL APPROACH

The primal algorithm described here is applicable to the generalized minimum cost flow problem in which the signs of arc costs and arc gains are unrestricted. Although the procedure for finding the initial basic solution is specialized for the network with a single output with fixed external flow, the algorithms for implementing the primal simplex procedure can operate on more general structures with more than one external flow. The only requirement is that some initial basic feasible solution be provided. The algorithm is similar to the primal algorithm for the pure problem and has the following steps:

1. Start with an initial primal basic feasible solution. Assign dual variables to satisfy complementary slackness.
2. Find an arc that violates dual feasibility. Let this be the entering arc. If there are none, stop with the optimum solution.
3. Find the flow-augmenting trail that includes the entering arc and basic arcs. Augment the flow through the trail by as much as possible. Determine the arc to leave the basis.
4. Change the basis by deleting the leaving arc and inserting the entering arc. Modify the dual variables to satisfy complementary slackness for the new basis.

Before stating the algorithm in detail, we specifically consider each step of this rough outline and develop some concepts that arise in connection with this algorithm.

Initial Basic Solution

The particular model considered in this example has a single external flow at the sink node. Thus an artificial starting procedure requires only a single artificial arc from the source node to the sink node. Subroutine ARTIFG assumes that such an arc has been provided in the input data with the index KST. The subroutine sets the flow on this arc to F_t. Setting all other arc flows to zero results an initial basic solution.

It still remains to determine a basis network with associated π_i values that satisfy complementary slackness. We assume for this that a directed path exists from the source node to each node in the network. Thus DSHRTG will give a spanning tree rooted at the source. The tree is defined by back pointers. The algorithm also provides the associated π_i values. The back pointer and π_i value for the sink node must be changed to reflect the fact that the artificial arc is in the initial basis. Calling TREINT provides the triple pointer representation.

In some applications of the primal algorithm, the initial basis is provided by some other previous computation. In these cases, ARTIFG will not be required.

ARTIFG ALGORITHM

Purpose: To provide an initial basic artificial solution for the network-with-gains minimum cost flow problem.

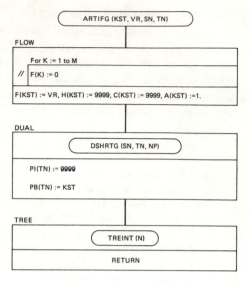

1. (FLOW) Set all arc flows to zero. Set the artificial flow to the required sink flow. Set the cost and the capacity on the artificial arc to very large values.
2. (DUAL) Use DSHRTG to obtain an initial tree and associated π values. Reset the π value and back pointer of the sink node to represent the artificial arc.
3. (TREE) Use TREINT to construct a triple pointer representation of the basis.

Determining the Entering Arc

A candidate for the entering arc is any arc that violates the dual feasibility condition for the network with gains.

$$\pi_i + h_k - a_k \pi_j \geqslant 0 \qquad \text{for arc } k(i,j)$$

Thus

$$\pi_{i_E} + h_{k_E} - a_{k_E} \pi_{j_E} < 0$$

for the entering arc $k_E(i_E, j_E)$.

There are many ways to select the entering arc. However, for our present purposes, we will use algorithm SELECG, which differs from the SELECT algorithm of Chapter 7 by a single change.

$$D := PI(I) + H(K) - A(K)PI(J)$$

replaces

$$D := PI(I) + H(K) - PI(J)$$

in the single place it occurs.

Since the arithmetic for the gains algorithm is performed with "real" numbers rather than with integer numbers, it is important that tests for equality and inequality be made with appropriate concern for round-off errors. For example, in SELECG the test for $D = 0$ should be replaced by

$$-\varepsilon \leqslant D \leqslant \varepsilon$$

where ε is some small quantity. Our flow charts will not show such adjustments, but they should appear in any computer implementation of the algorithm.

The Flow-Augmenting Trail

Once an arc is chosen to enter the basis for the primal algorithm, the flow-augmenting trail must be found so that we may determine the amount of flow change on each arc of the trail for each unit of flow in the entering arc. The entering arc will be in one of six general types of arrangements with respect to the current basis network. These arrangement types are illustrated in Figure 10.10. For simplicity, we discuss Figure 10.10 assuming the gain of the entering arc is positive.

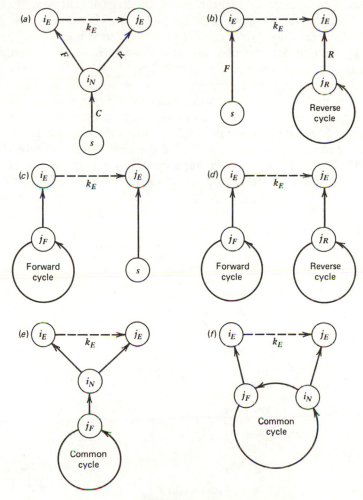

Figure 10.10
Possible Arrangements of Flow-Augmenting Trails

Consider first Figure 10.10a where the entering arc is incident to two nodes on the tree component of the basis network. There is only one tree component that is rooted at the source node s. The tree shown in Figure 10.10a has three parts, the forward part labeled F, the reverse part R, and the common part C. For the pure network problem, an increase in the flow in the entering arc k_E requires equal increases in the arcs in F and equal decreases in the arcs of R. No flow change in part C is required. For the gains problem, however, the flow changes in F and R are not necessarily equal. Hence, there may be a required change of flow in part C. For example, consider Figure 10.11 where an increase of one unit of flow in the entering arc, arc 4, is desired. The corresponding flow changes in the arcs, the Δ_k, are shown in the figure.

Note that in certain special realizations of the arrangement type illustrated in Figure 10.10a, the reverse part may not exist. For instance, when j_E is an element of F, there is no reverse part. If j_E is the source node, then the reverse part is equivalent to the common part and consists only of the source node. Other special cases can also easily be formulated.

In the arrangement of Figure 10.10b, there is no common part. Instead, flow is increased in the forward part and decreased in the reverse part. The cycle in the reverse part of the trail acts as the flow-absorbing cycle. Notice that j_E can be a member of the reverse cycle that reduces the reverse part to a reverse cycle alone. Similar observations can be made for the remainder of Figure 10.10.

The arrangement of Figure 10.10c has a cycle in the forward part of the trail, and there is no common part. The cycle acts as a flow-generating cycle.

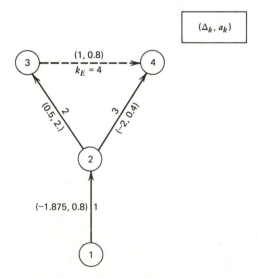

Figure 10.11
Flow Changes in a Flow-Augmenting Trail

Figure 10.10*d* includes a flow-generating cycle in the forward part of the trail and a flow-absorbing cycle in the reverse part of the trail.

Figures 10.10*e* and 10.10*f* both include a common cycle that can be considered simultaneously both flow generating and flow absorbing. They differ in that Figure 10.10*e* also has common arcs on the backpaths from i_E and j_E to the cycle while Figure 10.10*f* does not.

The development of the relationships that describe the flow change required for any of these arrangements requires some additional definitions. Consider a node i in the basis network. We define the arcs in the directed path from node i to i_E to be M_{Fi} and the arcs in the directed path from node i to j_E to be M_{Ri}. We define the node gain γ_i, in the context of this generalized primal minimum cost algorithm, to be the total flow change into node i required by a unit of flow in arc k_E. Once γ_i is computed, we can compute the flow change in arc $k(i,j)$ as a function of the flow change in arc k_E. Let Δ be the flow increase in arc k_E. The corresponding flow increase in arc $k(i,j)$ is

$$\Delta_k = \frac{\gamma_j \Delta}{a_k} \tag{20}$$

The computation of γ_i depends on the location of the node in the basis network. There are six possible locations for a node. We state the definition of γ_i in each case.

1. On the forward part but not on a cycle.

$$\gamma_i = \frac{1}{\displaystyle\prod_{k \in M_{Fi}} a_k} \tag{21}$$

2. On the reverse part but not on a cycle.

$$\gamma_i = \frac{-a_{k_E}}{\displaystyle\prod_{k \in M_{Ri}} a_k} \tag{22}$$

3. On the common part but not on a cycle.

$$\gamma_i = \frac{1}{\displaystyle\prod_{k \in M_{Fi}} a_k} - \frac{a_{kE}}{\displaystyle\prod_{k \in M_{Ri}} a_k} \tag{23}$$

4. On the forward part and on a cycle. We define β_F as the cycle gain on the forward path.

$$\gamma_i = \frac{\beta_F}{\beta_F - 1} \frac{1}{\displaystyle\prod_{k \in M_{Fi}} a_k} \tag{24}$$

5.   On the reverse part and on a cycle. Then β_R is the cycle gain.

$$\gamma_i = -\frac{\beta_R}{\beta_R - 1} \frac{a_{k_E}}{\displaystyle\prod_{k \in M_{Ri}} a_k} \tag{25}$$

6.   On the common part and on a cycle. Then β_c is the cycle gain.

$$\gamma_i = \frac{\beta_c}{\beta_c - 1} \left(\frac{1}{\displaystyle\prod_{k \in M_{Fi}} a_k} - \frac{a_{k_E}}{\displaystyle\prod_{k \in M_{Ri}} a_k} \right) \tag{26}$$

In any case where M_{Ri} or M_{Fi} have no elements the product terms

$$\prod_{k \in M_{Ri}} a_k$$

or

$$\prod_{k \in M_{Fi}} a_k$$

take on a value of 1.

We illustrate the use of these equations with the examples of Figure 10.12, which are in one-to-one correspondence to the possible types of flow-augmenting trails illustrated in Figure 10.10. Table 10.1 details the appropriate equation governing the computation of γ_i for each node in each of Figures 10.12a to 10.12f.

Direct use of equation (21) through (26) would first require catagorization of every node. This would be computationally inefficient since the evaluation of the γ_i's may be computed as an integral part of a single pass through the augmentation trail. This process is described and implemented by the PATHPG algorithm.

Equations (21) through (26) correctly determine the node gains in the presence of positive and negative arc gains and flow-generating and flow-absorbing cycles. Figure 10.10 requires some reinterpretation under these conditions. For instance, in Figure 10.10a, a negative gain for arc k_E implies an increase in flow in all parts of the trail: forward, reverse, and common. As flow is increased in k_E, additional flow is required from node i_E. Because of the negative gain, flow is simultaneously required from node j_E in the amount $-a_{k_E}$. Note that this amount is positive because a_{k_E} is negative. In the case of Figure 10.10d, a negative a_{k_E} (and no other negative gains in the trail) implies that both the forward and reverse cycles must act as flow-generating cycles. Similar observations may be made about the other cases of Figure 10.10.

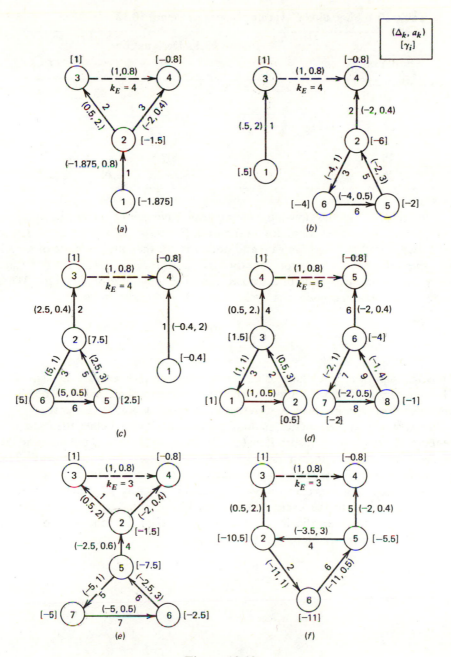

Figure 10.12
Computational Illustrations of Figure 10.10

321

Table 10.1   Equation Governing Nodes in Figure 10.12

Equation	Figure 10.12 Designation					
	a	b	c	d	e	f
21	3	1,3	3	4	3	3
22	4	4	1,4	5	4	4
23	1,2				2	
24			2,5,6	1,2,3		
25		2,5,6		6,7,8		
26					5,6,7	2,5,6

Negative arcs on the flow-augmenting trail have the effect of changing the sign of γ on two successive nodes on the trail. For instance, assume arc $k(i,j)$ on the forward trail has negative gain and node i is not the junction node of a cycle nor the intersection node of the forward and reverse parts. In this case, $\gamma_i = \gamma_j / a_k$. A negative value of a_k causes γ_i and γ_j to be of different signs. When node i is the junction node of a cycle and node j is a path node

$$\gamma_i = \frac{\beta \gamma_j}{(\beta - 1) a_k}$$

Thus the sign of a_k, the value of β, and the sign of γ_j determine the sign of γ_i. The sign of γ_i indicates the direction of flow change through node i. A positive γ indicates flow is to increase. A negative γ indicates flow is to decrease.

The sign of γ at the junction node of a cycle indicates whether the cycle is to generate flow ($\gamma > 0$) or absorb flow ($\gamma < 0$). Every cycle with β not equal to one or zero can act as either a flow generator or absorber. A cycle with $\beta > 1$ generates flow by increasing flow into the cycle and absorbs flow by decreasing flow into the cycle. A cycle with $\beta < 1$ (including negative β) generates flow by decreasing flow into the cycle and absorbs flow by increasing flow into the cycle. All this is automatically accounted for through the computation of the node gains γ using equations (21) through (26).

PATHPG ALGORITHM

Purpose: To find the flow-augmenting TRAIL for a generalized primal minimum cost flow algorithm and compute γ_i for each node on the trail.

1. (INITFOR) Initiate search for forward path. Set node check indicators ICHK to zero. Put i_E into the node list. If k_E is a forward arc, set γ_{i_E} to 1. If k_E is a mirror arc, set γ_i to $a(-k_E)$.

2. (FORWARD) Identify the arcs and nodes in the forward and common paths. Find the back pointer to the current node. If the back pointer is zero, we have found the source node and the forward (or common) path has no cycle; go to step 3. If the back pointer is not zero, find the next node in the path (IJ), update the gain factor (GAM), and put the next arc (K) and its origin node (IJ) on the arc and node lists. Check if the next node (IJ) has been encountered before. If not, assign its node gain, and check the node, and repeat step 2. If it has been encountered, a cycle has been found. Compute the cycle gain, indicate the junction node and go to step 3.

3. (INITREV) Initiate search for the reverse path. Put node j_E on the node list. If node j_E has already been encountered, it must be on the forward path so we indicate it is the intersection node and go to step 5. If j_E hasn't been encountered, compute γ_{j_E}, check node j_E, and go to step 4.

4. (REVERSE) Search for the reverse path. Find the back pointer to the current node. If it is zero, there is no common path or reverse cycle; go to step 6. If it is not zero, find its origin node. Put the arc and node on the corresponding lists. Check if the node has been encountered previously. If it

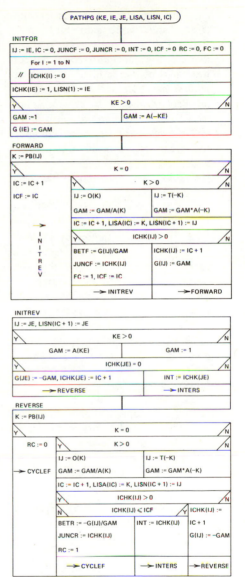

has not, mark the node as encountered, assign node gain, and repeat step 4. If it has been encountered, see if it has been encountered in the forward part of the path. If so, an intersection has been found. Indicate the intersection and go to step 5. Otherwise there is no common path and a reverse cycle has been found. Compute the cycle gain, indicate the junction node, and go to step 6.

5. (INTERS) An intersection has been found and the common part of LISN and LISA has been identified. Go down this list INT to ICF and recompute the node gain factors by subtracting the gain factor of the reverse path. If there is no common cycle (the situation of Figure 10.10*a*), go to step 7. If there is a cycle but the intersection node is not on the cycle (Figure 10.10*e*), go to step 6. If the intersection node is on the cycle (Figure 10.10*f*) and not all the cycle nodes have had their γ_i's adjusted to reflect that they are on a common cycle, adjust the remaining cycle nodes' γ_i's.

6. (CYCLEF) Adjust the gain factors for the nodes on the forward cycle to account for the cycle gain.

7. (CYCLER) Adjust the gain factors for the nodes on the reverse cycle to account for the cycle gain.

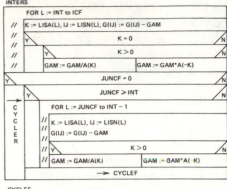

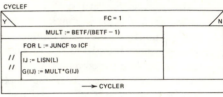

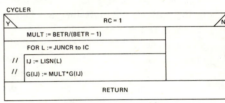

As an example of the use of PATHPG, let us consider its application to the network of Figure 10.12*f*. At the conclusion of steps INITIAL and FORWARD, the following information has been obtained.

$$\text{LISN} = [3, 2, 5, 6]$$
$$\text{LISA} = [1, 4, 6, 2]$$
$$\text{ICHK} = [0, 2, 1, 0, 3, 4]$$
$$\text{G} = [0, 0.5, 1.0, 0, 0.167, 0.333]$$
$$\text{BETF} = 1.5, \text{JUNCF} = 2, \text{FC} = 1, \text{and ICF} = 4.$$

Notice that we have simply followed the back pointers from node $i_E = 3$ until a node is encountered that has been visited before, that is, node 2. The list ICHK serves two purposes: to mark nodes that have been visited and to indicate the position in LISN in which each node is currently stored. LISN contains the nodes in the order that they are visited by PATHPG, and LISA contains the back pointers of the nodes in LISN. These back pointers comprise the list of arcs in the trail. ICF contains the position in LISN at which the last node encountered by step FORWARD is stored and JUNCF contains the position in LISN of the junction node for the forward cycle. For this example, LISN(ICF)=LISN(4)=6, which is the last node found by FORWARD, and LISN (JUNCF)=LISN(2)=2, which is junction node for the example's forward cycle. The G list gives the value of γ for each node.

At the completion of steps INITREV and REVERSE, the lists have been updated to

$$LISN = [3, 2, 5, 6, 4, 5]$$

$$LISA = [1, 4, 6, 2, 5]$$

$$ICHK = [0, 2, 1, 5, 3, 4]$$

$$G = [0, 0.5, 1.0, -0.8, 0.167, 0.333]$$

and, since ICHK(5)<ICF, we conclude that no reverse cycle exists and that a common part or intersection exists. INT has been set to a value of 3, which identifies node 5 as the intersection node between the forward and reverse parts; that is, LISN(INT)=LISN(3)= 5.

At the completion of step INTERS, the list G has been changed to

$$G = [0, -3.5, 1.0, -0.8, -1.83, -3.67]$$

Finally, control is passed to step CYCLEF where G(2), G(5), and G(6) are multiplied by MULT=3 to yield the final G list.

$$G = [0, -10.5, 1.0, -0.8, -5.5, -11.0]$$

A Basis Change Method

At each iteration, subroutine MFLOG will determine the arc that limits the flow change on the flow-augmenting trail and that arc will leave the basis. In order to properly perform the change of basis and subsequently modify the π values to reflect the new basis, we must recognize four different possibilities.

1. When the entering arc limits the maximum flow change in the flow-augmenting trail, the entering arc k_E is also the leaving arc k_L. In this case, neither the basis nor the π values change.
2. The leaving arc is in the reverse part. Two examples of this case are shown in Figure 10.13. In Figure 10.13a, the leaving arc is not part of a cycle. In Figure 10.13b, the leaving arc is a cycle arc. The cycle is broken and the nodes are connected to the source or a flow-generating cycle through the entering arc.
3. The leaving arc is in the forward part. Two examples of this case are shown in Figure 10.14. Here the entering arc must be reversed in order to form the new basis network. As in case 2 above, the effect of the leaving arc depends on whether it is in the cycle part of the forward path.
4. The leaving arc is in the common part. Three examples are shown in Figure 10.15. In this case, the entering arc will always form a cycle.

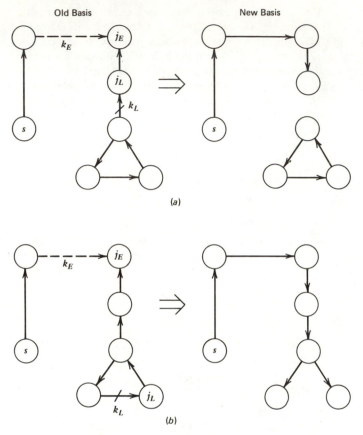

Figure 10.13
Basis Change When the Leaving Arc Is in the Reverse Part

In all cases, the change in the basis network is accomplished by first determining the direction of the entering arc (k_E or $-k_E$), then deleting arc k_L, reversing all the arcs from terminal node of k_E to the terminal node of k_L, and then inserting the arc k_E. Except for the first step, these are exactly the operations performed by the TRECHG algorithm and indeed, it may be used to accomplish the basis change operation for our generalized primal algorithm.

The method used to determine the new dual variables, the π, depends again on where the leaving arc is located. If it is located on the reverse path, the value of π_{i_E} does not change with the change in basis. The value of π_{j_E} becomes $(\pi_{i_E} + h_{k_E})/a_{k_E}$. The set of nodes and arcs in the tree rooted at j_E is found with the ROOTG algorithm and the π values are set using equation (29) of Chapter 9. A similar process is used when the leaving arc is in the forward path, except in

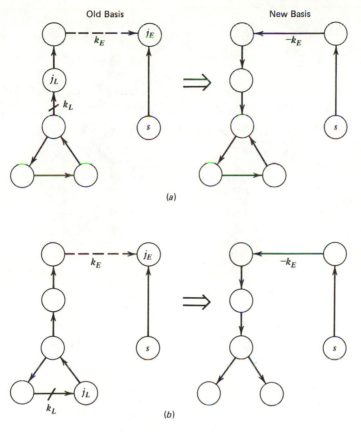

Figure 10.14

Basis Change When the Leaving Arc Is in the Forward Part

this case π_{j_E} remains fixed and the π's are changed in the tree rooted at node i_E.

When the leaving arc is located on the common path, we cause arc k_E to enter the basis. With this basis, we compute the cycle gain and the unit cost of passing flow through the cycle starting at node i_E. We then compute the value of π_{i_E} using the cycle gain, β, and equation (31) of Chapter 9. With π_{i_E} computed, the π values are computed for the nodes in the network rooted at i_E using equation (29) of Chapter 9. The primal network-with-gains algorithm is implemented in algorithm PGAINS.

Figure 10.16 presents an example application of PGAINS. Notice that the selected initial basic feasible solution of Figure 10.16b contains two flow-augmenting cycles and that the required flow of 10 units at node 10, the sink node, is supplied by the

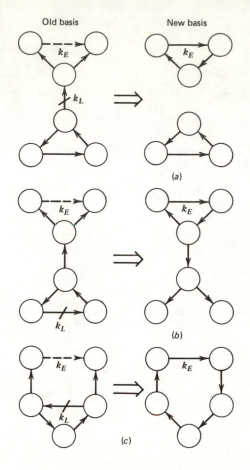

Figure 10.15
Basis Change When the Leaving Arc Is in the Common Part

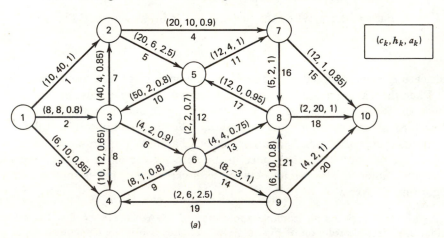

Figure 10.16
Example Application of PGAINS

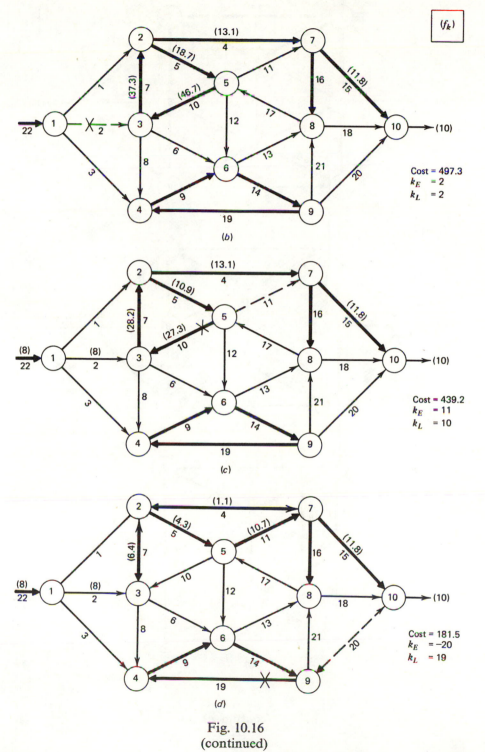

Fig. 10.16
(continued)

329

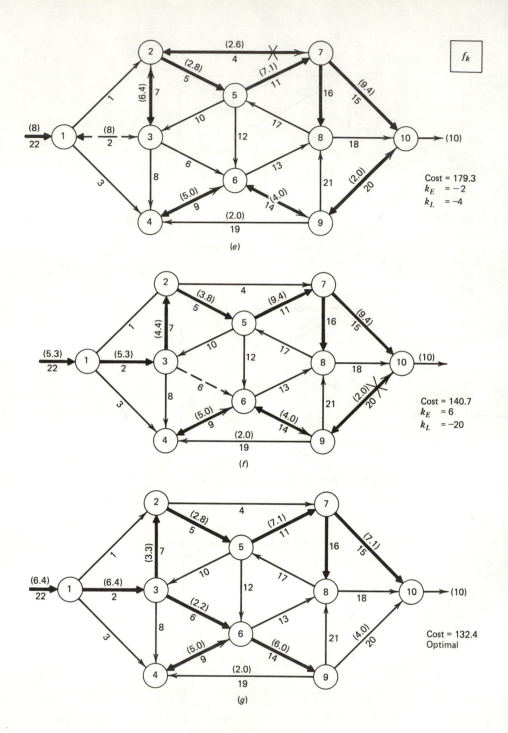

Figure 10.16
(continued)

330

flow-augmenting trail of arcs 5, 10, 7, 4, and 15. Further, node 1, the source node, is isolated and forms a tree with no arcs.

In each of Figures 10.16*b* to 10.16*f*, the entering arc is indicated by a dashed line and the leaving arc is marked by an *X*. Both the flow at the completion of each basis change and the resulting total cost of that flow are also given at each step. The basis arcs are indicated by the heavier lines. the large arrow heads indicate the direction of the arc in the basis. Figure 10.16*c* is of particular interest since it illustrates the formation of a new-generating cycle.

Figure 10.16*d* illustrates the case of the entering arc connecting two cycles. One acts as a flow-generating cycle and the other as a flow-absorbing cycle.

For those interested in the details of the algorithm, we describe the process by which PGAINS determines the location of the leaving arc in the trail. Recall that LISN is a list of nodes on the trail. Algorithm PATHPG creates this list. The nodes on the forward part are listed first, then the nodes on the common part are listed, and finally the nodes on the reverse part are listed. Three specific indices that are determined in PATHPG are:

IC: the number of nodes in the list
INT: the index of the first node in the common part. If there is no common part, this variable is zero.
ICF: index of the last node in the common part. If there is no common part, this is the index of the last node in the forward part.

With these indices, we can determine the location of the leaving arc in the trail. This affects how the entering arc should appear in the basis.

Algorithm MFLOG finds the variable ILC which is the index of terminal node of the leaving arc. Comparing ILC to the INT and ICF ascertains the location of the leaving arc. First if INT = 0, there is no common part and

$$ILC \leqslant ICF \Rightarrow \text{forward part}$$
$$ILC > ICF \Rightarrow \text{reverse part}$$

When INT > 0, then there is a common part.

$$ILC < INT \Rightarrow \text{forward part}$$
$$ILC > ICF \Rightarrow \text{reverse part}$$
$$INT \leqslant ILC \leqslant ICF \Rightarrow \text{common part}$$

These tests are implemented in the CHANGE portion of PGAINS.

PGAINS ALGORITHM

Purpose: To perform the primal algorithm for the generalized minimum cost flow problem.

1. (INITIAL) Set the iteration counter (IST) to 1.
2. (SELECT) Select an arc that violates dual feasibility. Let this arc be k_E. If there are no such arcs (IFIN = 1), stop with optimal solution. Otherwise go to step 3.
3. (LEAVE) Find the origin (i_E) and terminal (j_E) nodes of the entering arc. Find the maximum flow increase in the entering arc (MFE). Find the flow-augmenting trail that includes k_E. Find the maximum flow increase in this trail (MF) and the associated limiting arc (k_L). ILC is the index of this arc in LISA. If the maximum flow increase in k_E is less than the maximum flow increase in the trail, let the leaving arc be k_E. Change MF to MFE.
4. (FLOW) If MF > 0, increase the flow in the flow-augmenting trail by an amount determined by MF. Also change the flow in k_E. Go to step 5. If MF = 0, some arc in the flow-augmenting trail must be inadmissible. This is a degenerate iteration. Go to step 5.
5. (CHANGE) If $k_L = k_E$, no basis change is necessary; go back to step 2. Else, determine the location of the leaving arc: forward, common, or reverse. If k_L is in the forward part, reverse the entering arc and go to step 7. If k_L is in the common part, go to step 6. If k_L is in the reverse part, go to step 7.
6. (COMMON) Change the basis network by deleting k_L and inserting k_E. Compute the cycle gain and unit cost. Compute π_{i_E} and go to step 8.

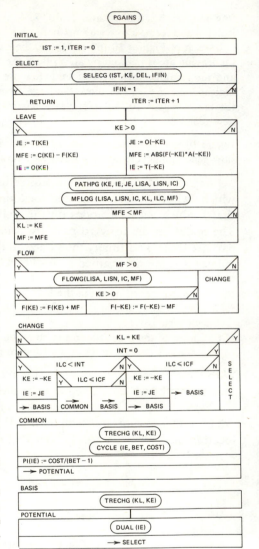

7. (BASIS) Change the basis by deleting k_L and inserting k_E.
8. (POTENTIAL) Change the π values for the part of the network rooted at j_E. Return to step 2.

10.8 HISTORICAL PERSPECTIVE

The first computational procedures for the generalized network problem related to the generalized transportation problem. Here the network has a bipartite structure and arcs are uncapacitated. Dantzig (1963) and Charnes and Cooper (1961) describe early attempts at adapting the simplex method to this special problem. Later authors—Eisemann (1964), Lourie (1964), and Balas and Ivanescu (1964)—provide the implementation of primal procedures related to the stepping-stone method of the pure transportation problem. Eisemann (1964) also addresses the capacitated transportation problem. Balas (1966) provides a dual approach for solution. All these computational techniques utilize the matrix representation of the transportation problem and are based on the bipartite nature of the network. Although we treat in this chapter the more general capacitated network problem, many of the computational problems are the same as those encountered for the transportation problem. In particular, Eisemann (1964) and Balas and Ivanescu (1964) provide excellent descriptions of the different forms taken by bases and the problems of pivoting.

Jewell (1962) introduces a procedure similar in concept to the out-of-kilter algorithm to solve the generalized minimum cost flow problem on a general network with capacities. Minieka (1972) modifies Jewell's algorithm to guarantee finite termination.

Johnson (1966) suggests the use of the three-pointer representation for representing the basis in a generalized network problem. Maurras (1972) and Glover, Klingman, and Stutz (1973) utilize a three-pointer representation to implement primal algorithms for the generalized min-cost flow problem. Glover and Klingman (1973) provide additional computational simplifications for the generalized transportation problems. Jensen and Bhaumik (1977) describe a dual incremental approach.

Although the maximum flow problem is not explicitly considered in these chapters, there are papers dealing specifically with the generalized maximum flow problem. Onarga (1967) provides an iterative algorithm to maximize flow at the sink with a given input flow. Grinold (1973) uses a dual simplex approach similar to INCREMG specialized to the maximum flow problem. Jarvis and Jezior (1972) describe a primal–dual algorithm similar to the out-of-kilter algorithm to solve the problem for an acyclic network.

Charnes and Raike (1966) provide a shortest path algorithm for the generalized network very similar to DSHRTG. A proof of its optimality is provided.

An algorithm for the longest path in a generalized acyclic network is also provided. The paper also discusses making all arc costs greater than zero with a transformation that uses a feasible dual solution.

Glover, Klingman, and Napier (1972) extend the method of Charnes and Raike (1966) to obtain basic dual feasible solutions to generalized networks having positive costs, positive gains no greater than one, and rank $m-1$. Hultz et al. (1976) provide an improved dual feasible start and make computational comparisons on pure network test problems. They find that for some problems the new start results in improved computational performance over the out-of-kilter algorithm but that primal methods are superior.

EXERCISES

1. Find the shortest path subnetwork for the network given below.

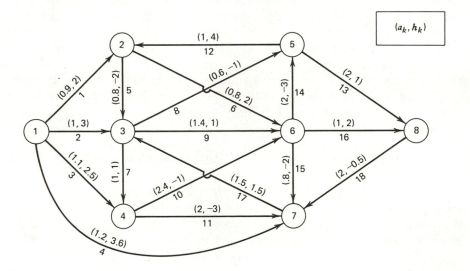

2. Modify PSHRTG to use equation (6) to update the values of the π_i's for nodes on a cycle.
3. Why is the change of basis from Figure 10.4*a* to Figure 10.4*b* a degenerate iteration?
4. Verify that the optimal objective function value to the example problem of Figure 10.4 is -2.07 by summing the costs of the flows on arcs 11, 15, 16, and 17.
5. Solve the problem of Figure 10.4 if a_{11} is set to a value of 1.0 instead of its current value.
6. In step 7 (CYCLE) of algorithm DULSPG, arcs that would form a cycle with $\beta < 1$ are not candidates for the entering arc. Why?

7. Use DULSPG to obtain the new optimal solution for the network of Figure 10.4 if arcs 11 and 15 are declared inadmissible.

8. What must be done before one can use algorithm INCREMG if an initial solution in which all flows are set to zero does *not* form an intermediate optimum for $F_t = 0$?

9. Use INCREMG to find the optimal flows for the following generalized minimum cost flow network.

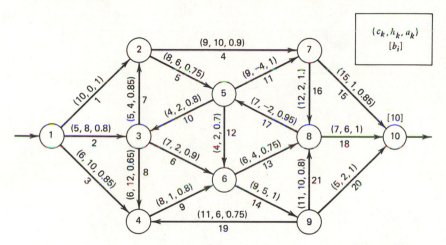

10. Use the methods of algorithm PATHPG to evaluate γ_i for each node on the flow-augmenting trail for each of the networks given below. The entering arc is indicated by the dotted line.

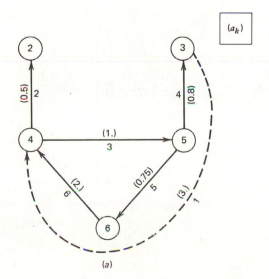

(a)

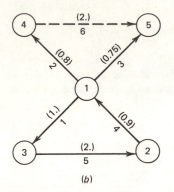

(b)

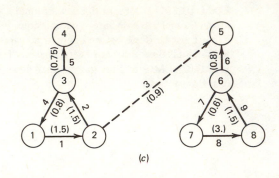

(c)

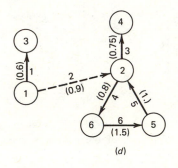

(d)

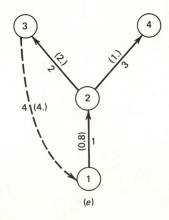

(e)

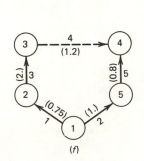

(f)

11. Refer to the cycle of Figure 9.26. Show that the representation of the cycle in either forward or reverse orientation will yield the same value of the node potentials. First compute π_1 using the forward arc orientation of the cycle $[1, 2, 3, 4]$ as in Figure 9.26. Then compute π_1 using the mirror arc orientation of the cycle $[-4, -3, -2, -1]$.

12. Solve the problem of Figure 10.9 using PGAINS.

13. Solve the problem of Figure 10.16 using INCREMG.

14. Solve the problems of Section 2.8 for the optimal flows.

15. Obtain the optimal flows for exercise 10 of Chapter 2.

16. Determine the maximum flow that can be provided to node 7 in the network given below, and give the arc flows that may be used to achieve that flow.

(a_k, c_k)

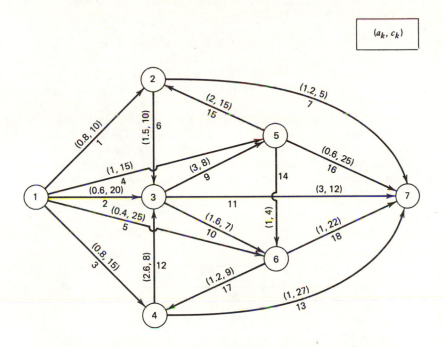

17. For each case in problem 10, replace the gain of arc 1 by its negative and compute the value of γ_i for each node.

18. Given a network with undirected arcs, a node cover is a subset of arcs such that each node is incident to at least one arc. The node covering problem is to select a cover with the fewest arcs.

    (a) For the network shown below, set up this problem as a linear integer program.

    (b) Show the network with all arc gains equal to -1 that solves the problem of part (a) with the integrality constraints relaxed.

    (c) Why does the network solution not guarantee an integer value?

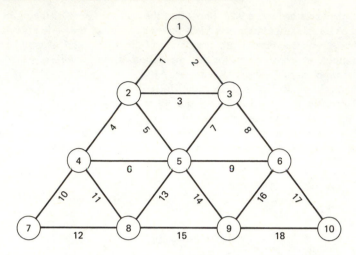

19. The maximum matching problem is to choose a maximum subset of arcs such that no two arcs selected are incident to the same node. This too is an integer programming problem. Drop the integrality requirement and solve the relaxation to the matching problem as a network flow problem using negative gains on each arc.

20. Set up the transportation problem below using only arcs with negative gains and only positive external flows.

Cost Matrix

	4	5	6	7	Supplies
1	8	3	9	X	20
2	7	X	9	15	30
3	5	10	X	20	50
Demands	10	15	35	40	

The X's imply disallowed transmission.

CHAPTER 11
THE CONVEX MINIMUM
COST FLOW PROBLEM

The previous chapters were restricted to consideration of minimum cost flow problems in which arc cost was a linear function of arc flow. In this chapter, we generalize the procedures used for the linear problem to solve the case in which an arc cost is a convex function of arc flow. The difference is important because solutions to the convex case may be *substantially* different from solutions that are obtained when the convexity is ignored and arc costs are modeled as a simple linear function of arc flow. Also an ability to solve the convex problem will allow the accurate and efficient solution of several important new classes of practical problems.

11.1 CONVEX COST FUNCTIONS

In the preceding chapters, we required arc flows to be nonnegative. In this chapter, we drop this restriction and allow the lower bound on flow to become a negative number. Thus the feasibility condition is

$$\underline{c}_k \leqslant f_k \leqslant c_k$$

where $\underline{c}_k$ may be a negative value. Arc directions will still be indicated by arrows; however, these are for reference purposes. Thus for an arc $k(i,j)$ when $f_k > 0$ the flow is from node i to node j. When $f_k < 0$, the flow is from node j to node i.

We use the functional form $h_k(f_k)$ for the cost function on arc k. Figure 11.1 illustrates several different functions that are convex. A function is convex if a straight line drawn between any two points on the function lies entirely on or above the function. This is illustrated in Figure 11.2. This statement may be

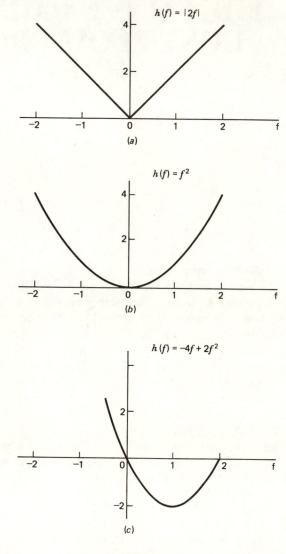

Figure 11.1
Example Convex Cost Functions

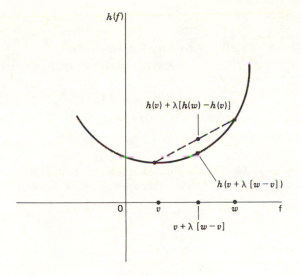

Figure 11.2
A Convex Function

written symbolically as

$h(f)$ is convex if and only if:

$$h(v) + \lambda[h(w) - h(v)] \geqslant h(v + \lambda[w - v]) \tag{1}$$

for all λ, $0 \leqslant \lambda \leqslant 1$

where v and w are any flow values such that $w > v$. The quantity on the left of this inequality defines the line between $h(v)$ and $h(w)$. The quantity on the right is the value of the function at corresponding points. This definition is useful visually but rarely useful in an analytic determination of convexity.

For a convex function, we require $h(f)$ to be continuous. We do not require that $h(f)$ be everywhere differentiable; however, the function will have a well-defined first derivative at all but a finite number of points. An alternative definition of convexity under these conditions is that the derivative of h be monotonically nondecreasing as f increases. Using h' to be the first derivative we find that

$h(f)$ is convex if and only if it is continuous

and $h'(w) \geqslant h'(v)$ \tag{2}

for all $w > v$ at which the derivative is defined.

This definition does not require that h' be continuous but in cases in which h' is continuous and everywhere defined, the second derivative can also be used to test convexity. When h and h' are both continuous, then the function is convex if and only if

$$h''(f) \geqslant 0 \qquad \text{for all } f \text{ for which the} \tag{3}$$
$$\text{second derivative is defined}$$

Note that, in cases where arc flows are bounded by $\underline{c}_k$ and c_k, the specifications above need only hold for flows in the range $(\underline{c}_k, c_k)$.

EXAMPLE

Suppose $h(f)$ is represented by the absolute value function as in Figure 11.1a and the per unit cost of flow is equal to 2 for flows in either the positive or negative direction. Clearly,

$$h(f) = |2f|$$

or

$$h(f) = \begin{cases} -2f & \text{for } f < 0 \\ 2f & \text{for } f \geqslant 0 \end{cases}$$

Computing first derivatives, we have

$$h'(f) = \begin{cases} -2 & f < 0 \\ 2 & f > 0 \end{cases}$$

at $f = 0$ the first derivative is not defined. This function fulfills condition (2), so the function is convex.

The negative of the cost function described above is

$$h(f) = -|2f|$$

Here

$$h'(f) = \begin{cases} 2 & f < 0 \\ -2 & f > 0 \end{cases}$$

Condition (2) fails so the function is not convex. Note that condition (3) is not applicable in this case since h' is not continuous. Applying condition (3) would lead to erroneous results since $h'' = 0$ for all f, $(f \neq 0)$.

EXAMPLE

Suppose the following quadratic cost function applies.

$$h(f) = f^2$$

Therefore

$$h'(f) = 2f \quad \text{and} \quad h''(f) = 2$$

Since h' is increasing and continuous both condition (2) and condition (3) indicate convexity. Similarly it is easy to show that $h(f) = -f^2$ is not convex.

11.2 THE CHARACTER OF SOLUTIONS

One important reason to consider the use of convex cost functions as a modeling tool is the very different nature of the solutions of a convex problem as compared to those of the linear problem. Recall in the linear problem the optimum solution is always a basic solution. Thus the flows on a large portion of the arcs (all the nonbasic arcs) are either at capacity or zero.

However, in a problem with strictly convex cost functions, the arc bounds are frequently not important to the solution and most arc flows are strictly between upper and lower bounds. Consider a situation in which there are no arc upper bounds as illustrated in Figures 11.3 and 11.4, which compare the solutions to a power distribution problem for linear and quadratic cost functions, respectively. The solutions are very different, indeed.

For planning applications, it is often true that a solution similar to the convex case is more satisfactory to decision makers than a solution similar to the linear case. This may be due to the uncertainty inherent in planning activities that do not favor an all-or-nothing solution. As we will see in a later section, we can explicitly represent certain kinds of uncertainty using convex costs.

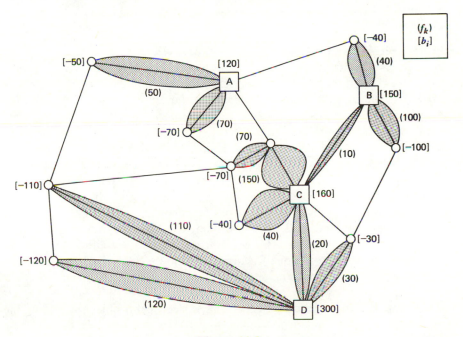

Figure 11.3
A Power Distribution Problem with Linear Costs

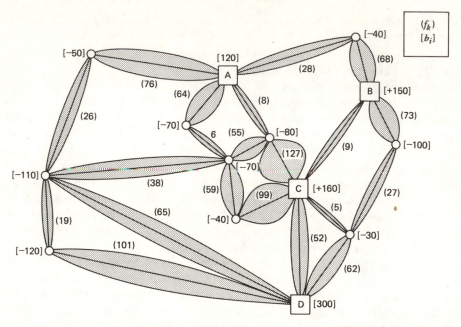

Figure 11.4
A Power Distribution Problem with Quadratic Costs

11.3   PHYSICAL NETWORK FLOW PROBLEMS

Most engineering disciplines describe problems that involve flow of some physical quantity in a network. In electricity, we are concerned with the flow of electrical current in wires. In pipeline analysis, we are concerned with the flow of fluid in pipes. In a thermal system, we are concerned with the flow of heat through conduction paths.

For each of these physical systems, computational techniques have been derived to solve the network problem. The flow variables are determined for given inputs and network parameters. There is an equivalence between the solution techniques for these physical flow problems and the network flow programming problems given in this book. In particular, each of the problems noted above can be stated and solved as a network flow programming problem with convex arc cost functions. This equivalence allows the solution of physical flow problems with the techniques of this chapter. It also allows the solution of certain convex network programming problems with the techniques derived for the physical flow problems.

To illustrate the equivalence, we consider the electrical problem of finding the currents in a resistive, direct current network. It is not possible to use these techniques in a network that involves inductance or capacitance and alternating currents. The pipeline and heat flow problems are analogous to the dc electrical problem.

Consider the example network of Figure 11.5. The resistors shown in the figure are assumed linear (the nonlinear case will be considered later). A linear resistor R has the characteristic that if a current I is passed through it, a voltage v will appear across its terminals such that

$$v = IR.$$

The quantites v, I, and R are measured, respectively, in units of volts, amperes, and ohms. The battery shown in the figure presents a constant voltage E between its terminals. This voltage is independent of the current passing through the battery.

This network is to be solved for the currents in the resistors and the voltages at the nodes. The solution can be obtained using Kirchoff's laws.

1. The sum of the currents entering a node must equal the sum of the currents leaving the node. This is equivalent to conservation of flow.
2. The voltage across every resistor must equal the difference in the potentials between the nodes that terminate the resistor.

To use these laws, we define node potentials V_i and branch currents I_k on the diagram. The directions given to the currents I_k are arbitrary and used for reference. Note that each I_k may be positive or negative and is not bounded in magnitude.

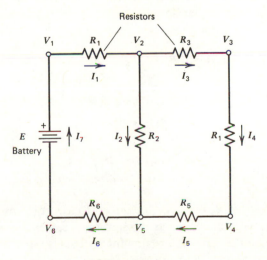

Figure 11.5
A DC Resistive Network

For the example problem, Kirchoff's current law is written

$$
\begin{aligned}
I_7 - I_1 \qquad\qquad\qquad\qquad &= 0 \\
I_1 - I_2 - I_3 \qquad\qquad\qquad &= 0 \\
I_3 - I_4 \qquad\qquad &= 0 \\
I_4 - I_5 \quad &= 0 \\
I_2 \qquad\quad + I_5 - I_6 &= 0 \\
-I_7 \qquad\qquad\quad + I_6 &= 0
\end{aligned}
$$

These equations are the same as the conservation-of-flow equations for the network programming problem. One of these equations is redundant. Kirchoff's voltage law yields

$$
\begin{aligned}
-V_1 \qquad\qquad\qquad + V_6 &= -E \\
V_1 - V_2 \qquad\qquad\qquad &= I_1 R_1 \\
V_2 \qquad\quad - V_5 \quad &= I_2 R_2 \\
V_2 - V_3 \qquad\qquad &= I_3 R_3 \\
V_3 - V_4 \qquad &= I_4 R_4 \\
V_4 - V_5 \qquad &= I_5 R_5 \\
V_5 - V_6 &= I_6 R_6
\end{aligned}
$$

Since the node potentials all appear as differences in these equations, one may arbitrarily be set to zero. We choose V_6. We now have 7 unknown branch currents and 5 unknown node voltages and 12 linear equations in these variables. Thus, we can solve for the currents and voltages using standard procedures for solving linear equations.

EXAMPLE
Let $E = 10$, $R_1 = 1$, $R_2 = 3$, $R_3 = 1$, $R_4 = 2$, $R_5 = 1$, and $R_6 = 2$. Solving the set of equations yields

$$
\begin{aligned}
I_1 &= I_6 = I_7 = 2.12 \\
I_3 &= I_4 = I_5 = 0.91 \\
I_2 &= 1.21 \\
V_1 &= 10, V_2 = 7.88, V_3 = 6.97 \\
V_4 &= 5.15, V_5 = 4.24, V_6 = 0
\end{aligned}
$$

Procedures to find and solve the set of linear equations defined by a dc electrical circuit with linear resistors are, of course, well known. There are cases, however, in which the resistors are nonlinear. This is represented by a nonlinear relation between voltage and current. Nonlinear resistors result in nonlinear voltage equations. These are more difficult to solve but techniques such as the modified Newton–Raphson techniques are available.

We now describe an alternative method for solving for the currents of a dc network that represents the problem as a convex minimum cost network problem. This method uses the concept of the *content* of a resistive element. The content of a two-terminal resistive element in a dc electric circuit depends on its voltage characteristic. Consider an element k as in Figure 11.6. We arbitrarily establish a positive direction for current through the element as passing from node i to node j. The value of this current is I_k. When the current passes from node j to node i, I_k has a negative value. The voltage across the element is defined as a function of I_k and can be written

$$v_k(I) = V_i - V_j$$

The function $v_k(I)$ is called the *voltage characteristic*. The form of this function depends on the particular kind of element under consideration. Several are listed in Table 11.1.

For the linear resistive element, the voltage characteristic is a linear function

$$v_k(I) = IR$$

Table 11.1   Voltage Characteristics for Various Elements

Elements	Symbol	Voltage Characteristic
Linear resistor	R	$v(I) = IR$
Diode		$I \geqslant 0 \quad v(I) = IR_1$ $I < 0 \quad v(I) = IR_2$ $R_2 \gg R_1$
Battery	E	$v(I) = -E$
Nonlinear resistor	R, k	$I \geqslant 0 \quad v(I) = RI^k$ $I < 0 \quad v(I) = R(-I)^k$

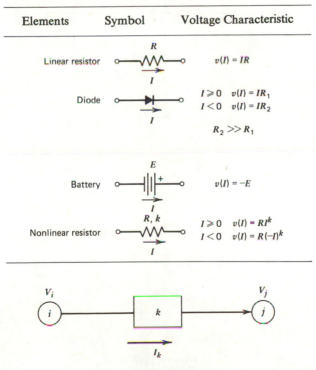

Figure 11.6
A Resistive Element

where the parameter R is the value of the resistance. Table 11.1 identifies several other resistive elements that are nonlinear. Note that it is possible for the voltage characteristic to have different definitions for positive and negative currents.

The content of an element for current I_k is defined to be the area under the voltage characteristic between 0 and I_k as illustrated in Figure 11.7a.

$$G_k(I_k) = \int_0^{I_k} v(I)\, dI \tag{4}$$

For negative values of I_k, the content of the element is the area indicated in Figure 11.7b. Formula (4) still holds as a definition of the content function. However, the equivalent form shown in equation (5) is also appropriate when I_k

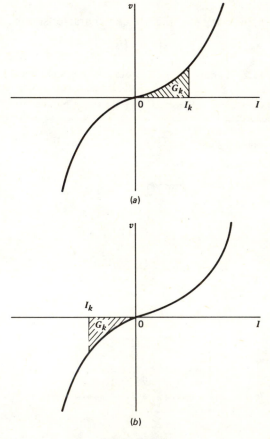

Figure 11.7
The Content of an Element (*a*) for Positive Current (*b*) for Negative Current

is negative.

$$G_k(I_k) = -\int_{I_k}^0 v(I)\,dI \qquad (5)$$

In applying equations (4) and (5), care should be taken to use the proper functional form for $v(I)$ since the form may differ for positive and negative currents. Table 11.2 lists the content functions for the elements considered in Table 11.1.

The content of an electrical network is defined to be the sum of the contents of its components.

$$G_{NET} = \sum_{k \in M} G_k$$

where M is the set of elements in the network. It has been shown (cf. Section 11.7) that the solution for the currents of the electrical network is a stationary point of the network content function when the currents satisfy the conservation-of-flow constraints at the nodes.

To illustrate this consider the simple battery and linear resistance circuit of Figure 11.8. From Table 11.2, the contents of the elements are

$$G_B = -10I_B$$
$$G_R = 2I_R^2$$

The network content is

$$G_{NET} = -10I_B + 2I_R^2 \qquad (6)$$

Conservation of flow requires that

$$I_B = I_R = I \qquad (7)$$

Table 2   CONTENT FUNCTION FOR VARIOUS ELEMENTS

Element	Content Function
Linear resistor	$G(I) = I^2 R/2$
Diode	$I \geqslant 0 : G(I) = I^2 R_1/2$
	$I < 0 : G(I) = I^2 R_2/2$
Battery	$G(I) = -EI$
Nonlinear resistor	$I \geqslant 0 : G(I) = RI^{k+1}/(k+1)$
	$I < 0 : G(I) = R(-I)^{k+1}/k+1$

Substituting equation (7) in equation (6) and setting the derivative to zero to find the stationary point of the content function, we obtain

$$\frac{dG_{NET}}{dI} = -10 + 4I = 0$$

or

$$I = 2.5 \text{ amps}$$

This is the current that flows in the circuit of Figure 11.8.

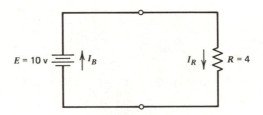

Figure 11.8
Example Electrical Network

When the function $v_k(I)$ is monotonically nondecreasing, as are all the examples of Table 11.1, the content function defined by equation (4) is a convex function. Thus an electrical network consisting only of elements of this type must have a convex content function. This guarantees that a stationary point is also a global minimum point. Thus, the techniques of this chapter that find the minimum cost flows in a network with convex arc cost functions can also be used to find the minimum content flows and thus solve the physical network problem.

The network programming approach may be a practical scheme for solving physical network problems. The data is simple, only requiring the physical network description and parametric descriptions of components. It is not necessary to derive the Kirchoff equations. Nonlinear components are handled with little more difficulty than linear ones. Diodes and current-limiting components are easily modeled with lower and upper bounds on arcs. Current sources and sinks are modeled as external node flows. The network programming formulation relates arc flow to current and node potential (π) to node voltage. Thus the solution by network programming yields both currents and voltages.

An analogous problem is the case of determining fluid flow in a pipe network. Here f_k is the flow rate in pipe k and π_i is the head at node i. The head loss in a pipe $k(i,j)$ is given by the Hazen–Williams formula

$$\pi_i - \pi_j = Kf^x$$

where K is a constant that depends on the length, width, and material of the pipe and x is an exponent that assumes the value 1.85 or 2 depending on the approximation used. In this case, the content of pipe k is

$$G_k = \begin{cases} \dfrac{Kf^{x+1}}{x+1} & \text{for } f \geqslant 0 \\[2em] \dfrac{K(-f)^{x+1}}{x+1} & \text{for } f < 0 \end{cases}$$

This becomes a term in the objective function for solution by the convex network flow programming approach.

For thermal systems, the thermal resistance to heat transfer through a solid is $R = L/kA$ where L is the length of the heat transmission path, A is the cross-sectional area, and k is the coefficient of thermal conductivity. Here f_k corresponds to the heat flow rate in heat units per unit time and π_i represents the temperature in degrees.

11.4   COST FUNCTIONS THAT DEPEND ON RANDOM VARIABLES

All the problems previously considered were deterministic; that is, they allowed no direct consideration of random variables. Certain practical problems that involve risk possess deterministic equivalents that have convex arc cost functions.

Consider an arc with a cost function $h(f, x)$ that depends both on flow and on some random variable x. If x has a probability density function $g(x)$, the expected cost as a function of flow is

$$\bar{h}(f) = E[h(f, x)] = \int_{-\infty}^{\infty} h(f, x) g(x) \, dx \tag{8}$$

A sufficient condition for $\bar{h}(f)$ to be convex is that it possess a continuous first derivative and that its second derivative be everywhere positive; that is

$$\frac{d^2\bar{h}}{df^2} = \int_{-\infty}^{\infty} \frac{\partial^2 h}{\partial f^2} g(x) \, dx \geqslant 0$$

Since $g(x)$ is a density function and is everywhere nonnegative, the convexity condition is satisfied if $h(f, x)$ is a convex function of f for all values of x where $g(x)$ is strictly positive. A more general sufficient condition does not require second derivatives to exist at all points, but only requires that the partial derivative of h with respect to f be a monotonically nondecreasing function of f for all x.

EXAMPLE

Consider an arc whose flow represents the number of workers available in a shop. Let x be a random variable that represents the work requirement in terms of the number of workers. Let a penalty $D(x-f)^2$ be charged for differences between the number of workers available and the number of workers required. If $g(x)$ is the probability density function of x with mean μ and variance σ^2, the expected value of the cost on arc k is

$$\bar{h}(f) = \int_\infty^\infty D(x-f)^2 g(x)\,dx$$

$$= \int_\infty^\infty D[x-\mu-(f-\mu)]^2 g(x)\,dx$$

$$= D\left[\int_\infty^\infty (x-\mu)^2 g(x)\,dx \right.$$

$$- 2(f-\mu)\int_\infty^\infty (x-\mu)g(x)\,dx$$

$$\left. + (f-\mu)^2 \int_\infty^\infty g(x)\,dx \right]$$

$$= D\left[\sigma^2 + (f-\mu)^2 \right]$$

The cost function includes a constant term that depends on the variance of the distribution and a quadratic function of flow that is convex. Note that the expected cost function does not depend on the form of the density function.

Another form of cost function that frequently arises in practice is

$$h(f,x) = \begin{cases} h_1(f,x) & x \leqslant f \\ h_2(f,x) & x > f \end{cases} \tag{9}$$

where, again, x is a random variable with probability density $g(x)$. Computing the expected cost requires that the integral be broken into two parts.

$$\bar{h}(f) = \int_{-\infty}^f h_1(f,x)g(x)\,dx + \int_f^\infty h_2(f,x)g(x)\,dx \tag{10}$$

Assuming first and second derivatives exist at all points, the second derivative of the expected cost is

$$\frac{d^2\bar{h}(f)}{df^2} = \int_{-\infty}^f \frac{\partial^2 h_1}{\partial f^2} g(x)\,dx$$

$$+ \int_f^\infty \frac{\partial^2 h_2}{\partial f^2} g(x)\,dx$$

$$+ g(f)\left[\frac{\partial h_1}{\partial f} - \frac{\partial h_2}{\partial f} \right]_{x=f}$$

The notation in the last term indicates that the first derivatives are to be evaluated at $x=f$. Since $g(x)$ is always positive, a sufficient condition for the expected cost function to be convex is

1. h_1 and h_2 are convex for all x and f
2. $\partial h_1/\partial f > \partial h_2/\partial f$ at $x=f$

EXAMPLE

Consider a situation when one must allocate a flow to a certain arc to meet a demand x. No cost is incurred if the flow provided is greater than the demand but, if flow is less than demand, a penalty of $D(x-f)$ is incurred. The positive quantity D is a penalty for each unit of demand not met. Thus

$$h(f,x)=\begin{cases} 0 & \text{if } x \leqslant f \\ D(x-f) & \text{if } x \geqslant f \end{cases}$$

$$h_1(f,x)=0 \qquad h_2(f,x)=D(x-f)$$

$$\frac{\partial h_1}{\partial f}=0 \qquad \frac{\partial h_2}{\partial f}=-D$$

$$\frac{\partial^2 h_1}{\partial f^2}=0 \qquad \frac{\partial^2 h_2}{\partial f^2}=0$$

Note that both h_1 and h_2 are convex (linear) for all x. Also condition (2) is met, since $0 > -D$. Therefore, the theory indicates that this cost function is convex. The expected cost function is

$$\bar{h}(f)=\int_f^\infty D(x-f)g(x)\,dx$$

Suppose the random variable has a uniform density function

$$g(x)=\begin{cases} \dfrac{1}{3} & 0 \leqslant x \leqslant 3 \\ 0 & \text{otherwise} \end{cases}$$

Clearly,

$$\bar{h}(f)=\int_f^3 \frac{D(x-f)}{3}\,dx \qquad 0 \leqslant f \leqslant 3$$

$$= \frac{3D}{2} - Df + \frac{Df^2}{6}$$

Evaluating the second derivative will indicate that this indeed is a convex function.

EXAMPLE

A simple transportation problem involving one source and two destinations is given in Figure 11.9. Products are manufactured at node 1 at a cost of $10 each and shipped to

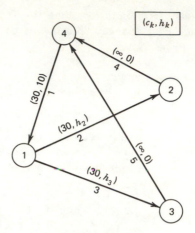

Figure 11.9
A Stochastic Network Problem

the destinations. The product's selling price at node 2 is \$15 each when a demand exists for it. However, the amount of demand is a random variable x_2, with a uniform probability distribution given by

$$g_2(x_2) = \begin{cases} .1 & 10 \leqslant x_2 \leqslant 20 \\ 0 & \text{otherwise} \end{cases}$$

The product can be sold at node 3 for \$14 each; however, the amount of demand at node 3 is also a random variable x_3, with probability density

$$g_3(x_3) = \begin{cases} .005\,x_3 & 0 \leqslant x_3 \leqslant 20 \\ 0 & \text{otherwise} \end{cases}$$

Any of the perishable product shipped to nodes 2 or 3 but not sold is lost. Any amount not exceeding 30 units may be produced. The problem is to find the production number and the amount to ship to each destination that will maximize the expected profit.

The revenue at node 2 depends on both the amount shipped f_2 and the demand x_2. If the demand is less than the supply, then the amount sold is equal to the demand.

$$r_2(f_2, x_2) = 15x_2 \qquad \text{for } x_2 \leqslant f_2$$

If the demand exceeds the supply, the entire supply is sold.

$$r_2(f_2, x_2) = 15f_2 \qquad \text{for } x_2 \geqslant f_2$$

In a similar manner at node 3

$$r_3(f_3, x_3) = \begin{cases} 14x_3 & \text{for } x_3 \leqslant f_3 \\ 14f_3 & \text{for } x_3 \geqslant f_3 \end{cases}$$

The expected revenue at node 2 is

$$\bar{r}_2(f_2) = \begin{cases} 15f_2 & 0 \leqslant f_2 < 10 \\ -.75(f_2)^2 + 30f_2 - 75 & 10 \leqslant f_2 \leqslant 20 \\ 225 & f_2 > 20 \end{cases}$$

A similar analysis for node 3 yields

$$\bar{r}_3(f_3) = \begin{cases} -.0117(f_3)^3 + 14f_3 & 0 \leqslant f_3 \leqslant 20 \\ 186.4 & f_3 > 20 \end{cases}$$

These functions are continuous and concave. The profit function for the problem is

$$\text{Profit} = -10f_1 + \bar{r}_2(f_2) + \bar{r}_3(f_3)$$

To solve this problem using a minimum cost algorithm, we assign the cost functions to arcs 1, 2 and 3 as

$$h_2(f_2) = -\bar{r}_2(f_2)$$
$$h_3(f_3) = -\bar{r}_3(f_3)$$
$$h_1(f_1) = 10f_1$$

Since $\bar{r}_2$ and $\bar{r}_3$ are concave functions, h_2 and h_3 are convex functions (the negative of a concave function is convex). The important point to note here is that we have been able to write the stochastic network problem as an equivalent deterministic convex minimum cost flow problem.

11.5   PIECEWISE LINEAR APPROXIMATION

One approach to solving the separable convex problem is to make a piecewise linear approximation to the convex cost function at each arc, define a new variable, that is, an arc with a linear cost function, for each piece, and solve the resulting linear problem. This is illustrated in Figure 11.10. In this example, the original cost function is

$$h(f) = f^2$$
$$-2 \leqslant f \leqslant 2$$

This function may be approximated by dividing the range of f into four pieces each of length 1 and defining new variables f_1, f_2, f_3, and f_4 to represent the pieces. The approximate arc cost function formed is

$$h_p(f) = -3f_1 - 1f_2 + 1f_3 + 3f_4 + 4$$
$$f = f_1 + f_2 + f_3 + f_4 - 2$$

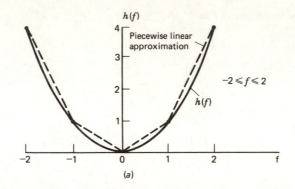

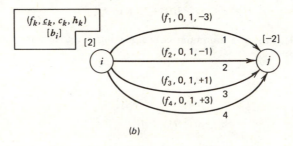

Figure 11.10
Piecewise Linear Approximation to a Convex Cost Function

Notice that this function is appropriate only if we add the additional condition that the variables increase in order; that is

$$\text{if } f_2 > 0 \quad \text{then } f_1 = 1$$
$$\text{if } f_3 > 0 \quad \text{then } f_2 = 1$$
$$\text{if } f_4 > 0 \quad \text{then } f_3 = 1$$

This condition is satisfied automatically for the minimum cost problem because, when $h(f)$ is convex, $h_1 < h_2 < h_3 < h_4$. Thus it always is optimal to increase f_1 to its maximum before f_2 is changed from zero. Similar statements hold for the other pieces.

For the case when $h(f)$ is concave, it will always be true that $h_1 > h_2 > h_3 > h_4$, and the conditions above will not be satisfied automatically. This leads to significant computational problems, which are considered in Chapter 12.

The network structure for the example case appears in Figure 11.10b. Note that the single arc has been replaced by four arcs. Node external flows have been added to account for the negative lower bound of the original arc. The constant term in h_p has been dropped. Of course, this does not affect the optimum flows but does affect the total cost of the optimum solution. The advantage of this transformation is that the problem can now be solved with linear techniques of Chapter 7. The disadvantage is that the problem is expanded in size by a factor equal to the number of pieces in the approximation. A further disadvantage is that the solution is a rough approximation with $h_p(f)= h(f)$ at only five points.

The accuracy of approximation is in the control of the modeler. More pieces in the approximation yield both a more accurate representation and a larger problem.

In general terms, consider the convex function on an arc $k(i,j)$.

$$h(f) \qquad \underline{c} \leqslant f \leqslant c$$

Define the breakpoints on the f axis $a_0, a_1, a_2, a_3, \ldots, a_p$ where $\underline{c} = a_0$ and $c = a_p$. Let $h(a_\ell)$ be the cost at the break point a_l. The parameters on the ℓth arc are

$$
\begin{aligned}
&\text{Lower bound: } \underline{c}_\ell = 0 \\
&\text{Upper bound: } c_\ell = a_\ell - a_{\ell-1} \qquad \ell = 1, \ldots, p \\
&\text{Cost } h_\ell = \frac{h(a_\ell) - h(a_{\ell-1})}{a_\ell - a_{\ell-1}}
\end{aligned}
\qquad (11)
$$

Changes on the external flows on the common terminal nodes of the arcs, $\ell = 1, \ldots, p$, are

$$
\begin{aligned}
\Delta b_i &= -\underline{c} \\
\Delta b_j &= \underline{c}
\end{aligned}
$$

These results follow directly from the discussion of the transformation for zero lower bounds given in Chapter 3, Section 3.2. The constant term dropped from the objective by the transformation is $h(\underline{c})$.

11.6 IMPLICIT PIECEWISE APPROXIMATION

The method of the last section suffers from two primary disadvantages.

1. The expansion in size of the network with the associated expansion in computer time and space cost.
2. The inaccuracy of the solution inherent in the approximation.

In this section, we describe a procedure that does much to overcome both limitations. The procedure is not as computationally efficient as the linear pure minimum cost algorithms of Chapter 7 but, of course, we are solving a more difficult problem.

As an example, consider the network of Figure 11.11. Each arc has the convex quadratic cost function

$$h_k = u_k f_k^2$$

We begin our procedure by assuming that a feasible flow F is defined that satisfies all node external flows. We define a step size Δ and allow flows to be changed only in multiples of Δ. A marginal cost network is constructed that describes the effects of changing arc flows. For each arc in the original network, the marginal network includes both a forward and a mirror arc.

Costs for the marginal network are defined as follows.

For forward arcs

$$
\begin{aligned}
h_k &= \frac{h_k(f_k + \Delta) - h_k(f_k)}{\Delta} &&\text{if } f_k + \Delta \leqslant c_k \\
h_k &= R &&\text{if } f_k + \Delta > c_k
\end{aligned}
\tag{12}
$$

For mirror arcs

$$
\begin{aligned}
h_{-k} &= \frac{h_k(f_k - \Delta) - h_k(f_k)}{\Delta} &&\text{if } f_k - \Delta \geqslant \underline{c}_k \\
h_{-k} &= R &&\text{if } f_k - \Delta < \underline{c}_k
\end{aligned}
\tag{13}
$$

where R is a large number.

The cost in the forward arc is the unit cost change caused by increasing f_k by an amount Δ. The cost in the mirror arc is the unit cost change by decreasing f_k

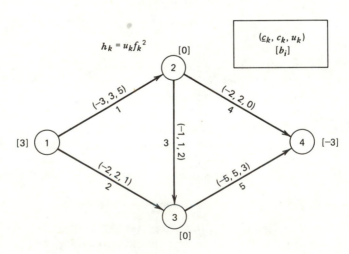

Figure 11.11
Example Network with Quadratic Cost Functions

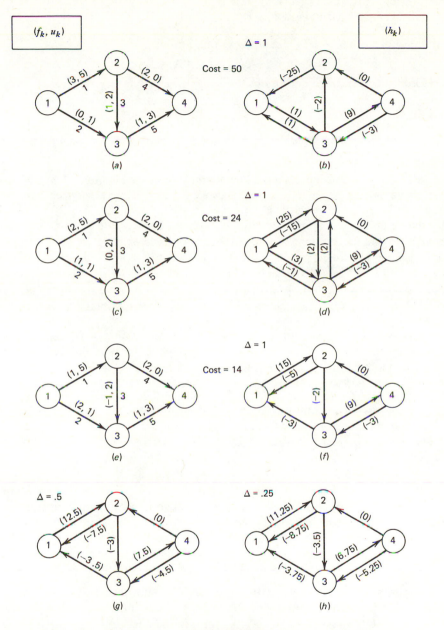

Figure 11.12
Example Application of Implicit Piecewise Approximation

by an amount Δ. Since $h_k(f)$ is a convex function

$$h_k(f_k) - h_k(f_k - \Delta) \leqslant h_k(f_k + \Delta) - h_k(f_k)$$

Therefore from the above definitions

$$h_k + h_{-k} \geqslant 0$$

When a flow change would result in a violation of either the upper or the lower bounds, h_k, or h_{-k}, respectively, is set to a very large value to prevent the flow change.

Figure 11.12a shows an assignment of flows to the problem of Figure 11.11, and Figure 11.12b shows the corresponding marginal network for $\Delta = 1$. Only arcs which have *not* been assigned a cost of R are shown in the figure.

A cycle of arcs in the marginal network implies a cycle in the original network on which flows may be changed without affecting node external flows. If a cycle in the marginal network has a negative total cost, then changing the flow around this cycle will result in a decrease in cost. Thus, if there exist no negative cycles, the current flow solution is optimal for the current value of Δ. If there exists a negative cycle, the flow can be changed on the cycle to obtain a less costly solution.

For the example problem, in Figure 11.12b the cycle, $M_c = \{-1, 2, -3\}$, has a cost of -26. changing the flow on this cycle by an amount 1 yields the new solution shown in Figure 11.12c with corresponding marginal network in Figure 11.12d. The new marginal network still has a negative cycle, $M_c = \{-1, 2, -3\}$, with cost of -10. Changing the flow on this cycle yields Figures 11.12e and 11.12f. Review of Figure 11.12f indicates that no negative cycles remain. This then is the optimal solution for $\Delta = 1$.

When the optimal solution is found for given Δ, we next make the determination whether the step size is small enough. The question relates to the desired accuracy of the solution, because for the convex problem Δ may be made as small as desired. If integer flows are desired, the optimum is obtained when $\Delta = 1$. If a smaller step size is desired, we reduce Δ by a factor of 2, recompute the admissibility and costs of the marginal network arcs, and continue the process of searching for negative cycles.

For the example problem, we continue with the marginal network of Figure 11.12g for $\Delta = .5$. There are still no negative cycles and the process continues with $\Delta = .25$. The marginal network in Figure 11.12h shows that the flows are optimal for this case as well.

The algorithm that implements this procedure is subroutine CONVEX. The algorithm begins with an initial feasible flow and an initial spanning tree. A subroutine HFUN must be provided to compute the cost functions for the arcs of the network. The HFUN shown below computes marginal costs for a quadratic function. Other functions will require a different version of HFUN. Notice that HFUN is called for each arc and that the costs for the forward and mirror arcs are stored in the lists HP and HM, respectively. After the proper values for the node potentials are assigned, PSHRTM is used to identify negative cycles in the marginal network.

CONVEX ALGORITHM

Purpose: To find the minimum cost flow when the arc costs are convex functions of arc flow. The functions are defined by subroutine HFUN. It is assumed that an initial feasible flow is defined by the vector **F** and that an initial spanning tree is defined.

1. (INITIAL) Set the flow step to the maximum value.
2. (COST) Determine the flow increase and flow decrease marginal costs for each arc. Determine the corresponding dual variables π.
3. (CYCLE) Find a negative cost cycle in the marginal network. If there are none, stop with the optimal solution for the current step size. Go to step 7. If a negative cycle is found, go to step 4.
4. (PATH) Use the back pointers to find the arcs on the cycle.
5. (FLOW) Change the flow on the cycle arcs by an amount Δ. Recompute the flow increase and flow decrease marginal costs for arcs whose flows have changed.
6. (DUAL) Reform the tree that existed before the cycle was formed. (KL is passed from PSHRTM.) Find the set of nodes in the tree rooted at the cycle node. Recompute the π values for these nodes. Return to step 3.
7. (DELCHG) Set the new flow step size as half of the previous value. If the new flow step size is smaller than the minimum desired value, stop. Otherwise, return to step 2.

HFUN ALGORITHM

Purpose: To calculate the forward and reverse marginal costs for changing flow by an amount DELF. The case shown computes these costs for a quadratic objective with constant term

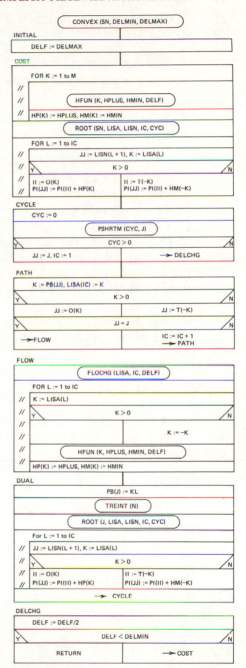

$H(K)$. For other convex functions a different version of HFUN should be supplied.

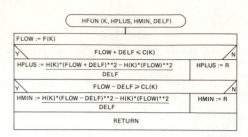

Notes: HPLUS is the marginal cost for the forward arc K.

HMIN is the marginal cost for the mirror arc $(-K)$.

FLOW is the flow on arc K.

DELF is the step size for flow change.

The reader will recall that SMNSPM is used by PSHRTM to determine the most favorable arc to enter the currently nonoptimal shortest path basis. Because the arc costs are convex functions for this particular application, a new version of SMNSPM called algorithm SMNCNVX is provided. As for strictly linear costs, only one of the forward arc, mirror arc pair may be favorable for consideration as the entering arc. However, when the arc cost is a convex function, d_k is not necessarily equal to $-d_{-k}$. This is the single reason that algorithm SMNCNVX is required.

One additional new algorithm, ORGSCNV, a modified version of ORIGS, is also required. Because of the presence of convex arc cost functions, the simple modification present in ORIGS to drive all arc lower bounds to zero is no longer necessary.

SMNCNVX ALGORITHM

(A version of SMNSPM)

Purpose: To find the arc that has the most negative value of d_k for arcs with convex arc cost functions.

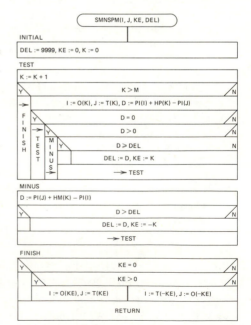

1. (INITIAL) Set DEL = 9999, $k_E = 0$, and $k = 0$.

2. (TEST) Increment the arc index and check if all arcs have been considered. If so, return. Else compute d_k for the forward arc. If $d_k = 0$, go to the next arc. If $d_k > 0$, check the mirror arc. If $d_k \geqslant$ DEL, go to the next arc. Else, set DEL = d_k and $k_E = k$ and go to the next arc.

3. (MINUS) Compute d_{-k}. If $d_{-k} >$ DEL, go to the next arc. Else, set DEL = d_{-k} and $k_E = -k$ and go to the next arc.

4. (FINISH) If $k_E = 0$, return. Set I equal to the origin node of k_E. Set J equal to the terminal node of k_E.

ORGSCNV ALGORITHM
(A version of ORIGS)
Purpose: To accept an arc data item for arcs with convex arc cost functions and store it in an arc list ordered by ascending origin node.
ORGSCNV

1. (INITIAL) If first call to ORIGS, set all node pointers to 1. Else go to step 2.
2. (MOVE) Increase M by one. Increase all node pointers greater than I by one. Move all arcs above the new entry one index higher on the list.
3. (ARC) Insert arc in the last position alloted to node I.

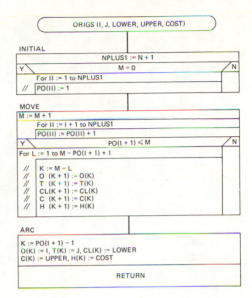

11.7 HISTORICAL PERSPECTIVE

The results of this chapter are based on several important precursors. Charnes and Cooper (1958) suggest a simplex type algorithm for convex network problems using an arbitrary step size for piecewise linear approximations. Beale (1959) describes a primal approach for uncapacitated, convex cost, transportation problems. Minty (1960) introduces monotone networks for which each arc has a monotonically increasing characteristic. This is equivalent to the marginal cost function of this chapter. Minty (1961) describes how the theory of monotone networks can be used to solve nonlinear electrical network problems and Minty (1962) extends the application to convex cost network flow programming problems. Minty's approach is a primal–dual method.

Menon (1965) and Weintraub (1974) describe primal nonbasic procedures for the convex problem. Both search for a cycle on which a flow change will result in a reduced cost. Menon suggests an exhaustive search of all cycles while Weintraub uses an assignment algorithm to find the best cycle. Hu (1966) uses a flow incremental approach. Collins et al. (1977) describe a convex simplex approach using a basis tree. Flow is changed on a cycle by an amount that produces the greatest reduction in cost.

The approach used in this chapter that solves the convex problem with successively smaller step sizes has a predecessor in Edmonds and Karp (1972), who suggest a similar scaling approach for linear network problems. Zadeh (1973) shows that for certain network problems scaling is necessary to obtain an algebraic rather than exponential increase in computational cost with problem size.

The principal early work relating physical network problems to convex optimization is Millar (1951). His paper describes the concept of content used in this chapter. Dennis (1959) describes electrical equivalents to various network programming problems. Later authors,—Birkhoff and Diaz (1956), Minty (1961), Stern (1969), and Collins (1977)—describe the application of mathematical programming techniques to the solution of physical network problems.

There have been several papers on stochastic networks of the type considered in this chapter. These particularly relate to the stochastic transportation problem in which demands are known only with uncertainty. Authors writing on this problem are Elmaghraby (1960), Williams (1963), Swarc (1964), and Wilson (1972).

EXERCISES

1. Use CONVEX to solve the problem of Figure 11.12. Use the same initial solution and show all intermediate trees with node potentials and d_{KE}'s from PSHRTM.
2. Are the divisions by Δ in equations (12) and (13) necessary? Why or why not?
3. Describe, in detail, a general technique to provide the initial spanning tree and flows for a network with convex cost functions prior to applying algorithm CONVEX.
4. Solve the problem of Figure 11.11 by approximating each arc, where required, by four arcs with strictly linear arc costs.
5. Solve the problem of exercise 2.9, Chapter 2, if the costs of cooling for plants A and B are changed to $10X_A^2$ and $15X_B^2$, respectively. Compare this solution with the solution to exercise 2.9, Chapter 2.
6. Solve for the voltages and currents in Figure 11.5 using algorithm CONVEX. Assume $E = 10$, $R_1 = 1$, $R_2 = 3$, $R_3 = 1$, $R_4 = 2$, $R_5 = 1$, and $R_6 = 2$. Compare your results with those obtained through the use of Kirchoff's laws.
7. The pressure drop in a pipe is given by the Hazen–Williams equation

$$h = aQ^{1.852}$$

where h is the pressure drop, Q is the flow in the pipe, and a is a constant that depends on the diameter, length, and physical characteristics of the pipe. In terms of network flows, the Hazen–Williams equation can be written for a pipe $k(i,j)$ as

$$\pi_i - \pi_j = a_k f_k^{1.852}$$

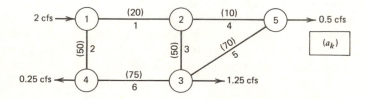

Here π_i is interpreted as the head at node i and f_k as the flow in pipe k. The accompanying figure shows a pipeline network with the values of a_k shown adjacent to the figure. Inflows and outflows are also shown. Find the solution for pipe flows and heads. Assume the head at node 3 is fixed at 100 ft.

8. The accompanying figure shows a water distribution system with a reservoir. The reservoir has the effect of maintaining a head of 200 ft. at node 1. The flow out of the reservoir is a variable to be determined as well as the flows and heads in the remainder of the system.

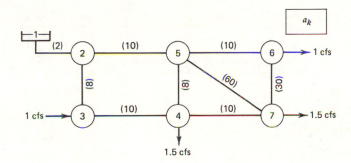

9. You desire to set up a system to provide some social service. The accompanying figure shows a representation of the geographical area for which you want to provide the service. It has been divided into cells. The number in each cell is the average hourly demand for a service to be provided the system. The actual demand is a Poisson process (i.e., the number of requests for service in an hour has a Poisson distribution with mean λ given by the numbers in the cell). The service can be provided at either cell 2 or cell 8. A facility exists at each location, which can be modeled as a single channel queue. The service time is a random variable that has an exponential distribution. The service rate at cell 2 is $\mu_2 = 20/\text{hr}$ and at cell 8 is $\mu_8 = 40/\text{hr}$. You want to make an assignment of the demand to service centers in order to minimize time of travel plus waiting and service time. Assume all travel is rectilinear and travel time is 10 minutes per mile. Set up the network model to solve this problem.

1 8	2 10	3 15	4 7
5 12	6 8	7 2	8 10

2 miles

— 4 miles —

10. Solve the simple transportation problem of Figure 11.9 using CONVEX.

11. Find the feasible flows that minimize the total cost on the given network.

Arc	$h(k)$
1	$3f_1$
2	$2f_2$
3	$3f_3$
4	f_4^2
5	$f_5 + f_5^2$
6	$5f_6$
7	f_7^2
8	$2f_8$
9	f_9

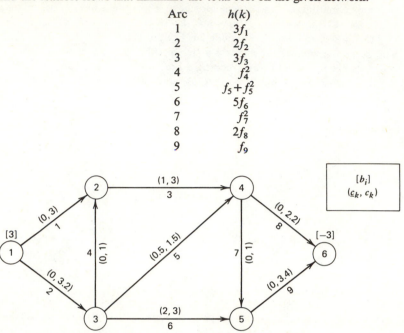

12. You are in charge of providing labor for a manufacturing shop for the next six months. Currently, there are 20 people working for the shop. Each worker can manufacture five units of product per month. The cost of hiring and training a new worker is $500. The cost of laying off a worker is $1000. At the end of the six-month period you want 20 persons working in the shop. The labor cost per employee is $1000 per month. Your product is perishable so that it can only be sold in the month it is produced. Demand in each month is a random variable with a uniform distribution; its range is shown in the table below. If demand is lower than production capacity, some workers are idle. If demand is higher than labor cost, sales are lost with a penalty of $20 per lost sale. How many workers should be hired or laid off at the beginning of each month to minimize the expected cost? You may also lay-off workers at the end of month 6.

Month	Minimum Demand	Maximum Demand
1	50	150
2	100	200
3	75	175
4	50	150
5	200	250
6	100	200

CHAPTER 12
CONCAVE COSTS

Given our success at dealing with convex arc cost functions as a simple variant of the linear cost case, one would hope that we could handle concave arc cost functions as easily. Unfortunately, this is not the case. A simple change from convex to concave cost curves produces a major change not only in the solution characteristics but also in the procedures necessary to obtain the solution. While solutions to the convex case may be achieved for several thousands of arcs, the presence of fewer than 100 arcs with concave costs may result in unacceptable computer costs to obtain an optimum solution.

There are several reasons for this result. For the convex problem and the linear problem, a local minimum is also a global minimum. Thus, tests that indicate a local minimum are also sufficient to indicate a global minimum. For the concave problem, there may be many local minima. A solution procedure must in some way search over all these local minima to locate the global minimum. Such a search requirement presents a formidable computational problem. Indeed, this problem is one of the class of problems that has been identified as *NP complete*. No algorithm bounded in polynomial computation time has been found for *any* problem in this class and most researchers doubt that one exists. All algorithms that do solve the problem can only be bounded by exponential functions, that is, those in which computational effort increases as k^{m_c}, where m_c is the number of arcs with concave cost functions and k is some constant.

We describe in this chapter an implicit enumeration algorithm adapted from those used to solve $0-1$ integer programming problems. This algorithm is $o(2^{m_c})$ in computational cost. Although this is certainly not a desirable bound, the algorithm can be used to solve small but important practical network models with concave arc cost functions.

12.1 APPLICATIONS

A concave function is illustrated in Figure 12.1. When one connects any two points on a concave function with a straight line, the line lies entirely on or below the function. When the function is continuous and differentiable, concavity is indicated when the second derivative is negative at all feasible points. Two important instances of a concave cost are illustrated in Figure 12.2a and 12.2b. An arc exhibiting the cost function of Figure 12.2a reflects economies of scale. Here cost increases with increasing flow but the marginal cost decreases. This figure might represent the cost of purchasing some raw material that provides price discounts for larger quantities. It might also represent the construction cost of a facility where the flow is the size of the facility. Construction costs often exhibit economies of scale.

Figure 12.2b shows a more idealized cost function with cost of zero for zero flow. When flow exceeds zero, a fixed cost is incurred, h_F. In addition, a linear cost h_V is charged for flow. This cost curve could represent the construction and operating cost of a road. With zero flow, the road is not built; hence, cost is zero. With flow greater than zero, the road must be built. The fixed cost associated with right-of-way, design, and other costs that are independent of size is expended. As the size of the road, flow, is increased, costs that do depend on size are represented by the linear portion of the curve.

The curve of Figure 12.2b could also represent the cost of an inventory replenishment decision. Here the flow is the amount of product ordered. The fixed cost is the reorder or setup cost. The variable cost is the cost of the product.

In this chapter, we consider *explicitly* only cost curves like that given in Figure 12.2b. It is possible to use this type of cost curve to represent more complex concave cost curves by using multiple arcs and a piecewise linear approximation

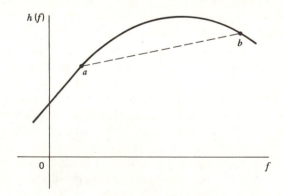

Figure 12.1
A Concave Function

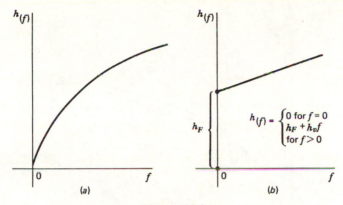

Figure 12.2
Example Concave Cost Functions
(*a*) Economies of Scale (*b*) Fixed Charge

as in Figure 12.3. The example problem requires three arcs to provide the approximation. Note that the model does allow flows that will obtain the cost on the dotted line on Figure 12.3. For example, $f_2 = c_1/2$, $f_1 = 0$ is a feasible solution. Such flows will never be present in an optimum solution, however, because there exists another feasible flow with less cost, $f_1 = c_1/2$, $f_2 = 0$. Note that the piecewise model allows a total capacity of $c_1 + c_2$ for the sum of the two flows where the original arc capacity was c_2. To eliminate this possibility, we add a third arc in series with the combination with capacity c_2.

The modeling capability provided by concave cost curves allows a number of interesting new applications. One is the facility location problem as illustrated in Figure 12.4. Here the nodes on the right represent customers. The external flows at each of these nodes are negative and represent the demand of each customer. Arcs coming into these nodes are transportation links with linear costs and capacities. The arcs leaving the source represent potential sites for facilities that are to provide for the demand. The values of h_F and h_v are the fixed and variable costs of construction and operation of each site. The capacity of the arc is the proposed size of the facility. A minimum cost flow algorithm would select the sites at which facilities should be built to minimize cost. Transportation arcs with fixed costs can also be included to represent cost of installing a transportation link.

Network design problems that require the selection of arcs to be included in a network can often be described using arcs with concave costs. Such problems might arise in the design of transportation or pipeline networks.

Production control models in which flow represents the quantity produced are more accurately modeled with concave costs with the production setup or inventory reordered cost modeled as a fixed charge.

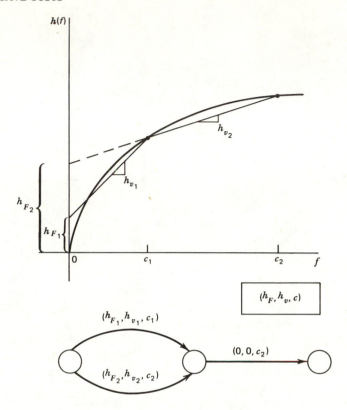

Figure 12.3
Piecewise Linear Approximation for a Concave Function

Indeed, a large number of discrete decision problems can be modeled using concave cost network models. The analyst should be aware, however, of the great difficulty that will be experienced with solving these models. The algorithms described in this chapter provide a general procedure for solving fixed charge concave problems. It is difficult to define precisely the size of problem for which it will efficiently provide solutions. The computational cost of solution depends to a great extent on factors other than the number of arcs with concave costs. However, solutions should be possible if this number is less than 50. Numbers between 50 and 100 would present considerably more difficulty. With more than 100 such arcs, it is not likely that an optimum solution will be obtained.

Many problems to which these techniques may be applied admit more efficient specialized solution approaches. When size of the problem makes the algorithms of this chapter impractical, the analyst should seek procedures that take advantage of the special structure of the decision problem.

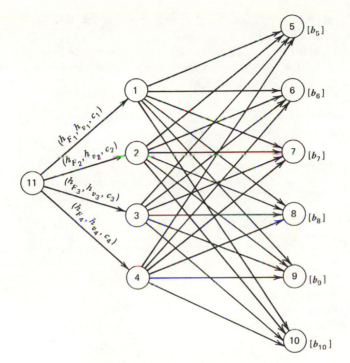

Figure 12.4
Example Facility Location Problem

12.2 NOTATION

Given a network $D=[N,M]$, we identify a subset K_c of m_c arcs with concave cost functions, that is, $K_c=\{k_c(1),k_c(2),\ldots,k_c(m_c)\}$, where $k_c(\ell)$ is the arc index in the original network description. Since the analyst will often desire to change this set without changing the data for the original network, this method of defining a set of concave arcs in the original network allows a welcome flexibility for applied problems.

For each concave arc $k_c(\ell)$, we identify a fixed cost $h_F(\ell)$ and variable cost $h_v(\ell)$. Thus the cost function implied by these numbers for arc $k=k_c(\ell)$ is

$$h_k(f_k) = \begin{cases} 0 & f_k=0 \\ h_F(\ell)+h_v(\ell)\cdot f_k & 0<f_k \leqslant c_k \end{cases}$$

where a finite arc capacity c_k must be defined for each arc. Conceptually, the procedure described here may be applied to either pure or generalized network problems with linear or convex cost. However, for convenience in exposition and programming, we assume the original network is defined as a pure linear minimum cost flow problem. The principal effect of this is that the flow variables may be assumed to be integer.

In order to read the data required for the concave problem, we provide the READCC subroutine, which is designed to be called immediately after execution is returned from subroutine READ.

READCC ALGORITHM
Purpose: To read in parameters associated with a concave minimum cost flow problem.

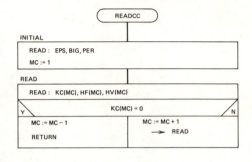

1. (INITIAL) Read in program parameters:

    EPS: a small number used to test for 0.
    BIG: a large number used to test for ∞.
    PER: the accepted deviation from the optimum objective allowed.
    ZB: the objective for some known feasible solution for the concave problem. If none is known, let ZB = BIG

2. (READ) Read in the parameters of arcs with concave costs:

    KC: the index of the arc in the original network.
    HF: the fixed charge associated with the arc
    HV: the variable charge associated with the arc.

    This step repeats until a blank is encountered and then returns.

12.3 EXHAUSTIVE ENUMERATION
In the course of the algorithm, we associate the variable x_i with the ith concave arc. We say that the ith arc is "closed" with no flow allowed if $x_i = 0$. We say that the ith arc is "open" with flow allowed between 0 and c_k if $x_i = 1$. Let **X** be the vector $[x_1, x_2, x_3, \ldots]$.

With these variables defined, the concave cost problem can be written as a $0-1$ integer programming problem.

$$z = \sum_{i=1}^{m_c} h_F(i)x_i + \sum_{k=1}^{m} h_k f_k \tag{1}$$

where for concave arcs

$$h_k = h_v(i) \qquad \text{for } k = k_c(i)$$

$$\text{st. } \mathbf{Af} = \mathbf{b}$$

$$0 \leqslant f_k \leqslant c_k \qquad k \in M$$

$$f_k \leqslant x_i c_k \qquad \text{for } i = 1, \dots, m_c \quad \text{and} \quad k = k_c(i)$$

$$x_i = 0, 1 \qquad \text{for } i - 1, \dots, m_c$$

The constraints that do not involve the $\mathbf{X}$ variables represent the conservation-of-flow and bounding constraints of the original network problem. The constraints $f_k \leqslant x_i c_k$ for the concave arcs force the flow variable f_k to be equal to zero when $x_i = 0$, and establish the original bound on f_k if $x_i = 1$.

One solution technique for this problem is exhaustive enumeration of the possible realizations of $\mathbf{X}$. A particular solution to the problem is an assignment of 1 or 0 to each x_i. Given any realization of $\mathbf{X}$, a linear network problem remains whose optimal objective function value can be obtained by the procedures of Chapters 7 and 8.

If one tries all possible assignments and evaluates the solution to each, choosing the solution with the smallest cost will obtain the optimum. Of course, there are 2^{m_c} possible assignments, which is a large but finite number for moderate values of m_c.

Exhaustive enumeration is not a conceptually difficult process. For a problem of three variables, the set of $2^3 = 8$ possible solutions is obtained by listing all possible binary numbers consisting of three digits.

$$(x_1, x_2, x_3)$$
$$(0 \quad 0 \quad 0 \ )$$
$$(0 \quad 0 \quad 1 \ )$$
$$(0 \quad 1 \quad 0 \ )$$
$$(0 \quad 1 \quad 1 \ )$$
$$(1 \quad 0 \quad 0 \ )$$
$$(1 \quad 0 \quad 1 \ )$$
$$(1 \quad 1 \quad 0 \ )$$
$$(1 \quad 1 \quad 1 \ )$$

In our computer implementation of this process, we perform enumeration by constructing a binary tree as illustrated in Figure 12.5, where the circles are called *vertices* and the line segments are called *branches*.

The tree is divided into levels as shown in the figure. Each vertex represents a set of realizations of the X vector. The single vertex at level 0 represents the set of all realizations (8 in the example case). The vertices at level 1 are formed by making a binary decision on one of the X variables (x_1 in the illustration). The branch to the left represents all realizations with $x_1 = 1$. We represent this by the symbol 1 in vertex 9. The branch to the right represents all realizations with $x_1 = 0$. The symbol $\bar{1}$ in vertex 2 represents this decision. Subsequent levels describe additional binary decisions that separate the realization set into smaller and smaller subsets. Finally the vertices at the bottom of the tree represent the eight individual realizations of the problem.

Each vertex of the tree may be identified by a sequence of numbers indicating the binary decisions made to reach the vertex. The numbers in this sequence represent the fixed variables. The appearance of i in the sequence reveals that x_i has been fixed to the value 1. The appearance of $\bar{i}$ reveals that x_i has been fixed to the value 0. Numbers not in the sequence identify the free variables, those that have not yet been assigned a value. For example, vertex 6 of Figure 12.5 may be identified by the sequence $\bar{1}2$, showing that $x_1 = 0$, $x_2 = 1$ and that x_3 is free.

We call the process of creating two new vertices from a single vertex the *separation* process. If a sequence $\{s_\ell\}$ identifies a vertex of the tree at level ℓ, the two new vertices at level $\ell + 1$ are assigned the sequences $\{s_\ell, i\}$ and $\{s_\ell, \bar{i}_\ell\}$. It is clear that this process *separates* the set of solutions defined by $\{s_\ell\}$ into two mutually exclusive subsets that together comprise the entire set defined by s_ℓ.

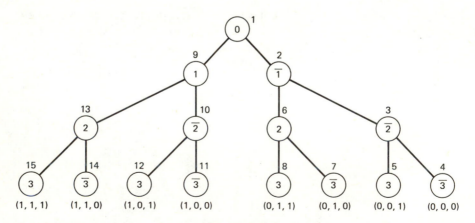

Figure 12.5
A Binary Tree with Variables Fixed in Natural Order

There is no requirement in the subsequent discussion that the variables be explored in numerical order or that the order be the same for all branches of the tree. Figure 12.6 illustrates one of the several alternatives.

The computational procedures explore the enumeration tree one path at a time (all paths originate at level 0) and do not explicitly store the tree in computer memory. Thus, we need a notation that describes where we are in the dynamic enumeration process. This is accomplished by using the underline symbol, together with the sequence notation. The symbol i appearing in a sequence indicates that $x_i = 1$ and that the alternative branches where $x_i = 0$ have already been explored for this part of the tree. The symbol $\bar{i}$ indicates that $x_i = 0$ and that the alternative paths for $x_i = 1$ have already been explored for this part of the tree. To explore the tree in the order defined by the vertex numbers of Figure 12.6 would require the following sequences.

$$0, 3, 32, 321, 32\bar{1}, 3\underline{2}, 3\underline{2}1, 3\underline{2}\bar{1}, \underline{3}, \underline{3}\bar{1}, \underline{3}\bar{1}2,$$

$$\underline{3}\bar{1}\underline{2}, \underline{3}\underline{\bar{1}}, \underline{3}\underline{\bar{1}}\bar{2}, \underline{3}\underline{\bar{1}}\underline{2}$$

The sequence and underline notation allows the tree to be searched in an arbitrary order with the provision that a path be extended to as high a level as possible before another path is explored. This is called a *backtrack* procedure because the path is extended as far as possible and then the process backtracks to the last vertex originating as yet unexplored branches. The exhaustive enumeration algorithm allows the tree to be searched in the manner just described.

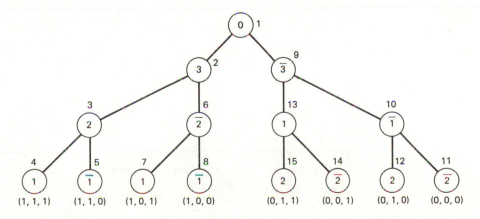

Figure 12.6
A Binary Tree with the Variables Fixed in an Arbitrary Order

EXHAUSTIVE ENUMERATION ALGORITHM

1. (INITIAL) Start with an empty sequence. Let $Z_B = \infty$ and $\mathbf{X}_B = 0$ with all variables free.

2. (SEPARATE) If there are no free variables, go to step 3. Else, choose a free variable i and determine if it should be set to 0 or 1. If it is to be set to 1, add i to the sequence. If it is to be set to 0, add $\bar{i}$ to the sequence. Repeat step 2.

3. (EVALUATE) All the variables are fixed and a complete solution is defined. Identify the $\mathbf{X}$ defined by the sequence. Evaluate the objective Z for $\mathbf{X}$; go to step 4.

4. (REPLACE) If $Z < Z_B$ replace Z_B by Z and replace $\mathbf{X}_B$ by $\mathbf{X}$. Go to step 5.

5. (BACKTRACK) Start at the end of the sequence (to the right) and delete all underlined symbols until a symbol is reached that is not underlined. Underline and change the state of the symbol (i.e., i becomes $\bar{i}$, $\bar{i}$ becomes $\underline{i}$.) Go to step 2. If all symbols in the sequence are underlined, stop; the enumeration is complete. The optimum solution is $\mathbf{X}_B$ with value Z_B.

This algorithm uses $\mathbf{X}_B$ to store the "best solution found so far" in the process. We call this solution the *incumbent solution*. The value of Z_B is the objective value for the incumbent solution. A solution remains incumbent until a better solution is found. The incumbent solution at the termination of the algorithm is the *optimum solution*. We leave the making of a flowchart of this algorithm to the student (see exercise 1).

12.4   IMPLICIT ENUMERATION

Because of the large number of possible solutions for even a relatively small problem (over 1,000,000 for $m_c = 20$), exhaustive enumeration is not practical in most cases. For this reason, we would like to devise tests that allow the program to "fathom" tree vertices before they are enumerated. To fathom a vertex is to reject it from consideration in the search for a solution without actually evaluating the vertex. Fathomed vertices are, in a sense, enumerated without actually being generated. Thus we call this procedure *implicit enumeration*.

We consider in this section a fathoming test that uses a lower bound on the objective function. Assume we have obtained in some manner a feasible solution $\mathbf{X}_B$ with value Z_B. The current vertex s_t represents some set of feasible solutions that could be enumerated and identified by setting the free variables to every possible combination. By some method, we obtain a *lower bound* Z_{LB} on all solutions in this set. We are sure that all solutions in the set have a value larger than Z_{LB}. If it happens that

$$Z_{LB} \geqslant Z_B$$

then all the solutions in the set can be judged nonoptimal or at least no better than the solution currently in hand. All the vertices in the tree that are descendants of the vertex under consideration can be implicitly enumerated. If on the other hand

$$Z_{LB} < Z_B$$

we cannot implicitly enumerate the vertices.

The efficiency of this test depends on two things.

1. The quality of the lower bound (the larger the bound the better).
2. The sensitivity of the objective function value to the decision variables in $\mathbf{X}$.

The quality of the lower bound is usually related to the computational difficulty in obtaining it. For many problems, a poor lower bound can be found very easily and a good lower bound can be found only with considerable computational effort. The analyst must select a good compromise between goodness of bound and computational efficiency.

The sensitivity of the objective function's value to changes in the values of the decision variables is a characteristic of the problem being solved. Problems for which the optimum value of Z is significantly less than values for other decisions will enhance the effectiveness of the bounding test. If there are many near optimal solutions, the bounding test will not be as effective. The effectiveness of the fathoming tests is important because of the size of the tree. When $m_c = 20$, the exhaustive enumeration tree has over 2×10^6 vertices. If the lower bound test is 90% effective, there are still 2×10^5 vertices to be evaluated.

Thus, the implicit enumeration procedures will allow the analyst to solve significant problems, but they will never have the capacity to solve the size of problems easily handled by linear or convex network codes. Where the latter can handle thousands of variables, the solution of a problem with $m_c = 100$ will probably be very difficult. This is the reason for the great difference between the difficulty of the concave and the linear or convex network flow programming problems.

We will strengthen the elimination test by using a parameter that specifies the allowed percent deviation from the optimum, PER. The bounding test then becomes:

$$\text{If } Z_{LB} \geqslant Z_B - \text{PER } |Z_B|/100$$

then eliminate the vertex under consideration.

This test may result in the discarding of the optimum solution. We are assured, however, that the resultant solution is within a percentage, PER, of the optimum. This is often a useful compromise that obtains a "good" though not necessarily optimal solution with acceptable computational cost. A possible strategy is to solve the problem with successively smaller values of PER until the optimum solution is assured (PER $= 0$) or the computer budget runs out.

12.5   LOWER BOUND

At an intermediate stage of the enumeration process, certain of the $0-1$ decision variables are set to 1 and certain of them are set to 0. The remainder of the variables are free. For convenience, define W^+ as the set of variables set to 1, W^- as the set of variables set to zero, and W^0 as the set of free variables. The mathematical program defining the concave network problem at this intermediate stage becomes

$$\text{Min. } \sum_{i \in E^+} h_F(i) + \sum_{i \in W^o} h_F(i)x_i + \sum_{i=1}^{m} h_k f_k \qquad (2)$$

where

$$h_k = h_v(i) \qquad \text{for } k = k_c(i)$$

$$\text{st.} \quad \mathbf{Af} = \mathbf{b}$$

$$0 \leqslant f_k \leqslant c_k \qquad \text{for } k \in M$$

$$f_k \leqslant x_i c_k \qquad \text{for } i \in W^0, k = k_c(i)$$

$$f_k = 0 \qquad \text{for } i \in W^-, k = k_c(i)$$

$$x_i = 0, 1 \qquad \text{for } i \in W^0$$

The flows are forced to zero on the closed arcs. The sum of the fixed charges for the open arcs appears as a constant term in the objective.

To obtain a lower bound to all solutions associated with W^0, we first drop the condition that x_i be integer and replace it with

$$0 \leqslant x_i \leqslant 1 \qquad \text{for } i \in W^0$$

The new problem is said to be a relaxation of the problem of (2) because the integrality condition has been relaxed. Since the feasible region is enlarged with a relaxation, the optimum objective for the relaxation must be a lower bound to the solution of the integer problem.

Now note that, for the relaxed problem, the optimum solution will always yield

$$x_i = \frac{f_k}{c_k} \qquad \text{for } i \in W^0, \, k = k_c(i)$$

This is true since the objective coefficient for x_i, $h_F(i)$, is positive and x_i only appears in its simple upper bound constraint and in the single constraint that bounds f_k. Therefore, we can replace x_i by using the relation: $x_i = f_k / c_k$. Thus the coefficient for f_k may be reexpressed as

$$h_k = \left[\frac{h_F(i)}{c_k} + h_v \right]$$

Using this result, we find that the mathematical statement of the relaxed problem is

$$Z_{LB} = \text{Min.} \sum_{i \in W^+} h_F(i) + \sum_{i=1}^{m} h_k f_k \qquad (3)$$

where

$$h_k = h_v(i) \qquad \text{for } k = k_c(i), \, i \in W_i^+$$

$$h_k = \frac{h_F(i)}{c_k} + h_v(i) \qquad \text{for } k = k_c(i), \, i \in W_i^0$$

$$\text{st.} \quad \mathbf{Af} = \mathbf{b}$$

$$0 \leqslant f_k \leqslant c_k \qquad \text{for } k \in M$$

$$f_k = 0 \qquad \text{for } k = k_c(i), \, i \in W_i^-$$

This is a linear network flow problem that can easily be solved using the linear network flow optimization procedures of Chapters 7 and 8.

It is instructive to note how the linearized cost for f_k in the relaxation relates to the concave cost function it replaces. This is shown in Figure 12.7. The linear cost function is only accurate at the end points, $f_k = 0$ and $f_k = c_k$. At all other points, the linear function is an underestimate of the concave function. This is an additional justification for the fact that the solution of the relaxed problem provides a lower bound to the solution of the original mixed integer programming problem.

In the algorithms described below, we use a slightly different form of the relaxed problem that replaces the restriction that

$$f_k = 0 \qquad \text{for } k = k_c(i), \, i \in W_i^-$$

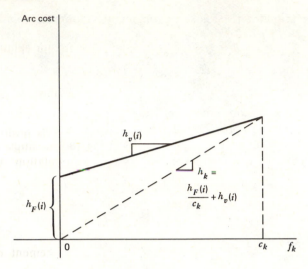

Figure 12.7
Linear Underestimate for an Arc with a Concave Cost Function

with the specification

$$\left.\begin{array}{c} h_k = R \\ 0 \leqslant f_k \leqslant c_k \end{array}\right\} \quad \text{for } k = k_c(i), \, i \in W_k^-$$

where R is a large positive number.

With a large cost on closed arcs, the flow on these arcs will be nonzero only if such a flow is required in order to obtain a feasible solution. The algorithms test the flows on closed arcs. If one is greater than zero, an infeasibility indicator is set and the tree vertex is fathomed. The advantage to this procedure is that only arc costs are changed during the enumeration process. Since the constraint set is never changed, once a feasible solution is obtained for the network flow problem it will never be lost. Subsequent solution of the linear problem can then be found using a primal algorithm.

The optimal solution to a relaxed problem can be rounded to a feasible solution of the original problem by letting

$$\begin{array}{ll} x_i = 1 & \text{if } f_k > 0 \\ x_i = 0 & \text{if } f_k = 0 \end{array} \quad \text{for } i \in W^0 \quad \text{and} \quad k = k_c(i)$$

The values of the fixed variables are already set to 0 or 1. We call this a rounded solution $\mathbf{X}_R$. The corresponding objective is called Z_R where

$$Z_R = \sum_{i=1}^{m_c} h_F(i) x_i + \sum_{i=1}^{m} h_k f_k$$

where

$$h_k = h_v(i) \qquad \text{for } k = k_c(i)$$

A rounded solution is used to replace an incumbent best solution when $Z_R < Z_B$.

12.6 THE IMPLICIT ENUMERATION ALGORITHM

The solution of the concave problem is found using several new subroutines. The major algorithmic steps are accomplished with the ENUMER subroutine. The algorithms use notation that is slightly modified from that of the previous sections. A vector **W** is defined that assumes the values

$$W_i > 0 \qquad \text{if } i \text{ is fixed to 1}$$
$$W_i = 0 \qquad \text{if } i \text{ is free}$$
$$W_i < 0 \qquad \text{if } i \text{ is fixed to 0}$$

Thus, **W** replaces the sets W^+, W^-, and W^0 used previously. The numerical value used for W_i is the level number at which the variable is fixed.

ENUMER ALGORITHM

Purpose: To control the operations of the implicit enumeration procedure that solves the fixed charge concave minimum cost network flow problem.

1. (INITIAL) Call SETUP to make the initial linear approximations for all arcs. Set the level indicator L to 0.

2. (FATHOM) Solve the linear approximation associated with the tree vertex. If the lower bound is larger than the objective for the incumbent best solution, go to step 4. If the lower bound is less than the incumbent, find the rounded solution with the associated objective value. If Z_R is not better than the incumbent Z_B, go to step 3. If it is better, replace the incumbent. If the rounded objective equals the relaxed objective, no improvement can be obtained by continuing. Go to step 4. Otherwise, go to step 3.

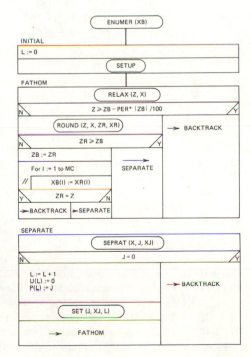

3. (SEPARATE) Find the fractional free variable on which to separate. If there is no separation variable, go to step 4. If there is a separation variable, advance to the next level of the tree. Add the variable to the sequence. Set the underline to 0. Set the variable to the proper value. Return to step 2.

4. (BACKTRACK) If the underline for level L is 0, find the variable J for step L, set the underline for this level to 1, find the new value for J, and go to step 5. If the underline is equal to 1, backtrack one level in the tree. If the level zero is encountered, stop; the process is completed. If not, repeat step 4.

5. (CHANGE) Make all variables fixed at or below level L into free variables. Set variable J to the value determined in step 4. Return to step 2.

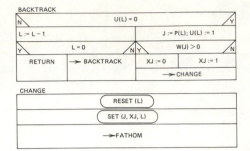

The tree sequence is defined by the pointer **P**, which has a value for each level. We use the vector **W** to keep track of the values assigned to the fixed variables. The vector **U** contains the underline information. If $U(L)=1$, the corresponding entry in $P(L)$ would be underlined in the sequence. If $U(L)=0$, it is not underlined. The variable L keeps track of the current tree level.

Other variables used are

BIG: a large number.
EPS: a small number.
Z: the objective for the relaxed problem.
X: the **X** solution for the relaxed problem which may be fractional for free variables.
ZB: the objective for the incumbent best solution.
XB: the $0-1$ vector for the incumbent best solution.
ZR: the objective for the rounded solution.
XR: the $0-1$ vector that is the rounded solution.
J: the separation variable.
XJ: the value J is to be set to, 0 or 1.

The subroutines used are

ENUMER: Controls the implicit enumeration tree search.
SETUP: Sets up the initial linear coefficients.
SET: Sets variable J to the value XJ.
RELAX: Solves the relaxed problem and computes the values of **X** and **Z**.
ROUND: Finds a rounded solution based on **X**.
SEPRAT: Chooses a separation variable and the value of the separation variable to be first explored.
RESET: Resets all the variables fixed at or below level L from fixed variables to free variables.

SETUP ALGORITHM

Purpose: To set up the initial linear approximations for the relaxed problem and to set all variables free.

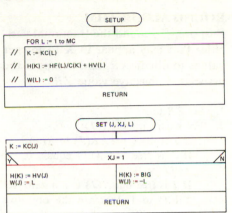

SET ALGORITHM

Purpose: To fix variable J to the value XJ at level L. Find the arc K associated with variable J. If $XJ=1$, set the linear cost for arc K equal to the variable cost of J. Set $W(J)$ to L. This indicates that variable J has been fixed at 1 at level L. If $XJ=0$, set the linear cost of arc K to a large value. Set $W(J)$ to $-L$. This indicates that variable J has been fixed at 0 at level L.

RELAX ALGORITHM

Purpose: To solve the relaxed problem, compute the values of X for the relaxed problem, and compute the relaxed objective value Z.

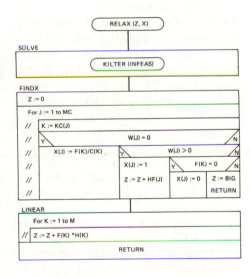

1. (SOLVE) Use the out-of-kilter algorithm to find the optimum solution to the relaxed problem. Note that the algorithm will initiate with the solution obtained at the previous iteration.
2. (FINDX) For each variable $X(J)$, find the corresponding arc in the original network. If the variable is free compute the fractional value of $X(J)$. If variable J is fixed to 1, set $X(J)=1$ and add the fixed charge to the objective. If variable J is fixed to 0 and the corresponding arc has nonzero flow, indicate infeasibility by setting Z to a large value and return. If variable J is fixed to 0 and the corresponding arc flow is zero, set $X(J)$ to 0.
3. (LINEAR) Add to Z the linear costs contributed by all the flow variables and return.

ROUND ALGORITHM

Purpose: To round the fractional variables that may appear in **X** to integer values to obtain XR, and compute the associated objective value ZR.

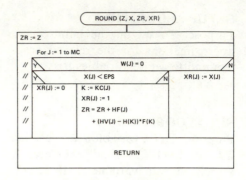

1. Set ZR to Z.
2. For each variable $X(J)$:
    If J is fixed, set $XR(J)$ to $X(J)$.
    If J is free but $X(J)$ equals 0, set $XR(J)$ to 0.
    If J is free but $X(J)$ is not zero, set $XR(J)$ to 1. Increase the objective by the difference between the concave cost and the linear underestimate for that arc.

    Repeat until all variables are considered; then return.

SEPRAT ALGORITHM

Purpose: Given a relaxed solution **X**, this procedure finds the free variable J that provides the greatest difference between the concave cost and the linearized cost. This is called the *separation variable*. It also determines the direction in which to branch ($XJ=0$ or 1)

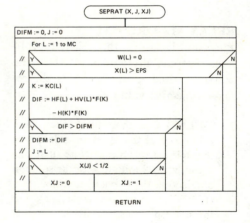

1. Set DIFM$=0$, $J=0$.
2. Do this process for each variable:
    If the variable is not free, go to the next variable.
    If the variable is zero, go to the next variable.
    Compute difference (DIF) between the concave cost function and the linearized cost function used for the relaxation.
    If this difference is greater than the value previously found (DIFM), make J the separation variable.
    If $X(J)$ is less than 1/2, branch first in the 0 direction. Otherwise, branch first in the 1 direction.
    Go on to the next variable unless finished. In that case, return.

EXAMPLE 385

RESET ALGORITHM

Purpose: To make free all variables at level L or higher.

For each variable J: If J is fixed at 1 or 0 at level L or higher (i.e., $W(J) \geqslant L$ or $W(J) \leqslant -L$) make variable J free by setting $H(K)$ to the linear under-estimate of the concave function and setting $W(J)$ to 0.

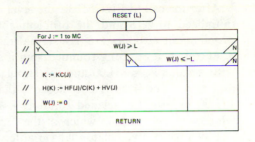

12.7 EXAMPLE

Let us refer to the network of Figure 12.8 and use it to illustrate the use of ENUMER. Prior to calling ENUMER, one must set Z_B to a large number, read the node and arc data for the network, and use READCC to stipulate those arcs having fixed charges associated with them. As may be observed from Figure 12.8 four arcs, arcs 1, 2, 3, and 4, possess fixed charges in this example.

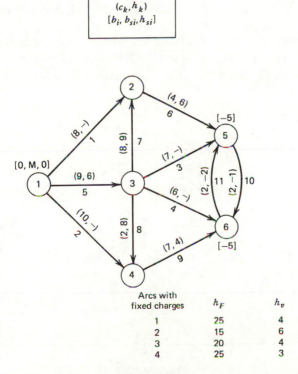

Arcs with fixed charges	h_F	h_v
1	25	4
2	15	6
3	20	4
4	25	3

Figure 12.8

Example Network Having Four Arcs with Concave Cost Functions

After the call to ENUMER, the level counter and the arrays **P**, **W**, and **U** are initialized to zero and the problem shown in Figure 12.9 is solved. In that problem, all variables are free so that relaxed coefficients are used for arcs 1, 2, 3, and 4. The resulting rounded solution becomes the first incumbent solution with a value $Z_B = 131.00$ and the stipulation that arcs 2 and 3 be fixed "open" and arcs 1 and 4 be fixed "closed"; that is, $XB = [0, 1, 1, 0]$. Noting that $ZR = 131.00$ exceeds the relaxed solution $Z = 115.05$, we proceed to call SEPRAT, which decides not only which of the four currently free variables is to be fixed but also which fixed state that variable will assume.

The method used in SEPRAT selects the separation variable on the basis of maximal difference between concave and linearized costs. There are many other techniques for choosing the separation variable. The one used was suggested by Soland (1974). The differences between the concave and linearized costs at vertex 0 are 0, 4.5, 11.45, and 0 for arcs 1, 2, 3, and 4, respectively. Since the flow on arc 3 is three units or less than half of the maximal flow on the arc, arc 3 is fixed "closed" at vertex 2. This decision is shown in Figure 12.10, which illustrates the implicit enumeration tree. Table 12.1 shows the sequence and the vectors **P**, **W**, and **U** developed by the algorithm for each tree vertex.

Solution of relaxed problem and subsequent rounding at vertex 2 yields no improvement in the incumbent solution and the implicit enumeration process proceeds to vertex 3 where arc 1 is also fixed closed. However, at this point, the optimal objective value for the associated relaxed problem exceeds that of the incumbent solution. Clearly, the remaining vertices in that part of the tree are fathomed; that is, the sequences $\overline{3}1\overline{2}$, $\overline{3}1\overline{24}$, $\overline{3}1\overline{2}4$, $\overline{3}12$, $\overline{3}124$, and $\overline{3}12\overline{4}$ need not be explicitly considered.

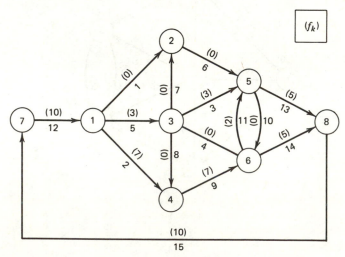

Figure 12.9
Equivalent Out-of-Kilter Network with the Initial Relaxation Solution

EXAMPLE   **387**

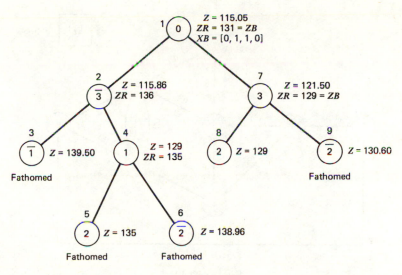

Figure 12.10
Branch-and-Bound Tree for the Network of Figure 12.8

Table 12.1   VECTORS DEVELOPED BY THE ENUMER ALGORITHM

Vertex	Sequence	P	W			U
1	0	0 0 0 0	0	0	0 0	0 0 0 0
2	$\bar{3}$	3 0 0 0	0	0	−1 0	0 0 0 0
3	$\bar{3}\,\bar{1}$	3 1 0 0	−1	0	−1 0	0 0 0 0
4	$\bar{3}\,1$	3 1 0 0	2	0	−1 0	0 1 0 0
5	$\bar{3}\,1\,2$	3 1 2 0	2	3	−1 0	0 1 0 0
6	$\bar{3}\,1\,\bar{2}$	3 1 2 0	2	−3	−1 0	0 1 1 0
7	$\underline{3}$	3 0 0 0	0	0	1 0	1 0 0 0
8	$\underline{3}\,2$	3 2 0 0	0	2	1 0	1 0 0 0
9	$\underline{3}\,\bar{2}$	· 3 2 0 0	0	−2	1 0	1 1 0 0

Since no sequence element has been underlined up to vertex 3, the backtrack procedure simply directs the search to vertex 4, which symbolizes the sequence $\bar{3}1$. Continuing the enumeration process, we explicitly evaluate a total of 9 out of the 31 possible vertices in the complete enumeration tree. At that point, we are assured that the optimal solution is XB=[0,1,1,0] with an objective function value of ZB=129. The optimal solution is given in Figure 12.11, where closed arcs 1 and 4 are not shown.

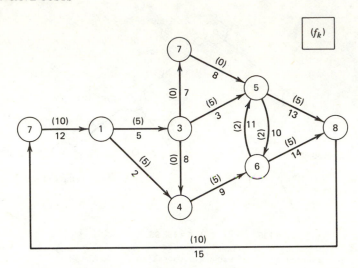

Figure 12.11
Optimal Solution for the Network of Figure 12.8

Notice that the optimal XB$=[0,1,1,0]$ was coincidentally encountered at vertex 0. However, the optimal flows and objective function value were not encountered until vertices 7 and 8. Indeed, it is only a result of the structure of this particular example problem that the flow solution from the relaxed problem at vertex 7 matched the optimal flows.

12.8 HISTORICAL PERSPECTIVE

The open literature in the general area of concave minimum cost flow problems is primarily confined to the consideration of special classes of problems, in particular facility location problems and fixed charge transportation problems. The combinatorial nature of the solution techniques usually proposed for these problems necessitates that the techniques take advantage of any structural characteristics possessed by a special class. Even so, the computational results reported in the literature suggest that the techniques are practical only for relatively small problems.

Facility location problems fall into the class of concave network flow problems because an important decision in location studies is the distribution policy for a commodity. The latter takes the form of the transportation model. Papers dealing with the facility location problem can be divided into two classes based on whether the facilities to be located are capacitated or not. Branch-and-bound optimization algorithms for the uncapacitated, fixed charge problem are described by Efroymson and Ray (1966), Spielberg (1969), and Khumawala (1972). The approaches in these papers are also appropriate when the facility cost is a piecewise linear concave function. Davis and Ray (1969) and Sá (1969) provide algorithms for the capacitated facility location problem that use the out-of-kilter algorithm to solve the relaxed problems of the branch-and-bound enumeration. Soland (1974) describes an algorithm that handles general concave arc costs for either facilities or transportation links. His approach is appropriate for both capacitated and uncapacitated problems. Resh and Barton (1970) describe a similar approach using the OKA to solve the relaxed problems. The algorithms of this chapter are essentially an extension of Soland's (1974) work to transshipment networks using only fixed charges to simplify the algorithms. A number of heuristic approaches have been suggested for location problems. Some of these are described in ReVelle et al. (1970).

The fixed charge transportation problem has the form of a transportation problem. The arcs in the problem are given a fixed charge in addition to a variable cost. This class also has relevance to the fixed charge transshipment problem because all such problems can be transformed into a fixed charge transportation problem—Malek–Zaverei and Frisch (1972). Murty (1968) describes an extreme point ranking optimization technique. Gray (1971) couples this with a branch-and-bound approach. Florian and Robillard (1971) transform the transshipment problem to the equivalent transportation problem and solve this with a branch-and-bound approach. Kennington and Unger (1973) show group theoretic results for this problem class. Kennington and Unger (1976) describe an advanced branch-and-bound code that uses a basic primal network code to solve the transportation problems arising from the relaxations. The latter paper provides references to a number of heuristic approaches.

Zangwill (1969) studies the general concave network problem for special classes of networks. Solutions are presented based on the fact that solutions to the concave problem must be extreme point solutions.

The general idea of branch and bound as used in this chapter can be found in a number of integer programming texts. Garfinkel and Nemhauser (1972) use the terminology of this chapter.

EXERCISES

1. Make a flowchart and code the exhaustive enumeration algorithm presented in Chapter 12.
2. Use exhaustive enumeration to solve the problem of Figure 12.8.
3. Use ENUMER to solve the problem of Figure 12.8 with PER = 10.
4. An arc has the following cost function.

$$h(k) = 100 - (f_k - 10)^2 \qquad 0 \leqslant f_k \leqslant 10$$

where

$$\underline{c}_k = 0 \quad \text{and} \quad c_k = 10$$

Use three arcs with fixed and linear costs and one additional arc with $c_k = 10$ to form a piecewise linear approximation to the concave cost functional given above.

5. Suppose that the cost function of arc 1 in Figure 12.8 is replaced by the cost function given in exercise 4. Solve the resulting minimum cost flow problem.

6. (a) Write the $0-1$ integer programming formulation for the given network.
   (b) Fix arc 3 to be closed and arc 5 to be open and write the $0-1$ integer programming subproblem thus specified.
   (c) Write the relaxed linear programming problem for the problem given in part (b).
   (d) Solve the problem of part (c).
   (e) Give the "rounded solution" for the solution found in part (d).

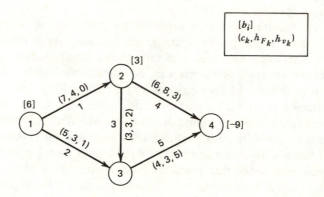

7. Suppose one or more arcs in a network flow problem have cost functions of the following form.

$$h(k) = h_{Fk} + h_{Vk} f_k^2 \qquad 0 < f_k < c_k$$

that is, arc k has a fixed charge and a *convex* cost function associated with flow on the arc. Outline a method that could be used to solve such a network flow programming problem.

8. Subroutine KILTER was used in subroutine RELAX only as a matter of convenience since KILTER requires neither a primal nor a dual feasible solution. Rather, KILTER requires only a flow solution that satisfies conservation of flow.
   (a) What modification would be required to use PRIMAL1 instead of KILTER?
   (b) What modifications would be required to use PRIMAL instead of KILTER?

9. Solve the Generous Electric Problem of Chapter 2 if arcs 1(1, 4), 6(1, 9), 14(2, 10), and 16(3, 5) have the following fixed costs attached to their use.

$$h_F(1) = 50, \qquad h_F(6) = 50, \qquad h_F(14) = 100, \quad \text{and} \quad h_F(16) = 30$$

REFERENCES

Balas, E., "The Dual Method for the Generalized Transportation Problem," *Management Science*, *12*, 555–568 (1966).

Balas, E., and P. L. Ivanescu, "On the Generalized Transportation Problem," *Management Science*, *11*, 188–202 (1964).

Barr, R. S., F. Glover, and D. Klingman, "An Improved Version of the Out-of-Kilter Method and a Comparative Study of Computer Codes," *Mathematical Programming*, *7*, 60–86 (1974).

Bartholdi, John J., L. Martin–Vega, and H. D. Ratliff, *Efficient Network Solutions to Parallel Processor Scheduling Problems*: *A Survey*, Dept. of Industrial and Systems Engineering, University of Florida, Gainesville Florida, Research Report No 76–16 (1976).

Bazaraa, M. S., and J. Jarvis, *Linear Programming and Network Flows*, John Wiley & Sons, New York (1977).

Bazaraa, M. S., and R. W. Langley, "A Dual Shortest Path Algorithm," *SIAM J. of Applied Mathematics*, *26*, 496–501 (1974).

Beale, E. M. L., "An Algorithm for Solving the Transportation Problem When the Shipping Cost over Each Route is Convex," *Naval Research Logistic Quarterly*, *6*, 43–56 (1959).

Belford, P. C., and H. D. Ratliff, "A Network-flow Model for Racially Balancing Schools," *Operations Research*, *20*, 619–628 (1972).

Bellman, R. E., "On a Routing Problem," *Quarterly of Applied Mathematics, 16,* 87–90 (1958).

Bellman, R., and S. E. Dreyfus, *Applied Dynamic Programming,* Princeton University Press, Princeton, N. J. (1962).

Bellmore, M., G. Bennington, and S. Lubore, "A Multivehicle Tanker Scheduling Problem," *Transportation Science, 5,* 36–47 (1971).

Bennington, G. E., "An Efficient Minimal Cost Flow Algorithm," *Management Science, 19,* 1042–1051 (1973).

Bennington, G. E., "Applying Network Analysis," *Industrial Engineering, 6,* 17–25 (1974).

Bhaumik, G., and P. A. Jensen, *Network Modeling of Multireservoir Water Distribution Systems,* Center for Research in Water Resources, U. of Texas at Austin, Report 107 (1974).

Birkhoff, G., and J. B. Diaz, "Non-linear Network Problems," *Quarterly of Applied Mathematics, 8,* (1956).

Bowman, Earl E., "Production Scheduling by the Transportation Method of Linear Programming," *Management Science, 12,* 778–784 (1956).

Bradley, G., "Survey of Deterministic Networks," *AIIE Transactions, 7,* 222–234 (1975).

Bradley, G., G. Brown, and G. Graves, "Design and Implementation of Large Scale Primal Transshipment Algorithms," *Management Science, 24,* 1–35 (1977).

Busacker, R. G., and P. J. Gowen, "A Procedure for Determining a Family of Minimal Cost Network Flow Patterns," ORO Technical Report 15, Operations Research Office, Johns Hopkins University (1961).

Busacker, R. G., and T. Saaty, *Finite Graphs and Networks,* McGraw–Hill, New York (1965).

Chapin, Ned, "New Format for Flowcharts," *Software Practice and Experience, 4,* 341–357 (1974).

Charnes, A., and W. W. Cooper, "The Stepping Stone Method of Explaining

Linear Programming Calculations in Transportation Problems," *Management Science, 1*, 49–69 (1954).

Charnes, A., and W. W. Cooper, "Nonlinear Network Flows and Convex Programming over Incidence-Matrices," *Naval Research Logistics Quarterly, 5*, 231–240 (1958).

Charnes, A., and W. W. Cooper, *Management Models and Industrial Applications of Linear Programming*, 2 Vols., John Wiley & Sons, New York (1961).

Charnes, A., and W. M. Raike, "One-Pass Algorithm for Some Generalized Network Problems," *Operations Research*, 14, 914–924 (1966).

Charnes, A., F. Glover, D. Karney, and J. Stutz, "Past, Present and Future of Large Scale Transshipment Computer Codes and Applications," University of Texas, CCS Report 131 (1973).

Clasen, R. J., "The Numerical Solution of Network Problems Using the Out-of-Kilter Algorithm," RAND Memorandum RM-5456-PR, Santa Monica, California (1968).

Collins, M., L. Cooper, R. Helgason, J. Kennington, and L. LeBlanc, "Solving the Pipe Network Analysis Problem Using Optimization Techniques," Southern Methodist University, Tech. Dept. 77009 (1977).

Cooper, L., "The Transportation-Location Problem," *Operations Research, 20*, 94–108 (1972).

Dantzig, G., "Application of the Simplex Method to a Transportation Problem," in *Activity Analysis of Production and Allocation*, Ed. T. C. Koopmans, John Wiley & Sons, New York (1951).

Dantzig, G. B., *Linear Programming and Extensions*, Princeton University Press, Princeton, N. J. (1963).

Dantzig, G. B., and D. R. Fulkerson, "Minimizing the Number of Tankers to Meet a Fixed Schedule," *Naval Research Logistics Quarterly, 1*, 217–22 (1954).

Dantzig, G. B., and D. R. Fulkerson, "On the Min-Cut Max-Flow Theorem of Networks," *Linear Inequalities and Related Systems*, Princeton University Press, Princeton N. J. (1956).

Davis, P. S., and L. Ray, "A Branch Bound Algorithm for the Capacitated

Facilities Location Problem," *Naval Research Logistics Quarterly*, *16*, 331–344 (1969).

Dennis, J. B., "A High-Speed Computer Technique for the Transportation Problem," *Journal of Association for Computing Machinery*, 132–153 (1958).

Dennis, J. B., *Mathematical Programming and Electrical Networks*, John Wiley & Sons, New York (1959).

Dijkstra, E. W., "A Note on Two Problems in Connection with Graphs," *Numerishe Mathematik*, *1*, 269–271 (1959).

Dorsey, R. C., T. J. Hodgson, and D. H. Ratliff, "A Network Approach to a Multi-Facility Production Scheduling Problem Without Backordering," *Management Science*, *21*, 813–822 (1975).

Dorsey, R. C., T. J. Hodgson, and H. D. Ratliff, "A Production Scheduling Problem with Batch Processing," *Operations Research*, *22*, (1974).

Dreyfus, S. E., "A Generalized Equipment Replacement Study," *Journal Society of Industrial and Applied Mathematics*, *8*, 425–435 (1960).

Dreyfus, S. E., "An Appraisal of Some Shortest Path Algorithms," *Operations Research*, *17*, 395–412 (1969).

Durbin, E. P., and D. M. Kroenke, "The Out of Kilter Algorithm: A Primer," RAND Memorandum RM-5472PR, Santa Monica, California (1967).

Edmonds, J., and R. M. Karp, "Theoretical Improvements in Algorithmic Efficiency for Network Flow Problems," *Journal of the Association for Computing Machinery*, *19*, 248–264 (1972).

Efroymson, M. A., and T. L. Ray, "A Branch and Bound Algorithm for Plant Location," *Operations Research*, *14*, 361–368 (1966).

Eisemann, K., "Simplified Treatment of Degeneracy in Transportation Problems," *Quarterly of Applied Mathematics*, 399–403 (1957).

Eisemann, K., "The Generalized Stepping Stone Method for the Machine Loading Model," *Management Science*, *11*, 154–176 (1964).

Elmaghraby, S. E., "Allocation Under Uncertainty when the Demand Has a Continuous Density Function," *Management Science*, *6*, (1960).

Elmaghraby, S. E., *Some Network Models in Operations Research*, Springer–Verlag, New York (1970).

Falk, J. E., and J. L. Horowitz, "Critical Path Problems with Concave Cost-Time Curves," *Management Science*, *19*, 446–455 (1972).

Flood, M. M., "On the Hitchcock Distribution Problem," *Pacific Journal of Mathematics*, *3*, 369–386 (1953).

Florian, M., and P. Robert, "A Direct Method to Locate Negative Cycles in a Graph," *Management Science*, 17, 307–311 (1971).

Florian, M., and P. Robillard, "An Implicit Enumeration Algorithm for the Concave Cost Network Flow Problem," *Management Science*, *18*, 184–193 (1971).

Ford, L. R., and D. R. Fulkerson, "Solving the Transportation Problem," *Management Science*, *3*, 24–32 (1956).

Ford, L. R., and D. R. Fulkerson, "A Simple Algorithm for Finding Maximal Network Flows and an Application to the Hitchcock Problem," *Canadian Journal of Mathematics*, *9*, 210–218 (1957).

Ford, L. R., and D. R. Fulkerson, "A Primal-Dual Algorithm for the Capacitated Hitchcock Program," *Naval Research Logistics Quarterly*, *4*, 47–54 (1957).

Ford, L. R., and D. R. Fulkerson, *Flows in Networks*, Princeton University Press, Princeton, N. J. (1962).

Francis, R., and J. White, *Facility Layout and Location*, Prentice-Hall, Englewood Cliffs, New Jersey (1974).

Frank, H., and I. T. Frisch, *Communication, Transmission and Transportation Networks*, Addison–Wesley, Reading, Massachusetts (1971).

Fulkerson, D. R., "A Network Flow Computation for Cost Curves," *Management Science*, *7*, 167–178 (1961).

Fulkerson, D. R., "An Out-of-Kilter Method for Solving Minimal Cost Flow Problems," *SIAM Journal of Applied Mathematics*, *9*, 18–27 (1961).

Fulkerson, D. R., "Flow Networks and Combinatorial Operations Research," *American Mathematical Monthly*, *73*, 115–138 (1966).

Fulkerson, D. R., and G. B. Dantzig, "Computation of Maximal Flows in Networks," *Naval Research Logistics Quarterly*, 2, 277–283 (1955).

Garfinkel, R. S., and G. L. Nemhauser, *Integer Programming*, John Wiley & Sons, New York (1972).

Gleyzal, A., "An Algorithm for Solving the Transportation Problem," *Journal of Research of the National Bureau of Standards*, 54, 213–216 (1955).

Glover, F., D. Karney, and D. Klingman, "The Augmented Predecessor Index Method for Locating Stepping-Stone Paths and Assigning Dual Prices in Distributions Problems," *Transportation Science*, 6, 171–180 (1972).

Glover, F., D. Karney, and D. Klingman, "Implementation and Computational Comparisons of Primal, Dual and Primal-Dual Computer Codes for Minimum Cost Network Flow Problems," *Networks*, 4, 191–212 (1974).

Glover, F., D. Karney, D. Klingman, and A. Napier, "A Computational Study on Start Procedures, Basis Change Criteria and Solution Algorithms for Transportation Problems," *Management Science*, 20, 793–819 (1974).

Glover, F., and D. Klingman, "On the Equivalence of Some Generalized Network Problems to Pure Network Problems," Research Report CS81, Center for Cybernetic Studies, The University of Texas, Austin (1972).

Glover, F., and D. Klingman, "A Note on Computational Simplifications in Solving Generalized Transportation Problems," *Transportation Science*, 7, 351–361 (1973).

Glover, F., and D. Klingman, "Network Applications in Industry and Government," University of Texas, Report CCS 247 (1975).

Glover, F., and D. Klingman, "New Advances in Solution of Large Scale Network and Network-Related Problems," *Colloquid Mathematica Societatis Janos Bolyai 12*, North Holland Publishing Co. (1975).

Glover, F., D. Klingman, and A. Napier, "Basic Dual Feasible Solutions for a Class of Generalized Networks," *Operations Research*, 20, (1972).

Glover, F., D. Klingman, and J. Stutz, "Extensions of the Augmented Predecessor Index Method to Generalized Network Problems," *Transportation Science*, 7, 377–384 (1973).

Glover, F., D. Klingman, and J. Stutz, "Augmented Threaded Index Method for Network Optimization," *INFOR*, *12*, 293–298 (1974).

Golden, B. L., and T. L. Magnanti, "Deterministic Network Optimization: A Bibliography," *Networks*, *7*, 149–183 (1977).

Graves, G. W., and R. D. McBride, "The Factorization Approach to Large-Scale Linear Programming," *Mathematical Programming*, *10*, 1 (1976).

Gray, P., "Exact Solution of the Fixed Charge Transportation Problem," *Operations Research*, *19*, 1529–1538 (1971).

Grinold, R. C., "Calculating Maximal Flows in a Network with Positive Gains," *Operations Research*, *21*, 528–541 (1973).

Halmos, P. R., and H. E. Vaughan, "The Marriage Problem" *American Journal of Mathemtics*, *72*, 214–215 (1950).

Hatch, R. S., "Bench Marks Comparing Transportation Codes based on Primal Simplex and Primal-Dual Algorithms," *Operations Research*, *23*, 1167–1172 (1975).

Hershdorfer, A. M., "Predicting the Equilibrium of Supply and Demand: Location Theory and Transportation Network Flow Models," *TRF Papers*, 131–143 (1966).

Hitchcock, F. L., " The Distribution of a Product from Several Sources to Numerous Localities," *Journal of Mathematics and Physics*, *20*, 224–230 (1941).

Hoffman, A. J., and S. Winograd, "Finding All Shortest Distances in a Directed Network," *IBM Journal of Research and Development*, *16*, 412–414 (1972).

Hu, T. C., "Minimum Convex Cost Flow in Networks," *Naval Research Logistics Quarterly*, *13*, 1–9, (1966).

Hu, T. C., "Laplace's Equation and Network Flows," *Operations Research*, *15*, 348–356 (1967).

Hu, T. C., *Integer Programming and Network Flows*, Addision–Wesley, Reading, Massachusetts (1969).

Hultz, J., D. Klingman, and R. Russell, "An Advanced Dual Basic Feasible

Solution for a Class of Capacitated Generalized Networks," *Operations Research*, *24*, 301–313 (1976).

Jarvis, J., and A. M. Jezior, "Maximal Flow with Gains through a Special Network," *Operations Research*, *20*, 678–688 (1972).

Jensen, P. A., and G. Bhaumik, "A Flow Augmentation Approach to the Network with Gains Minimum Cost Flow Problem," *Management Science*, *23*, 631–643 (1977).

Jewell, W. S., "Optimal Flow Through Networks With Gains," *Operations Research*, *10*, (1962).

Jewell, W. S., "Models for Traffic Assignment," *Transportation Research*, *1*, 31–46 (1967).

Johnson, E. L., "Programming in Networks and Graphs," ORC 65-1, University of California, Berkeley (1965).

Johnson, E. L., "Networks and Basic Solutions," *Operations Research*, *14*, 619–623 (1966).

Kantorovich, L., "On the Translocation of Masses," *Comptes rendas* (*Doklady*) *de l' academie des sciences de l' URSS*, *XXXVII*, (1942).

Kelley, J. E., Jr., "Critical Path Planning and Scheduling: Mathematical Basis," *Operations Research*, *9*, 296–320 (1961).

Kennington, J., and E. Unger, "A New Branch-and-Bound Algorithm for the Fixed-Charge Transportation Problem," *Management Science*, *22*, 1116–1126 (1976).

Khumawala, B. M., "An Efficient Branch and Bound algorithm for the Warehouse Location Problem," *Management Science*, *18*, B 718–B 731 (1972).

Klein, M., "A Primal Method for Minimal Cost Flows with Applications to the Assignment and Transportation Problems," *Management Science*, *14*, 205–220 (1967).

Knuth, D. E., *The Art of Computing, Volume 1*, Addison–Wesley, Reading, Mass. (1968).

Koopmans, T. C., "Optimum Utilization of the Transportation System," *Proc. of the International Statistical Conf., Washington D. C.* (1947), also *Econometric,* XVII, 136–146 (1949).

Kuhn, H. W., "The Hungarian Method for the Assignment Problem," *Naval Research Logistics Quarterly*, *2*, 83–97 (1955).

Kuhn, H. W., "Variants of the Hungarian Method for Assignment Problems," *Naval Research Logistics Quarterly*, *3*, 253–258 (1956).

Langley, R. W., J. L. Kennington, and C. M. Shetty, "Efficient Computational Devices for the Capacitated Transportation Problem," *Naval Research Logistics Quarterly*, *21*, 637–647 (1974).

Lawler, E. L., "On Scheduling Problems with Deferral Costs," *Management Science*, *11*, 280–288 (1964).

Lourie, Janice R., "Topology and Computation of the Generalized Transportation Problem," *Management Science*, *11*, 177–187 (1964).

Malek–Zavarei, M., and J. K. Aggarwal, "Optimal Flows in Networks with Gains and Losses," *Networks*, *1*, 355–365 (1972).

Malek–Zavarei, M., and I. T. Frisch, "On the Fixed-Cost Flow Problem," *International Journal of Control*, *16*, 897–902 (1972).

Maurras, J. F., "Optimization of the Flow Through Networks with Gains," *Mathematical Programming*, *3*, 135–144 (1972).

Menon, V. V., "The Minimum-Cost Flow Problem with Convex-Costs," *Naval Research Logistics Quarterly*, *12*, 163–172 (1965).

Millar, W., "Some General Theorems for Nonlinear Systems Posssessing Resistance," *Philosophy Mag.*, *13*, 1150–1160, (1951).

Minieka, E., "Optimal Flow in a Network with Gains," *INFOR*, *10*, (1972).

Minieka, E., *Optimization Algorithms for Network and Graphs*, Marcel Dekker, Inc., New York (1978).

Minty, G. J., "Monotone Networks," *Proceedings of the Royal Society A*, *257*, 194–212 (1960).

Minty, G. J., "Solving Steady-State Nonlinear Networks of 'Monotone' Elements," *Trans. Professional Group on Circuit Theory*, *CT-8*, 99–104 (1961).

Minty, G. J., "On an Algorithm for Solving Some Network Programming Problems," *Operations Research*, 10, 403–405 (1962).

Motzkin, T. S., "The Assignment Problem," *Proceedings of Symposia in Applied Mathematics, Vol VI—Numerical Analysis*, McGraw–Hill, New York (1956).

Munkres, J., "Algorithm for the Assignment and Transportation Problems," *Journal of the Society for Industrial and Applied Mathematics*, 5, 32–38 (1957).

Murty, K. G., "Solving the Fixed Charge Transportation Problem by Ranking the Extreme Points," *Operations Research*, 16, 268–279 (1968).

Nemhauser, G. L., "A Generalized Label Setting Algorithm For the Shortest Path Between Specified Nodes," *J. of Mathematical Analysis and Application*, 38, 328–334 (1972).

Onaga, K., "Optimum Flows in General Communication Networks," *J. of the Franklin Institute*, 283, 308–327 (1967).

Orden, A., "The Transshipment Problem," *Management Science*, 2, 276–285 (1956).

Phillips, D. T., and P. A. Jensen, "The Out-of-Kilter Algorithm," *Industrial Engineering*, 6, 36–44 (1974).

Potts, R. B., and R. M. Oliver, *Flows in Transportation Networks*, Academic Press, New York (1972).

Prager, W., "A Generalization of Hitchcock's Transportation Problem," *Journal of Mathematics and Physics*, 36, 99–106 (1957).

Ratliff, H. D., "Network Models for Production Scheduling Problems with Convex Cost and Batch Processing," Industrial and Systems Dept., U. of Florida, Research Report, 76–18 (1976).

Resh, P., and L. G. Barton, "A Nonconvex Transportation Algorithm," in *Applications of Mathematical Programming Techniques*, EML Beale, American Elsevier, New York (1970).

ReVelle, C., D. Marks, and J. C. Liebman, "An Analysis of Private and Public Sector Location Models," *Management Science*, 16, 692–707 (1970).

Sá, G., "Branch-and-Bound and Approximate Solutions to the Capacitated Plant-Location Problem," *Operations Research*, 17, 1005–1016 (1969).

Soland, R. M., "Optimal Facility Location with Concave Costs," *Operations Research*, *22*, 373–382 (1974).

Spielberg, K., "An Algorithm for the Simple Plant Location Problem with Some Side Conditions," *Operations Research*, *17*, 85–111 (1969).

Spira, P. M., "A New Algorithm for Finding All Shortest Paths in a Graph of Positive Arcs," *SIAM Journal of Computing*, *2*, 28–32 (1973).

Srinivasan, V., and G. L. Thompson, "Accelerated Algorithms for Labeling and Relabeling of Trees, with Applications to Distribution Problems," *Journal of the Association for Computing Machinery*, *19*, 712–726 (1972).

Srinivasan, V., and G. L. Thompson, "Coding Techniques for Transportation Algorithms," *Journal for the Association for Computing Machinery*, 20 (1973).

Srinivasan, V., and G. L. Thompson, "Benefit-Cost Analysis of Coding Techniques for the Primal Transportation Algorithm," *Journal of the Association for Computing Machinery*, *20*, 194–213 (1973).

Stanley, E. D., P. D. Honig, and L. Gainen, "Linear Programming in Bid Evaluation," *Naval Research Logistics Quarterly*, *1*, 48–54 (1954).

Stern, T. Z., "Some Network Interpretations of Systems Problems," *System Theory*, Inter-University Electronics Series, Vol. 8, 127–176, McGraw–Hill (1969).

Swarc, W., "The Transportation Problem with Stochastic Demand," *Management Science*, *11*, 33–50 (1964).

Texas Water Development Board, *Economic Optimization and Simulation for Management of Regional Water Resource Systems*, Report 179 (1972).

Texas Water Development Board, *Water Supply Allocation Model AL-IV* (1975).

Tomizawa, N., "On Some Techniques Useful for Solution of Transportation Network Problems," *Networks*, *1*, 173–194 (1972).

Truemper, K., "An Efficient Scaling Procedure for Gains Networks," *Networks*, *6*, 151–160 (1976).

Veinott, A. F., Jr., and H. M. Wagner, "Optimal Capacity Scheduling I and II," *Operations Research*, *10*, 578–646 (1962).

Waggener, H. A., and G. Suzuki, "Bid Evaluation for Procurement of Aviation Fuel at DFSC: A Case History," *Naval Research Logistic Quarterly*, *14*, 115–129 (1967).

Weintraub, A., "A Primal Algorithm to Solve Network Flow Problems with Convex Costs," *Management Science*, *21*, 87–97 (1974).

Whitting, P. D., and J. A. Hillier, "A Method for Finding the Shortest Route through a Road Network," *Operational Research Quarterly*, *II*, 37–40 (1960).

Williams, A. C., "A Stochastic Transportation Problem" *Operations Research*, *11*, 759–770 (1963).

Wilson, D., "An Apriori Bounded Model for Transportation Problems with Stochastic Demand and Integer Solutions," *AIIE Trans.*, *4*, 186–193 (1972).

Yen, J. Y., "On Hu's Decomposition Algorithm for Shortest Paths in a Network," *Operations Research*, *19*, 983–985 (1971).

Zadeh, N., "A Bad Network Problem for the Simplex Method and Other Minimum Cost Flow Algorithms," *Mathematical Programming*, *5*, 255–266 (1973).

Zangwill, W. I., "Minimum Concave Cost Flows in Certain Networks," *Management Science*, *14*, 429–450 (1968).

Zangwill, W. "A Backlogging Model and a Multi-Echelon Model of a Dynamic Economic Lot Size Production System—A Network Approach," *Management Science*, *15*, 506–527 (1969).

INDEX